Microbial Biochemistry

Publisher's Note

The *International Review of Biochemistry* remains a major force in the education of established scientists and advanced students of biochemistry throughout the world. It continues to present accurate, timely, and thorough reviews of key topics by distinguished authors charged with the responsibility of selecting and critically analyzing new facts and concepts important to the progress of biochemistry from the mass of information in their respective fields.

Following the successful format established by the earlier volumes in this series, new volumes of the *International Review of Biochemistry* will concentrate on current developments in the major areas of biochemical research and study. New volumes on a given subject generally appear at two-year intervals, or according to the demand created by new developments in the field. The scope of the series is flexible, however, so that future volumes may cover areas not included earlier.

University Park Press is honored to continue publication of the *International Review of Biochemistry* under its sole sponsorship beginning with Volume 13. The following is a list of volumes published and currently in preparation for the series:

Consultant Editors: H. L. Kornberg, Sc.D., F.R.S., Department of Biochemistry, University of Cambridge; and D. C. Philips, Ph.D., F.R.S., Laboratory of Molecular Biophysics, Department of Zoology, University of Oxford.

series:

INTERNATIONAL
REVIEW OF BIOCHEMISTRY

Volume 21

Microbial
Biochemistry

Edited by

J. R. Quayle, B.Sc., Ph.D., F.R.S.
Department of Microbiology
University of Sheffield
Sheffield, England

UNIVERSITY PARK PRESS
Baltimore

UNIVERSITY PARK PRESS
International Publishers in Science, Medicine, and Education
233 East Redwood Street
Baltimore, Maryland 21202

Typeset by Action Comp. Co., Inc.

Manufactured in the United States of America by Universal Lithographers, Inc., and The Optic Bindery Incorporated.

Library of Congress Cataloging in Publication Data

Main entry under title:

Microbial biochemistry.

(International review of biochemistry; v. 21) Includes index.
1. Microbiological chemistry.   I. Quayle, John Rodney.   II. Series.
[DNLM: 1. Microbiology.   W1 IN8296 v. 21/QW4.3 M626] QP501.B527
vol. 21 [QR84] 574.1′92′08s
ISBN 0-8391-1087-1   [576′.11′92]   78-12191

Consultant Editors' Note

The MTP *International Review of Biochemistry* was launched to provide a critical and continuing survey of progress in biochemical research. In order to embrace even barely adequately so vast a subject as "progress in biochemical research," twelve volumes were prepared. They range in subject matter from the classical preserves of biochemistry—the structure and function of macromolecules and energy transduction—through topics such as defense and recognition and cell differentiation, in which biochemical work is still a relatively new factor, to those territories that are shared by physiology and biochemistry. In dividing up so pervasive a discipline, we realized that biochemistry cannot be confined to twelve neat slices of biology, even if those slices are cut generously: every scientist who attempts to discern the molecular events that underlie the phenomena of life can legitimately parody the cry of *Le Bourgeois Gentilhomme*, "Par ma foi! Il y a plus de quarante ans que je dis de la Biochimie sans que j'en susse rien!" We therefore make no apologies for encroaching even further, in this second series, on areas in which the biochemical component has, until recently, not predominated.

However, we repeat our apology for being forced to omit again in the present collection of articles many important matters, and we also echo our hope that the authority and distinction of the contributions will compensate for our shortcomings of thematic selection. We certainly welcome criticism—we thank the many readers and reviewers who have so helpfully criticized our first series of volumes—and we solicit suggestions for future reviews.

It is a particular pleasure to thank the volume editors, the chapter authors, and the publishers for their ready cooperation in this venture. If it succeeds, the credit must go to them.

H. L. Kornberg
D. C. Phillips

Contents

Preface

Most reviews these days start with the author or editor presenting a nest of Chinese boxes to the reader and then, because of the wide-ranging title grandiloquently written on the outer case, the author in his introduction, with appropriate patter, proceeds to strip away successive cases until he is left with one that contains what he is really going to concern himself with. With the title of "Microbial Biochemistry" on the present outer case I too must extract a smaller box from within and explain to the reader the somewhat smaller print written on its lid.

The area of microbial biochemistry that forms the setting to this volume is the borderland between metabolism and energetics. With the main metabolic pathways now well charted, one of the primary thrusts in microbial biochemistry is toward a full understanding at the physicochemical level of the various mechanisms of energy transduction carried out by microbes. As the knowledge of cellular metabolism and energy transduction deepens, it becomes more feasible to calculate how much cell material should, or could, be made from a given quantity of substrate and/or energy source. This can then be compared with the values determined experimentally by the growth physiologist. The closeness of fit is a measure of the extent to which we understand the relationship between metabolism and energetics. It is therefore appropriate that the first chapter should be written by Professor Stouthamer on the search for correlation between theoretical and experimental growth yields. Readers may find it interesting to reflect on the progress that has been made since the pioneering studies of Monod, Gunsalus, Elsden, and their associates. Such correlations can only be based on the understanding of energy metabolism operating in the different modes of microbial life—aerobic, anaerobic, and photosynthetic; chapters 2, 3, and 4 describe the present state of play in these areas.

The development of the continuous culture chemostat over the last 20 years has provided an instrument to the growth physiologist as important as the spectrophotometer is to the biochemist. Much of the correlation between microbial growth behavior and biochemistry has only been made possible through applications of continuous culture technique, as will be evident throughout chapter 1. This book would therefore be incomplete without an account of some of the applications of the chemostat to microbial biochemistry. Professor Bull and Dr. Brown recount a selection of these in chapter 5 and in addition point the way to new and fascinating fields that the chemostat is now opening up. It is most important and timely that microbiologists and biochemists of all followings should be aware of the potentialities of continuous culture in mixed substrate and heterogeneous substrate systems and also of the new kinds of microbial biochemistry and population dynamics that are starting to emerge from the study of interacting microbial communities.

Chapters 1 through 5 encompass a central area lying between metabolism and energetics. Two topics have then been chosen where metabolism, energetics, and technological potential are combined. The first of these topics is dealt with in chapter 6 by Dr. Dalton, who describes the different modes of utilization of inorganic nitrogen by microbial cells. This is an area of great importance in growth yield studies and it abounds in hard and challenging biochemistry; the potential prize of grafting microbial nitrogen fixation into crop plants also hovers in the distance life a Holy Grail. The second of these topics is dealt with in a composite chapter by Professor Wolfe and Dr. Higgins on the microbial biochemistry of methane. This too abounds in hard and challenging biochemistry: here lie problems of metabolism and energetics that have resisted solution for three decades. The recent recognition of the great importance of methane in both terrestrial and celestial carbon cycles, and its central position in the technologies of waste recycling and possibly biomass production, have added a new urgency to fundamental researches into its microbiology and biochemistry.

Finally, bearing in mind that this volume is one of an international review series, it may be noted that it is written by authors from five different countries.

J. R. Quayle

International Review of Biochemistry
Microbial Biochemistry, Volume 21
Edited by J. R. Quayle
Copyright 1979   University Park Press   Baltimore

1
The Search for Correlation Between Theoretical and Experimental Growth Yields

A. H. STOUTHAMER
Department of Biology
Free University
Amsterdam, The Netherlands

With slight exaggeration one can say that the evolution of microorganisms has led to the existence of an enormous number of organisms with more or less the same cell structure but all differing in the nature of their energy-yielding mechanisms. One can distinguish differences in the number of substrates that can be used, the pathways by which these substrates are catabolized, the number of electron acceptors that can be used, the complexity of the respiratory chain, and the number of essential cellular components that can be formed from the ingredients of the growth medium. It is therefore evident that the study of the relationship between energy production and growth forms a central theme in microbiology.

During heterotrophic growth the utilization of a certain amount of substrate will give rise to the formation of a certain amount of new cell material. The yield of new cell material formed from a substrate will mainly depend on the amount of energy generated during catabolism and the amount of energy needed for the synthesis of new cell material. The purpose of yield studies in microorganisms is to find the relationship between substrate utilization, ATP generation, and the formation of new cell material. The interpretation of yield studies is easiest for microorganisms growing anaerobically, because in this case the ATP production from substrates that are degraded by known catabolic pathways can be directly calculated. Anaerobic experiments are therefore ideally suited to identifying the environmental factors that influence the relationship between energy production and growth. For growth with external hydrogen acceptors, yield studies are more difficult to interpret because in many cases the efficiency of oxidative phosphorylation is not known with certainty.

During the last few years the relationship between energy production and growth has been intensively studied. This chapter attempts to present an overall picture of this relationship integrated from the large number of papers on this subject. A fairly complete picture can be obtained about the various factors that influence molar growth yields. The evidence indicating that under some growth conditions there is a discrepancy between the rate of ATP production by catabolism and the rate at which ATP can be used for the formation of microbial cell material is presented. A speculative discussion on the various possible mechanisms that may adjust the rate of ATP production to the needs of anabolism is included. It is argued that the evolution of microorganisms has not led to mechanisms that result in an optimal growth yield under each condition of growth, but instead has led to great flexibility for growth under the maximum number of different conditions and to possibilities for rapid adaptation to changes in the environmental conditions.

THEORETICAL CALCULATIONS ON THE ATP REQUIREMENT FOR THE FORMATION OF MICROBIAL BIOMASS

Influence of the Carbon Source and Complexity of the Medium

In theoretical calculations of the ATP requirement for the formation of cell material, the macromolecular composition of the cells is taken as a base. Subsequently the ATP requirement for the formation of each cell constituent is calculated. Such calculations have been performed by Gunsalus and Shuster (1), Forrest and Walker (2), and Stouthamer (3). The last author used the detailed analysis of the cell composition of *Escherichia coli* by Morowitz (4) as a base. The ATP requirement for the formation of cell material of this composition from glucose and preformed monomers (amino acids and nucleic acid bases) and from glucose and simple inorganic salts is given in Table 1. The results show that, theoretically, 31.9 g cells can be formed per mol of ATP for growth with glucose and preformed monomers and 28.8 for growth with glucose and inorganic salts. It is remarkable that the difference between these values is relatively small, which is due to a small ATP requirement for monomer synthesis from glucose, as shown in Table 1. With other carbon sources the situation is completely different. The ATP requirement for the formation of cell material from pyruvate and preformed monomers and from pyruvate and inorganic salts is included in Table 1 for comparison. The results show that the nature of the carbon source has a strong influence on the ATP requirement for the formation of cell material. The amount of cell material formed per mol of ATP from pyruvate and inorganic salts is much smaller than the amount formed from glucose and inorganic salts. This difference is due to a larger ATP requirement for monomer formation and for transport processes during growth on pyruvate than during growth on glucose. Furthermore, during growth on

Table 1. ATP requirement for the formation of microbial cells from glucose or pyruvate in the presence or absence of amino acids and nucleic acid bases[a]

Macromolecule	Amount of macromolecule (g/100g cells)	ATP required ($10^4 \times$ mol/g cells formed)			
		A	B	C	D
Polysaccharide	16.6				
G6P formation		10.3	10.3	61.5	61.5
Polymerization		10.3	10.3	10.3	10.3
Protein	52.4				
Amino acid formation		0	13.5	0	148
Polymerization		191.4	191.4	191.4	191.4
Lipid	9.4	1.4	1.4	27	27
RNA	15.7				
Nucleoside mono- phosphate forma- tion		14.8	34.5	37	62
Polymerization		9.2	9.2	9.2	9.2
DNA	3.2				
Deoxynucleoside monophosphate formation		3.8	8.6	8	14
Polymerization		1.9	1.9	1.9	1.9
Turnover mRNA		13.9	13.9	13.9	13.9
Total		257.0	295.0	369.2	539.2
Transport of					
Carbon source		0	0	58.1	148
Amino acids		47.8	0	47.8	0
Ammonium ions		0	42.4	0	42.4
Potassium ions		1.9	1.9	1.9	1.9
Phosphate		7.7	7.7	7.7	7.7
Total ATP requirement		314.5	347.1	475.7	740.2
Gram cells per mol ATP		31.9	28.8	21.0	13.5

[a]Data from Stouthamer (3, 28, and unpublished results). Media: A) glucose, amino acids, and nucleic acid bases; B) glucose and inorganic salts; C) pyruvate, amino acids, and nucleic acid bases; D) pyruvate and inorganic salts.

pyruvate the effect of supplementation of the medium with amino acids and nucleic acid bases is much more pronounced than during growth on glucose. The results in Table 1 demonstrate that, theoretically, 21 g of cell material can be formed per mol of ATP from pyruvate and preformed monomers and 13.5 g of cell material from pyruvate and inorganic salts. The ATP requirement for the formation of cell material in a mineral salts medium with various carbon sources is given in Table 2. Again it is evident that the nature of the carbon source has a very profound influence on the ATP requirement for the formation of cell material. For growth on acetate, theoretically only 10 g of cell material can be formed per mol of ATP. For

Table 2. ATP requirement for the formation of cell material in mineral salts medium containing various carbon sources[a]

Synthesis of	ATP requirement ($10^4 \times$ mol/g cells) for growth on			
	Lactate	Malate	Acetate	CO_2
Polysaccharide	71	51	92	195
Protein	339	285	427	907
Lipid	27	25	50	172
RNA	85	70	101	⎱ 212
DNA	16	13	19	⎰
Transport	200	200	306	52
Total	738	644	995	1538
Gram cells per mol of ATP	13.4	15.4	10.0	6.5

[a]The data are simplified from Stouthamer (3) and Harder and van Dijken (5). The composition of the microbial cells is as given in Table 1.

autotrophic growth this value is only 6.5 g (5), which is due to the very large ATP requirement for monomer formation under these circumstances.

These calculations offer an explanation for two experimental observations. First, for aerobic batch cultures the Y_O values (g of dry weight per g atom oxygen taken up during growth) are smaller for growth on simple compounds than for growth on glucose (6). A Y_O value of 31.9 for growth of *Aerobacter aerogenes* on glucose (7) and a value of 3.9 for growth of *Pseudomonas oxalaticus* on formate (8) may be mentioned as extremes. In a recent review (9), $Y_{O_2}^{max}$ (g of dry weight per mol oxygen taken up, corrected for maintenance respiration) values for chemostat cultures of a number of organisms growing on various substrates have been listed. Many examples can be derived from this review that show that the $Y_{O_2}^{max}$ values for growth on simple compounds are smaller than for growth on glucose. Second, supplementation of a glucose–mineral salts medium with amino acids has scarcely an influence on the Y_O value (10). On the other hand, supplementation with amino acids strongly increases the Y_O value for cells growing on ethanol or acetate. This difference is explained by the higher ATP requirement for amino acid formation during growth on simple compounds than for formation during growth on glucose (compare Table 1). Similarly the results discussed below in "Anaerobic Chemostat Cultures" (see Table 7) show that the ATP requirement for the formation of cell material from citrate is larger than that for its formation from glucose.

ATP Requirement for Transport Processes

There are a number of uncertainties in the calculations treated in the previous section, especially in the amount of ATP required for transport processes. Glucose is transported by the phosphoenolpyruvate:sugar phosphotransferase system, by which glucose is converted into glucose 6-phosphate during transport (11); therefore no energy requirement is present for glucose transport. In the original calculations (3) the most recent views on transport processes for other compounds were taken into account (12, 13). It was considered that ATP or an energized state of the membrane is involved in transport, as in the concept of transport by chemiosmotic processes (14). According to the chemiosmotic hypothesis the hydrolysis of 1 molecule of intracellular ATP is associated with the extrusion of two protons, by which a membrane potential is generated. Accumulation of K^+ and NH_4^+ was attributed to electrogenic porters and consequently the uptake of 2 K^+ or NH_4^+ ions is associated with the hydrolysis of 1 mol of ATP. The uptake of phosphate was thought to require 1 mol of ATP (13, 14); the uptake of malate similarly requires 1 mol of ATP. However, the amount of ATP associated with the uptake of 1 mol of amino acid was at that time not known and it was assumed that 1 mol of ATP was required for the uptake of 1 mol of amino acid or carbon source. Agreement on the mechanism of transport has since been reached and it has been accepted that transport of substrates occurs by proton symport mechanisms predicted by the chemiosmotic hypothesis (15, 16). It has become clear that β-galactosides in *E. coli* are cotransported with protons, the stoichiometry of the process being 1 proton per molecule of β-galactoside transported (17, 18, 19). Uptake of lactate and of alanine was also associated with the uptake of 1 proton per molecule transported (20). Based on these observations it seems that the amount of ATP required for transport processes during growth has been overestimated in some previous calculations. However, Ramos and Kaback (21) have recently shown that the stoichiometry between the number of protons per mol of solute transported must be dependent on the extracellular pH. Furthermore, Collins et al. (20) selected mutants of *E. coli* during growth in the chemostat with alanine as the growth-limiting carbon/energy source, in which the stoichiometry was altered to 2 or even 4 protons per mol of alanine transported. Due to these changes in the stoichiometry, the mutants are capable of accumulating substrates to higher concentration ratios than the wild type. On the other hand, some doubt has arisen about the number of protons extruded per mol of ATP hydrolyzed (22). Because of this uncertainty, the lack of data about the stoichiometry of amino acid to protons during transport for all amino acids, and the possibility that different ratios exist for one substrate in different organisms and for one organism under different experimental conditions, no revision of the original calculations has yet been performed. Therefore, the amounts of ATP required for transport processes are only estimates (and probably overestimates). For this reason they are given separately in the previous tables, which will make correction easier when more definite data for the amounts of ATP required for transport processes become available.

The data in Table 1 indicate that on theoretical grounds 15 or 27% of the total ATP requirement for the formation of cell material during growth on mineral salts and either glucose or pyruvate, respectively, is used for transport processes.

It has been stressed before that these estimates are approximations. However, it seems a safe conclusion that the ATP requirement for transport processes during growth on glucose is much smaller than the requirement during growth on pyruvate. Recently, experimental evidence for the influence of the ATP requirement for transport processes on molar growth yields has been presented by Dijkhuizen and Harder (23). They observed molar growth yields of 3.8 g/mol and 3.4 g/mol for growth of *P. oxalaticus* with oxalate and formate, respectively. This was rather unexpected, because formate is assimilated by the ribulose diphosphate cycle and oxalate by the glycerate pathway. Therefore, the ATP requirement for the formation of cell material from formate is much higher than that from oxalate. That the molar growth yields were, nonetheless, so closely similar was explained by the assumption that the energy requirement for oxalate transport has a very great influence on the total energy budget of the organism during growth on that substrate. On the contrary, formate transport was not considered to be an active process requiring energy.

In a later section of this chapter an experimental estimate of the amount of ATP required for transport processes and membrane energization is given. The data indicate a much higher ATP requirement for membrane energization than that calculated on the assumptions mentioned above.

Influence of Cell Composition

The influence of cell composition on the ATP requirement for the formation of cell material is relatively small; in order to evaluate this, the ATP requirement for the formation of cellular macromolecules is given in Table 3. It is evident from these data that the ATP requirement for the formation of protein, RNA, and DNA is larger than the ATP requirement for the formation of polysaccharide and lipid. The composition of *A. aerogenes* differs considerably from that of *E. coli*, which was taken as the basis for

Table 3. ATP requirement for the formation of cellular macromolecules in a glucose–inorganic salts medium[a]

Macromolecule	ATP requirement $(10^4 \times \text{mol/g macromolecule})$
Polysaccharide	123.6
Protein	391.1
Lipid	14.8
RNA	373.2
DNA	330.0

[a] Results from Stouthamer (9). Calculated from the data in Table 1 (column B).

the calculations in the previous sections. A high protein content (75%) and a low polysaccharide content have been reported (24). On the basis of these data it can be calculated that, theoretically, about 25 g of cells per mol of ATP can be formed, a value that does not differ strongly from the value of 28.8 (Table 1, column B) calculated for the composition of *E. coli* as reported by Morowitz (4). It is well known that the composition of microbial cells is influenced by the growth rate and the growth condition (25, 26). The influence of these changes on the ATP requirement for the formation of cell material is extremely small, unless large amounts of storage material are formed.

The influence of growth conditions on the formation of energy storage compounds has been described extensively in a recent review by Dawes and Senior (27). The ATP requirement for the formation of storage materials has been treated by Stouthamer (28). When large amounts of storage materials are formed, the ATP requirement for the formation of cell material is much smaller than in the absence of the formation of storage material. For instance, 81 g of polysaccharide can be formed from glucose per mol of ATP, or 80 g of polymetaphosphate per mol of ATP. The formation of poly-β-hydroxybutyrate from glucose is even associated with net ATP formation and the formation of lipid from glucose scarcely needs ATP. It has indeed been observed that molar growth yields are higher under conditions in which storage materials are formed than under conditions in which they are not. Holme (29) has observed that glycogen deposition occurs in nitrogen-limited chemostat cultures of *E. coli*. The largest amounts of glycogen were deposited at low values of the specific growth rate. The same observations have been made for *A. aerogenes* (Stouthamer and Bettenhaussen, unpublished results). A large accumulation of glycogen coincided with a high value for the molar growth yield for glucose. These and other examples described by Herbert (25) suggest that the formation of storage compounds is largely controlled by the nitrogen content of the medium.

In oxygen-limited cultures of *Azotobacter beijerinckii* and of *Azotobacter chroococcum* the molar growth yield for glucose is much higher than in carbon-limited cultures (30, 31). This effect is explained by the observation that up to 50% of the bacterial dry weight in oxygen-limited chemostat cultures is poly-β-hydroxybutyrate. From these examples it is evident that the formation of storage materials indeed has a profound effect on the molar growth yield, which can be explained by a lower theoretical ATP requirement for formation of cell material under such conditions.

Influence of the Nitrogen Source

There is a large ATP requirement for nitrogen fixation. In cell-free extracts, utilization of 12–15 mol of ATP per mol of nitrogen fixed has been reported (32). Theoretical calculations of the amount of cell material that can be formed per mol of ATP for various assumed ATP requirements per mol of ammonia have been published by Stouthamer (9). During nitrogen fixation the assimilation of the ammonia formed occurs by the glutamine synthetase/ glutamate synthase pathway (for reviews see ref. 33 and Dalton, this volume). Assimilation of ammonia by this pathway requires 1 mol of ATP, whereas ammonia assimilation by glutamate dehydrogenase, which occurs during growth with excess ammonia, does not require ATP. This extra ATP requirement was not taken into account in the previous calculations (9). If this is done, the results described in Table 4 are obtained. During ammonia-limited growth or during growth with nitrate the assimilation of ammonia also occurs by the glutamine synthetase/glutamate synthase system. Theoretically, for ammonia-limited growth or growth with nitrate in a mineral salts medium, 23.1 g of cell material can be formed per mol of ATP, whereas 28.8 g of cell material can be formed with excess ammonia (Table 4).

It is evident that the nature of the nitrogen source in the medium has a very drastic influence on the ATP requirement for the formation of cell material. The amount of cell material that can be formed per mol of ATP is much lower during growth with molecular nitrogen than with ammonia. The experimentally observed molar growth yields are in accordance with these theoretical calculations (34, 35, 36). With growing cells of *Klebsiella pneumoniae* an ATP:N_2 ratio of 30 was observed (35), and for *Clostridium pasteurianum*, there was an ATP:N_2 ratio of about 20 (36). Similarly, in nongrowing cells of a derepressed nitrogen-fixing mutant of *K. pneumoniae* about 30 mol of ATP were required per mol of N_2 fixed (37).

Influence of the Carbon Assimilation Pathway of the Growth Substrate

The calculations given above indicate that the pathway of nitrogen assimilation has a profound influence on the ATP requirement for the formation of cell material. A similar influence is exerted by the carbon assimilation pathway. Different assimilation pathways have been found for growth with methane and methanol (5, 38, 39, Higgins and Wolfe, this volume). The

Table 4. Influence of the nitrogen source on the theoretical amount of cell material formed per mol of ATP[a]

Source	ATP requirement/mol N_2 fixed	Cell material (g)/ mol ATP
NH_3		28.8
NO_3'		23.1
N_2	12	11.1
	18	8.7
	24	7.1
	30	6.0

[a] For growth with molecular nitrogen various ATP:N_2 ratios are used. The value of 12 for the ATP:N_2 ratio is the minimal ratio that is possible based on a requirement of 4 mol ATP per electron pair transferred in the nitrogenase reaction (32).

theoretical amounts of cell material formed per mol of ATP for growth with methane or methanol using the different carbon assimilation pathways are shown in Table 5. It is evident that the influence of the carbon assimilation pathway on the ATP requirement for the formation of cell material is very great. In agreement with these theoretical calculations, it has been observed that the molar growth yield for methanol varies from 15.7 to 17.3 for a number of organisms using the ribulose monophosphate cycle and from 9.8 to 13.1 for a number of organisms using the serine pathway (40).

Calculation of Aerobic Growth Yields

The calculations of the amount of ATP required for the formation of microbial cell material do not give sufficient information for the description of yield studies under aerobic growth conditions. For aerobic experiments yields are generally expressed as Y_{sub} (g of dry weight per mol of substrate consumed) and Y_{O_2} (g of dry weight per mol of oxygen consumed). Therefore it seems very useful to have a method for calculating the theoretical values for these parameters from the theoretical calculations of the ATP requirement for the formation of microbial cell material given in the previous sections. As an example, an outline for such a calculation is given in Table 6. As a starting point the elementary formula of cells is used [compare Herbert, (41)]. Then the assimilation equation can be determined. In this equation the consumption of 0.25 "H_2" is included, in which "H_2" is used to indicate that reducing equivalents (e.g., $NADH_2$) are used during the formation of biomass. From the growth equation and the theoretical molar growth yield for glucose equations can be derived for the amount of $NADH_2$ or $FADH_2$ available for oxidation and the amount of ATP formed by substrate-level phosphorylation. The total amount of ATP produced is dependent on the number of phosphorylation sites present in the respiratory chain of the organism studied. In the example given in Table 6 the calculation is given for 2 or 3 sites in the respiratory chain, respectively. By multi-

Table 5.   Theoretical amount of cell material formed per mol of ATP for microorganisms using different carbon assimilation pathways for growth on methane and methanol[a]

Assimilation pathway	Cell material (g)/ mol ATP
Ribulose monophosphate cycle, fructose diphosphate aldolase variant	27.3
Ribulose monophosphate cycle, 2-keto-3-deoxy-6-phosphogluconate aldolase variant	19.4
Serine pathway	12.5
Ribulose diphosphate cycle	6.5

[a] Data from Harder and van Dijken (5).

Table 6.   Calculation of theoretical growth parameters for aerobic growth on glucose in mineral salts medium

The cell composition was taken as $C_6H_{10.84}N_{1.4}O_{3.07}$ (van Verseveld and Stouthamer, unpublished results). The molecular weight of cells is thus 152. Theoretically 1 mol of ATP allows the formation of 28.8 cells (Table 1). The symbols Y_{ATP}, Y_{glu}, and Y_{O_2} represent the theoretical amount of cell material formed per mol of ATP, glucose, or oxygen, respectively.

The assimilation equation is:

$$C_6H_{12}O_6 + 1.4\ NH_3 + 0.25\ "H_2" \rightarrow C_6H_{10.84}N_{1.4}O_{3.07} + 2.93\ H_2O$$

$Y_{glu} = x$, consequently $(x/152)$ mol of glucose are assimilated, and $1 - (x/152)$ are dissimilated. Per mol of glucose dissimilated, 10 mol of $NADH_2$, 2 mol of $FADH_2$ and 4 mol of ATP are formed. For assimilation $(0.25x/152)$ mol of $NADH_2$ are necessary. Available for oxidation are $10 - (10.25x/152)$ mol of $NADH_2$ and $2 - (2x/152)$ mol of $FADH_2$.

	Phosphorylation sites	
	2 sites	3 sites
ATP production		
From $NADH_2$	$20 - \dfrac{20.5x}{152}$	$30 - \dfrac{30.75x}{152}$
From $FADH_2$	$2 - \dfrac{2x}{152}$	$4 - \dfrac{4x}{152}$
Substrate phosphorylation	$4 - \dfrac{4x}{152}$	$4 - \dfrac{4x}{152}$
Total ATP production	$26 - \dfrac{26.5x}{152}$	$38 - \dfrac{38.75x}{152}$

Growth yield: $Y_{glu} = x = Y_{ATP} \times$ total ATP production

Conclusion			
	Y_{glu}	124.4	131.2
	Y_{O_2}	125.7	183.9

plying the total ATP production obtained in this way with the amount of cells that theoretically can be formed per mol of ATP (as given in Table 1), the theoretical value for Y_{glu} is obtained. By dividing this value by the oxygen uptake, we find the theoretical growth yield per mol of oxygen. From the example given in Table 6 it is evident that the theoretical value for Y_{glu} is scarcely influenced by the number of phosphorylation sites in the respiratory chain. On the other hand, the theoretical value for Y_{O_2} is very strongly influenced by the number of phosphorylation sites.

In Table 7 the theoretical values for Y_{glu} and Y_{O_2} are calculated for growth with various nitrogen sources. For this purpose the calculations on the theoretical ATP requirement for growth with various nitrogen sources given in Table 4 are used. A very important point in these calculations is that, in addition to a larger ATP requirement for the formation of cell material from nitrate and nitrogen, there is also a larger requirement for reducing equivalents under these conditions. Therefore there is also a smaller amount of ATP produced by oxidative phosphorylation per mol of substrate consumed during growth with nitrate and nitrogen than during growth with ammonia. The calculations show that the theoretical values for Y_{glu} are much smaller for growth with nitrate and nitrogen than for growth with ammonia. The theoretical values for Y_{O_2} for growth with molecular nitrogen are much smaller than for growth with ammonia and nitrate. The theoretical values for Y_{O_2} for growth with nitrate with 2 phosphorylation sites are larger than the similar value for growth with ammonia. However, with 3 phosphorylation sites the situation is the opposite. The most remarkable observation is that, especially with 3 phosphorylation sites, there is not much difference between the theoretical Y_{glu} values calculated on the basis of growth with nitrate and growth with molecular nitrogen.

These calculations, as those in Table 4, have been performed with various ATP:N_2 ratios. It has been shown that the high ATP:N_2 ratios observed with growing cells of *K. pneumoniae* (35) are due to the reductive

Table 7. Theoretical values for Y_{glu} and Y_{O_2} for growth with various nitrogen sources[a]

| Nitrogen sources | Phosphorylation sites | | | |
| | 2 sites | | 3 sites | |
	Y_{glu}	Y_{O_2}	Y_{glu}	Y_{O_2}
Ammonia	124.4	125.7	131.2	183.9
Nitrate	93.3	178.1	92.8	168.3
Nitrogen (ATP:N_2 = 12)	86.3	44.8	96.1	65.6
(ATP:N_2 = 18)	79.1	34.9	89.8	50.9
(ATP:N_2 = 24)	72.9	28.5	84.3	41.7
(ATP:N_2 = 30)	67.7	24.1	79.4	35.3

[a]The assimilation equations for growth with these nitrogen sources are:

$$C_6H_{12}O_6 + 1.4\ NH_3 + 0.25\ "H_2" \rightarrow C_6H_{10.84}N_{1.4}O_{3.07} + 2.93\ H_2O$$

$$C_6H_{12}O_6 + 1.4\ HNO_3 + 5.85\ "H_2" \rightarrow C_6H_{10.84}N_{1.4}O_{3.07} + 7.13\ H_2O$$

$$C_6H_{12}O_6 + 0.7\ N_2 + 2.35\ "H_2" \rightarrow C_6H_{10.84}N_{1.4}O_{3.07} + 2.93\ H_2O$$

The values are given for 2 or 3 sites of oxidative phosphorylation in the respiratory chain and for the ATP:N_2 ratios given in Table 4. The calculations are performed as outlined in Table 6.

formation of H_2 from H^+ by nitrogenase. For growing cells of *Azotobacter vinelandii* a more favorable $ATP:N_2$ ratio has been calculated (34). It has been shown that this organism is able to reoxidize the H_2 formed by nitrogenase (42), and a similar observation has been made for another nitrogen-fixing aerobic species of the genus *Methylosinus* (43). In this way reductant and ATP can be regenerated by oxidation of the H_2 evolved by nitrogenase, thereby increasing the efficiency of nitrogen fixation. Hydrogen evolution and reutilization have also been shown with bacteroids from various nitrogen-fixing nodules of plants (44). If reutilization of H_2 occurs with the same efficiency as outlined in a recent review by Aleem (45) for hydrogen-oxidizing bacteria, only 1–2 mol of ATP are lost per mol of H_2 evolved by nitrogenase, instead of 4 mol of ATP when no reutilization is possible. These data indicate that for these organisms $ATP:N_2$ ratios between 14 and 21 are expected for hydrogen-reoxidizing, nitrogen-fixing bacteria. At that $ATP:N_2$ ratio the Y_{glu} value for growth with molecular nitrogen is about the same as that for growth with nitrate. These theoretical calculations indicate that the processes of nitrogen fixation and nitrate assimilation are approximately equally expensive. It is a pity that lack of relevant experimental data as yet prevents verification of this prediction.

EXPERIMENTAL DETERMINATION OF THE ATP REQUIREMENT FOR THE FORMATION OF MICROBIAL BIOMASS

Anaerobic Batch Cultures

The experimental determination of the ATP requirement for the formation of microbial cell material started with the classic work of Bauchop and Elsden (46). These authors concluded that the amount of growth of a microorganism was directly proportional to the amount of ATP that could be obtained from the degradation of the energy source in the medium. They introduced the coefficient Y_{ATP}, which was defined as the amount of dry weight of organisms produced per mol of ATP. It is evident that Y_{ATP} can be determined only if the ATP yield during the degradation of the energy source is known. Only under anaerobic conditions can the ATP yield for substrate breakdown be calculated exactly, because the catabolic pathways for anaerobic breakdown of substrates are known. The subject has been reviewed many times during the last few years (2, 6, 9, 28, 47, 48, 49).

The molar growth yield, ATP yield, and Y_{ATP} for several organisms growing anaerobically on glucose are given in Table 8. Furthermore, some data are included on the influence of the complexity of the medium on Y_{ATP}. In Table 9 Y_{ATP} values for growth of a number of organisms growing on various substrates other than glucose are listed. Four conclusions can be drawn from the data in Tables 8 and 9. First, Y_{ATP} is not a constant for different microorganisms (9, 28, 49). For batch cultures, Y_{ATP} can vary

Table 8. Y$_{ATP}$ for batch cultures of a number of organisms growing anaerobically in various media on glucose

Organism	Medium[a]	Y (g/mol substrate)	ATP from substrate-level phosphorylation (mol/mol substrate)	Y$_{ATP}$	References
Streptococcus faecalis	c	20.0–37.5	2.0–3.0	10.9	(46, 75, 76, 91)
Streptococcus agalactiae	c	20.8	2.25	9.3	(77)
Streptococcus pyogenes	s	25.5	2.6	9.8	(78)
Lactobacillus plantarum	c	20.4	2	10.2	(79)
Lactobacillus casei	c	42.9	2.05	20.9	(51)
Bifidobacterium bifidum	c	37.4	2.85	13.1	(80)
Saccharomyces cerevisiae	c	18.8–22.3	2	10.2	(46)
Saccharomyces rosei	c	22.0–24.6	2	11.6	(81)
Zymomonas mobilis	c	8.5	1	8.5	(46, 82)
	s	6.5	1	6.5	(83, 84)
Zymomonas anaerobia	m	4.7	1	4.7	(50)
	c	5.9	1	5.9	(85)
Sarcina ventriculi	s	30.5	2.62	11.7	(86)
Aerobacter aerogenes	m	26.1	3	10.2	(7)
	c	47	2.6	18	(52)
Aerobacter cloacae	p,c	69.5	2.45	28.5	
	m	17.7–27.1	1.5–2.5	11.9	(87)
Escherichia coli	m	25.8	3.0	11.2	(6)
Ruminococcus flavefaciens	c	29.1	2.75	10.6	(88)
Proteus mirabilis	m	14.0	3	5.5	(54)
	c	38.3	3	12.6	
	p,c	48.5	2.6	18.6	(Stouthamer, unpublished results)
Actinomyces israeli	c	24.7	2.0	12.3	(89)
Clostridium perfringens	c	45	3.08	14.6	(56)
Streptococcus diacetilactis	p.d.	35.2	2.26	15.6	(59)
	c	43.8	2.04	21.5	
Streptococcus cremoris	p.d.	31.4	2.26	13.9	(59)
	c	38.5	2.04	18.9	

[a] Media: m = minimal; s = synthetic; p.d. = partially defined; c = complex; p = pH auxostat culture.

Table 9. Y_{ATP} values calculated from molar growth yields and ATP yields for several organisms growing anaerobically on various substrates

Organism	Substrate	Y (g/mol substrate)	ATP from substrate-level phosphorylation (mol/mol substrate)	Y_{ATP}	Reference
Streptococcus faecalis	Gluconate	17.6	1.8	10.4	(75)
	2-Ketogluconate	19.5	2.3	8.5	(90)
	Ribose	21.0	1.67	12.6	(46)
	Arginine	10.2	1	10.2	(46)
	Pyruvate	10.4	1	10.4	(91)
Aerobacter aerogenes	Fructose	26.7	3	10.7	(7)
	Mannitol	21.8	2.5	10.8	(92)
	Gluconate	21.4	2.5	11.0	(6)
Bifidobacterium bifidum	Lactose	52.8	5.08	10.4	(80)
	Galactose	27.8	2.80	9.9	(80)
Lactobacillus casei	Mannitol	27.8	2.35	11.8	(80)
	Mannitol	40.5	2.22	18.2	(51)
	Citrate	18.2	0.96	19.0	
Clostridium tetanomorphum	Glutamate	6.8	0.62	10.9	(93)
Clostridium aminobutyricum	γ-Aminobutyrate	7.6	0.5	15.2	(57)
	γ-Hydroxybutyrate	8.9	0.5	17.8	(57)
Clostridium glycolicum	Ethylene glycol	7.7	0.5	15.4	(58)
Clostridium kluyveri	Ethanol + acetate			9.2	(94)
	Crotonate	4.8	0.5	9.7	
Clostridium pasteurianum	Sucrose	73.1	6.64	11.0	(36)
Proteus rettgeri	Pyruvate	6.3	1.0	6.3	(64)
Veillonella alcalescens	Citrate	19.3	1.16	16.6	(60)
	Tartrate	13.1	0.94	13.9	(60)

from 4.7 for *Zymomonas mobilis* (50) to 20.9 for *Lactobacillus casei* (51). For a pH-auxostatculture of *A. aerogenes* a value of 28.5 for Y_{ATP} was found (52). Bauchop and Elsden (46) concluded that, under the cultural conditions that they used, the yield of *Streptococcus faecalis, Saccharomyces cerevisiae*, and *Z. mobilis* per mol of ATP was constant. Afterward the view became generalized that Y_{ATP} might be a general biological constant with a value of about 10.5. Forrest and Walker (2) determined an average of 10.6 $\pm$ 1.0 for Y_{ATP} out of 47 determinations for a large number of different organisms. Although they remarked that "the differences in Y_{ATP} shown for different organisms are real and usually greater than the experimental error of the determinations," their final conclusion was that "Y_{ATP} is a well defined biological constant, which can be used to predict yields of organisms and to assess the comparative efficiency of growth" (2). From the results described above it is evident that it is not justified to utilize the mean value of Y_{ATP} calculated from the results with other organisms for the prediction of the ATP yield for a process in an organism for which the ATP production is not known, because the value of Y_{ATP} can differ strongly from the mean value of 10.5 and, furthermore, Y_{ATP} can be strongly influenced by growth conditions. An example of erroneous conclusions resulting from the use of a Y_{ATP} value of 10.5 to calculate ATP yields is found in the work of Kapralek (53) with *Citrobacter freundii*. He concluded from a molar growth yield of 45 g/mol that during glucose fermentation 45/10.5, or 4.2, mol of ATP per mol of glucose were produced. However, the fermentation balance in growing cultures of enteric bacteria is 1 glucose → 1 ethanol + 1 acetate + 2 formate (7, 54). Consequently, the net ATP production is 3 mol per mol of glucose fermented, and Y_{ATP} for growth of *C. freundii* in complex medium with glucose is 15.

Second, the data in Table 8 demonstrate that for any one organism the highest value of Y_{ATP} is obtained for growth in a complex medium. A similar difference in carbon assimilation in simple and complex media has recently been observed for *Enterobacter aerogenes* and *Pseudomonas perfectomarinus* (55). High Y_{ATP} values have been found for a number of other organisms growing in complex media: *Lactobacillus casei* (51), *Clostridium perfringens* (56), *A. aerogenes* (52), *C. freundii* (53), *Clostridium aminobutyricum* (57), *Clostridium glycolicum* (58), *Streptococcus diacetilactis* and *Streptococcus cremoris* (59), and *Veillonella alcalescens* (60). Furthermore, high Y_{ATP} values have been measured for intraperiplasmic growth of *Bdellovibrio bacteriovorus* on *E. coli* as a substrate organism. A Y_{ATP} value of 18.5 was obtained for single growth cycle experiments and a value of 25.9 for multicycle experiments (61). During intraperiplasmic growth the Bdellovibrio did not need to synthesize monomers, because these are directly derived from the substrate organism (62). From these data it can be concluded that in general much higher values for Y_{ATP} are found for growth in complex media than in mineral salts media. In "Influence of the Carbon Source and Complexity of the Medium," above, it was found that theoretical calculations of the ATP requirement for the formation of cell material indicated that approximately similar amounts of cell material could be formed per mol of ATP in a glucose–mineral salts medium and in a glucose medium supplemented with amino acids and nucleic acid bases. This apparent discrepancy is discussed later in "Growth in Minimal Media."

Third, it is remarkable that the same Y_{ATP} value is found for growth of an organism on many different substrates (Tables 8 and 9). However, in most cases only growth on various sugars is compared, for which the theoretical ATP requirement for cell formation is the same. In other cases (e.g., *Lactobacillus casei*) growth on citrate has been studied in addition to growth on sugars. In these cases complex media were used for which the calculated ATP requirement is not very much different from that for growth on glucose in mineral salts medium (Table 1). For the determination of Y_{ATP} citrate, tartrate, and pyruvate are very suitable growth substrates. In a number of organisms that form succinate or propionate as an ultimate fermentation product additional ATP is formed by oxidative phosphorylation. This is due to the occurrence of anaerobic electron transport from $NADH_2$ to cytochrome b. The latter component is the hydrogen donor for the reduction of fumarate to succinate. This has been demonstrated in a number of organisms: *Vibrio succinogenes* (63), *Proteus* spp. (64, 65), *Propionibacterium* spp. (66, 67, 68), *Bacteroides* spp. (69, 70, 71), *V. alkalescens* (60, 72), and *Desulfovibrio gigas* (73). In these organisms it is difficult to determine the exact ATP production during fermentation of glucose or lactate, because the contribution of oxidative phosphorylation to the total ATP production is not known (for reviews see refs. 9, 28, and 74). However, during fermentation of citrate, pyruvate, and tartrate, in general, no succinate or propionate are formed, and with these substrates the ATP production is 1 mol per mol of substrate fermented. The Y_{ATP} value obtained in this way can then be used to obtain an estimate of the amount of ATP formed in association with the formation of succinate or propionate during fermentation of other substrates (60, 64).

Fourth, a very important general conclusion from the data in Tables 8 and 9 is that the experimental Y_{ATP} values are nearly always much smaller than those based on calculations of the theoretical ATP requirement for the formation of cell material (Tables 1 and 2). The only exceptions are the pH-auxostat cultures of *A. aerogenes* (52). Furthermore, the Y_{ATP} values found for growth of some organisms in complex media (*Lactobacillus casei*, *Cl. aminobutyricum*, *Cl. glycolicum*, and *V. alkalescens*) closely approach the theoretical values. The possible reasons for the discrepancy between experimental and theoretical Y_{ATP} values are discussed in the next two sections.

Anaerobic Chemostat Cultures

Growing bacteria, like all living organisms, require a certain amount of energy for maintenance processes. Maintenance energy is required for the turnover of cellular constituents, the preservation of the correct ionic composition and intracellular pH of the cell, and the maintenance of a pool of intracellular metabolites against a concentration gradient. In the experimental determination of Y_{ATP} values in batch cultures, no account is taken of the requirement for energy of maintenance. Therefore, it is possible that the occurrence of energy of maintenance might be one of the reasons for the discrepancy between the experimental and theoretical Y_{ATP} values. Furthermore, the great variation in Y_{ATP} among different microorganisms might be due to differences in the amount of energy required for maintenance purposes (49).

An equation that relates the molar growth yield and the specific growth rate has been derived by Pirt (95). In this derivation it is assumed that during growth the consumption of the energy source is partly growth-dependent and partly growth-independent, i.e.,

$$\frac{1}{Y_{glu}} = \frac{1}{Y_{glu}^{max}} + \frac{m_s}{\mu} \tag{1}$$

where Y_{glu} is the molar growth yield for glucose, m_s is the maintenance coefficient (mol of glucose per g of dry weight per hr), μ is the specific growth rate, and Y_{glu}^{max} is the molar growth yield for glucose corrected for energy of maintenance. The maintenance coefficient is normally determined by studying the influence of the specific growth rate on the molar growth yield in chemostat cultures. A double reciprocal plot of estimated values of Y_{glu} against the experimental μ values yields a straight line whose intercept is the reciprocal of Y_{glu}^{max} and whose slope is the maintenance coefficient (49, 95). In a number of cases no straight line was obtained (51, 95, 96), a behavior that was shown to be due to an influence of μ on the fermentation pattern and on the ATP yield of the organism (51, 96). Therefore, Y_{glu} and Y_{glu}^{max} were replaced by Y_{ATP} and Y_{ATP}^{max} (the growth yield per mol of ATP corrected for energy of maintenance). A double reciprocal plot of Y_{ATP} against μ did indeed yield a straight line with $L.$ $casei$ (51) and $Sa.$ $cerevisiae$ (96). By multiplication of the equation by μ, equation 2 was obtained (49),

$$q_{ATP} = \frac{\mu}{Y_{ATP}} = \frac{\mu}{Y_{ATP}^{max}} + m_e \tag{2}$$

where q_{ATP} is the specific rate of ATP production (mol of ATP per g of dry weight per hr) and m_e is the maintenance coefficient (mol of ATP per g of dry weight per hr).

As an example, the results of glucose-limited cultures of $A.$ $aerogenes$ are shown in Figure 1 (97). The specific rates of glucose consumption and acetate production are linear functions of μ. From these rates q_{ATP} can be calculated by using equation 3 (49, 54, 97), and

$$q_{ATP} = 2 q_{glu} + q_{ace} - 0.0139 \mu \tag{3}$$

where q_{glu} and q_{ace} are the specific rates of glucose consumption and acetate production, respectively. Because both q_{glu} and q_{ace} are linear functions of μ, it follows that q_{ATP} is also a linear function of μ as required by theory (equation 2). From the results in Figure 1a it is possible to calculate Y_{glu} and Y_{ATP} as a function of μ. The result (Figure 1b) shows that Y_{glu} and Y_{ATP} are indeed dependent on μ.

It seems a logical idea that this is due to the utilization of a relatively larger part of the energy source to provide energy for maintenance purposes at smaller μ values. Therefore, for the amount of energy used independently of growth rate the term "maintenance coefficient" was introduced (95). It might be, however, that the amount of energy source converted independently of growth is used for more purposes than for the true maintenance processes defined above. It has been concluded that this is indeed so (97), and hence the use of the term a "maintenance coefficient" for the mathematical parameters m_s and m_e in equations 1 and 2 may be misleading. Another important point is that in equations 1 and 2 the maintenance energy requirement is assumed to be a constant at all growth rates. There is evidence that this is not valid for all growth conditions (9, 98, 99). These points will be discussed below under "Growth in the Presence of Excess Energy Source."

Until now Y_{ATP}^{max} and m_e have been determined only in very few cases. The results are shown in Table 10. Three important conclusions can be drawn from these data. First, the results with $A.$ $aerogenes$ show that both Y_{ATP}^{max} and m_e are dependent on the growth condition. The same conclusion has been reached for m_e from the results of aerobic yield studies with $E.$ $coli$ (100, 101). Much lower Y_{ATP}^{max} values are obtained for $A.$ $aerogenes$ with glucose-limited chemostat cultures than with tryptophan-limited cultures. Because the opposite is true for the m_e values, we still find Y_{ATP} values that do not differ very much at the maximum specific rate. From the data in Table 10, Y_{ATP} values of 12.2 and 10.1 are obtained for a specific growth rate of 0.65 for glucose and tryptophan-limited cultures, respectively. These values are very close to those determined for batch cultures of this organism (Tables 8 and 9). The very high m_e value for tryptophan-limited cultures was the reason for concluding that m_e also included an energy component that was required for processes other than those of true maintenance (97).

Second, in accordance with the theoretical calculations of the ATP requirement for the formation of cell material from various substrates (Tables 1 and 2), the Y_{ATP}^{max} value found for citrate is lower than that found for glucose. Third, the Y_{ATP}^{max} for $Lactobacillus$ $casei$ is very close to the theoretical amount of cell material that can be formed per mol of ATP. The same applies to the Y_{ATP}^{max} for the tryptophan-limited culture of $A.$ $aerogenes$. However, from the latter observation no conclusions may be drawn, because m_e in that case is large and it is not certain that m_e is the same for all growth rates (99). In all other cases the experimental Y_{ATP}^{max} values are much smaller than the amount of cell material that theoretically can be

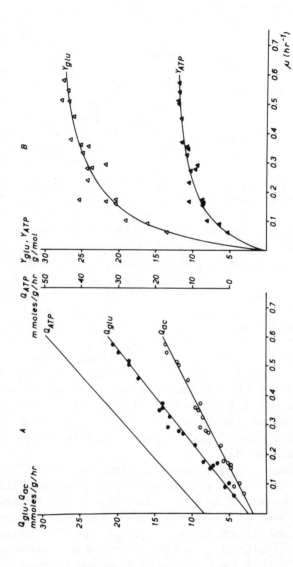

Figure 1.   Effect of the specific growth rate on the rates of glucose consumption and acetate and ATP production, and on the molar growth yields and Y_{ATP} by *Aerobacter aerogenes* during glucose-limited anaerobic growth. (A). Glucose consumption, ——●; acetate production, o——o; ATP production, ——. (B). Molar growth yield for glucose, ————△; Y_{ATP}, ▲——▲. Data from Stouthamer and Bettenhaussen (97). Reproduced from Stouthamer (192).

Table 10. Growth parameters of some organisms growing anaerobically in a chemostat

Organism	Growth-limiting factor	m_e	Y_{ATP}^{max}	References
Lactobacillus casei	glucose	1.5	24.3	(51)
Aerobacter aerogenes	glucose, minimal medium	6.8(5.4–8.0)	14.0(13.2–14.7)	(49, 52, 97)
	glucose, complex medium	2.3(0.4–4.9)	17.6(15.9–19.6)	
	tryptophan	38.7(33.9–41.7)	25.4(21.2–32.2)	
	citrate	2.2(0.4–3.9)	9.0(8.2–9.6)	(Stouthamer, unpub-lished results)
Escherichia coli	glucose	18.9(15.6–21.8)	10.3(8.9–10.7)	(100)
	glucose	6.9(5.1–8.7)	8.5(7.8–9.2)	(104)
Saccharomyces cerevisiae	glucose	0.5	11.0	(96)
	glucose	0.25	13.0	(105)
Saccharomyces cerevisiae (petite)	glucose	0.7	11.3	(105)
Candida parapsilosis	glucose	0.2	12.5	(105)

The numbers given in parentheses are the 95% confidence limits calculated by the model of de Kwaadsteniet, Jager, and Stouthamer (103).

formed per mol of ATP. Therefore the concept of maintenance energy offers no explanation per se for the discrepancy between theoretical and experimental growth yields and further explanation must be sought (97). The following equation for the relationship between theoretical and experimental Y_{ATP}^{max} values has been proposed (52):

$$q_{ATP} = \frac{\mu}{Y_{ATP}^{max}} + m_e = \frac{\mu}{(Y_{ATP}^{max})_{theor}} + m_g \cdot \mu + m_e \qquad (4)$$

where $(Y_{ATP}^{max})_{theor}$ is the theoretical Y_{ATP}^{max} value and m_g is the growth rate–dependent energy requirement for purposes other than formation of new cell material. It has generally been concluded that in aerobic yield studies there is also a discrepancy between experimental and theoretical growth yields (9, 101, 102). This is discussed in the next section.

Part of the results in Table 10 have been calculated with the aid of a stochastic model for heterotrophic growth of bacteria (103). This model allows the determination of the 95% confidence limits of Y_{ATP}^{max} and m_e and of Y_{ATP} at various μ values. The results show that the confidence limits of Y_{ATP}^{max} and m_e are rather large, whereas the interval estimates for Y_{ATP} are fairly precise. In many papers a statistical analysis of the data obtained from continuous culture experiments is either completely absent or quite insufficient. This is a serious point because the results (Table 10) show that the accuracy of these experiments generally has been overestimated.

Aerobic Cultures

In the presence of external hydrogen acceptors the energy source may be degraded further than under anaerobic conditions. Concomitantly, the ATP yield in the presence of hydrogen acceptors will be larger than in its absence. However, the total ATP yield during growth in the presence of hydrogen acceptors cannot be determined easily, because the P:$2e^-$ ratio for phosphorylation coupled to electron transport to the hydrogen acceptor is generally unknown. Therefore, for the interpretation of aerobic yield studies it is necessary to know either Y_{ATP} or Y_{ATP}^{max} to determine the P:$2e^-$ ratio, or conversely the P:$2e^-$ ratio must be known to determine Y_{ATP}^{max}. During aerobic growth the total ATP production may be given by equation 5:

$$q_{ATP} = \alpha (1-\beta) q_{sub} + q_{O_2} \cdot 2 \, P/O \qquad (5)$$

where α is the number of ATP molecules formed by substrate phosphorylation during the complete oxidation of the substrate, β is the part of the substrate that is assimilated, and q_{sub} and q_{O_2} are the specific rates of substrate or oxygen consumption, respectively. Insufficient information is obtained from an aerobic chemostat experiment to allow the determination of Y_{ATP}^{max}, m_e, and, P:O ratio with equations 2 and 5.

For aerobic chemostat experiments generally two parameters are given:

Y_{sub}^{max} (determined with equation 1) and $Y_{O_2}^{max}$. The latter is determined with equation 6:

$$\frac{1}{Y_{O_2}} = \frac{1}{Y_{O_2}^{max}} + \frac{m_O}{\mu} \tag{6}$$

where Y_{O_2} is the molar growth yield per mol of oxygen taken up during growth, m_O is the maintenance respiration rate (mol of oxygen per g of dry weight per hr) and $Y_{O_2}^{max}$ is the molar growth yield per mol of oxygen corrected for maintenance respiration. Y_{sub}^{max} and $Y_{O_2}^{max}$ values for chemostat cultures of a number of organisms growing on various substrates have been listed by Jones (106) and Stouthamer (9).

It must be stressed that Y_{sub}^{max} and $Y_{O_2}^{max}$ are not independent parameters, as illustrated in the data presented in Tables 6 and 7. This is further illustrated in Table 11 for gluconate-limited growth of *Paracoccus denitrificans* with nitrate as nitrogen source. (The results in the table have been taken from unpublished work by van Verseveld and Stouthamer.) The method for the calculation is exactly the same as outlined in Table 6. From the growth equation and Y_{glu}^{max} the O_2 uptake, CO_2 evolution, and $Y_{O_2}^{max}$ can be calculated and thus can be compared with the experimental values.

Table 11.   Calculation of growth parameters for gluconate-limited growth of *Paracoccus denitrificans* with nitrate as nitrogen source[a]

The growth equation is:

$$C_6H_{12}O_7 + 1.4\ HNO_3 + 6.85\ ``H_2" \rightarrow C_6H_{10.84}N_{1.14}O_{3.07} + 8.13\ H_2O$$

$Y_{glu}^{max} = 69.2$. Dissimilated: 54.4%; Assimilated 45.6%.

	Calculated	Observed
O_2 uptake (mol/mol gluconate)	1.43	1.52
CO_2 evolution (mol/mol gluconate)	3.26	3.03
$Y_{O_2}^{max}$	48.4	45.5

	Sites for oxidative phosphorylation	
ATP formation	2 sites	3 sites
1.77 $NADH_2$/mol gluconate	3.54	5.31
1.09 $FADH_2$/mol gluconate	1.09	2.18
Substrate-level phosphorylation	1.63	1.63
Total ATP formation	6.26	9.12
Y_{ATP}^{max}	11.0	7.6

[a] van Verseveld and Stouthamer, unpublished results.

It is evident that there is a very good agreement between the calculated and observed parameters. Assuming an unbranched respiratory chain (107) and 2 or 3 phosphorylation sites, the total ATP production can be calculated. In this way we arrive at Y_{ATP}^{max} values of 11.0 and 7.6, respectively, when 2 or 3 sites are present. This calculation illustrates the close relationship that exists between Y_{sub}^{max}, $Y_{O_2}^{max}$, and Y_{ATP}^{max}. Unfortunately this relationship between these parameters is absent in many of the published experimental yield data, which means either that not enough attention was given to obtaining a good mass balance or that the authors did not realize that in the experimental determination of these parameters large errors can be made.

From the data in Table 11 it is evident that without additional information it is not possible to make a choice between the two possibilities for Y_{ATP}^{max}. It is evident that other information is necessary to allow a choice between the two possibilities listed in Table 8. A number of possibilities enabling the determination of Y_{ATP}^{max} and the P:O ratio to be made from aerobic yield studies have been listed by Stouthamer (9).

The efficiency of oxidative phosphorylation has been a matter of great uncertainty for a great many years. However, some clarity is gradually starting to appear. There is a large variety in the composition of the respiratory chain and concomitantly in the efficiency of oxidative phosphorylation among various microorganisms (106, Jones, this volume). Furthermore, the number of phosphorylation sites is influenced by the growth condition. There is agreement now on the presence of 2 phosphorylation sites in *E. coli* (101, 108, 109, 110), in *A. aerogenes* (9, 97), *K. pneumoniae* (111), and *Bacillus megaterium* (112). Furthermore, there is agreement on the point that in some cytochrome *c*-containing bacteria (e.g., *Alcaligenes faecalis* and *Pa. denitrificans*) 3 sites of oxidative phosphorylation may be present (106). However, in *Pa. denitrificans* there is still disagreement about whether or not these 3 phosphorylation sites are expressed under all growth conditions.

It has been demonstrated that in heterotrophic cells phosphorylation site III is absent (113, 114, 115). This observation is in agreement with the results of aerobic yield studies (115, 116). With cell-free extracts (114) the presence of site III phosphorylation could be detected only from cells grown autotrophically. Yield studies with methanol-limited chemostat cultures [methanol in *Pa. denitrificans* is assimilated by the Calvin cycle (117)] similarly indicate that site III phosphorylation occurs in these cells (118). On the other hand, much higher growth yields for the same organisms have been obtained by Edwards, Spode, and Jones (119) and by Lawford (120). These authors concluded that 3 phosphorylation sites are present even in cells grown heterotrophically. These contradictory results might be reconciled by the possibility that the presence of phosphorylation site III is strongly dependent on the cultivation method.

With the available data on the efficiency of oxidative phosphorylation, it is possible to calculate Y_{ATP}^{max} values for aerobic chemostat cultures of the above-mentioned organisms. Y_{ATP}^{max} values calculated in this way are listed in Table 12. These data fortify the conclusions drawn from anaerobic yield studies: first, the experimental Y_{ATP}^{max} is smaller than the theoretical; and second, Y_{ATP}^{max} for growth on simple compounds is smaller than for growth on sugars. The latter conclusion had been drawn earlier from the measurement of Y_O values for batch cultures (6, 7).

POSSIBLE REASONS FOR THE DISCREPANCY BETWEEN EXPERIMENTAL AND THEORETICAL GROWTH YIELDS

Experimental Determination of the Amount of ATP Required for Membrane Energization

Although definitive agreement on the mechanism of oxidative phosphorylation has not yet been reached, most workers favor the idea of energy transmission by transmembrane gradients as proposed by Mitchell (14, 121). This leads to the scheme for energy generation and utilization shown in Figure 2. The energized membrane is the driving force for ATP synthesis, transport processes, and transhydrogenation (13, 14, 121). Under aerobic conditions the membrane is energized by respiration. During anaerobic growth, however, ATP is used for membrane energization. For *E. coli*, ATPase-negative mutants have been described (122, 123). In these mutants the membrane can be energized by respiration, but, due to the absence of the ATPase, the membrane potential cannot drive ATP synthesis. This offers an explanation for the absence of anaerobic growth of ATPase-negative mutants (124). Furthermore, the mutants do not show aerobic growth with substrates such as lactate and succinate, which do not yield

Table 12. Y_{ATP}^{max} values calculated for some organisms growing aerobically on various substrates[a]

Organism	Substrate	Y_{ATP}^{max}	References
Aerobacter aerogenes	glucose	12.4	(9)
Bacillus megaterium D440	glycerol	12.7	(112)
Bacillus megaterium M	glycerol	10.8	(112)
Paracoccus denitrificans	gluconate	9.9	(116)
	succinate	9.1	
Escherichia coli	glucose	13.9	(101)
	glycerol	12.7	(101)
	fumarate	10.1	(101)
	DL-lactate	9.5	(101)
	pyruvate	8.6	(101)
	acetate	7.1	(101)

[a] The calculation is based on the presence of 2 phosphorylation sites in the respiratory chain of these organisms.

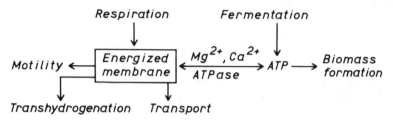

Figure 2. Scheme for the generation and utilization of ATP.

ATP by substrate-level phosphorylation. The molar growth yields for aerobic batch cultures of ATPase-negative mutants are lower than those for the wild-type (122, 125). A comparison of the growth parameters of glucose-limited anaerobic cultures of wild-type *E. coli* and of glucose-limited aerobic cultures of its ATPase-negative mutant has been performed by Stouthamer and Bettenhaussen (104). The results are shown in Table 13. The Y_{glu}^{max} value for aerobic growth of the mutant is much larger than that for anaerobic growth of the wild type. For aerobic growth of the wild type, a Y_{glu}^{max} value of 63.4 has been reported (100). The m_e values for the two growth conditions are about the same. It is obvious that the difference in Y_{ATP}^{max} between anaerobic cultures of wild-type *E. coli* and aerobic cultures of an ATPase-negative mutant is due to the fact that in the mutant no ATP has to be used for membrane energization. From the results in Table 13 it can be calculated that, in the mutant under aerobic conditions and in the wild type under anaerobic conditions, 57 mmol and 117 mmol of ATP, respectively, are required for the formation of 1 g of biomass. If we assume that under both conditions the efficiency of biomass formation is the same, the amount of ATP used for membrane energization during anaerobic growth of the wild type is 61 (117 − 57) mmol of ATP per g of biomass. Consequently, under anaerobic conditions 51% of the total energy production is used for membrane energization.

Table 13. Growth parameters for glucose-limited anaerobic cultures of wild-type *Escherichia coli* and for glucose-limited aerobic cultures of its ATPase-negative mutant[a]

| | Organism | |
Growth parameter	Wild type	ATPase negative
Y_{ATP}^{max}	8.5(7.8–9.2)	17.6(16.5–20.3)
Y_{glu}^{max}	20.8(18.9–23.0)	43.8(40.0–48.6)
m_e	6.9(5.1–8.7)	5.1(3.6–6.7)

[a]The data given are the point estimates and their 95% confidence limits calculated with the model of de Kwaadsteniet et al. (103). Data of Stouthamer and Bettenhaussen (104).

In the theoretical calculations of the amount of ATP required for biomass formation (Table 1), it was estimated that during growth in a glucose–mineral salts medium about 15% of the total ATP was required to drive transport processes (3). It is evident that the amount of ATP needed for membrane energization is much larger than this estimate. Two reasons may be mentioned for this difference. First, accumulated metabolites may leak away from the cell to the medium and may have to be taken up again. Second, in the chemiosmotic theory (13, 14, 121) the membrane is thought to be fully impermeable to protons. However, this is by no means the case, which is clearly shown by the determination of H^+:O ratios by oxygen pulse experiments. Because of the leakiness of the membrane to protons, the membrane potential will be lowered and this causes a higher energy requirement to maintain the energized state of the membrane. This second possibility seems a very important factor. The leakiness of the membrane to protons is discussed in more detail under "Dissipation of the Energized Membrane State."

The high energy requirement for membrane energization explains in large part the discrepancy between experimental and theoretical growth yields discussed in the previous sections. However, these results also have important consequences for the interpretation of aerobic growth yields. From the measurement of aerobic growth yields, a P:O ratio of 1.46 was calculated for wild-type *E. coli* (104). During the complete oxidation of glucose we expect an overall P:O ratio of 1.83 for complete oxidation of glucose when 2 phosphorylation sites are present (see "Aerobic Cultures," above). The agreement between these two values is reasonable. Assuming that the same amount of energy is needed for membrane energization under aerobic and anaerobic conditions, it was calculated that at $\mu = 0.38$ h^{-1}, q_{ATP} was 51.6 mmol of ATP per g of dry weight per h, of which the equivalent of 24.1 mmol of ATP per g of dry weight per h was used for membrane energization (104). Consequently, the actual ATP production was only 27.1 mmol of ATP per g of dry weight per h, of which 11.8 were produced by substrate-level phosphorylation and 15.3 by oxidative phosphorylation. Because the q_{O_2} in this experiment was 13.7 mmol of O_2 per g of dry weight per h, the actual P:O ratio was therefore only 0.56, whereas when the amount of energy needed for membrane energization was included the earlier mentioned value of 1.46 resulted. From these data it was concluded that the main task of respiration in *E. coli* is energization of the membrane (104). It is tempting to speculate that this conclusion also offers an explanation for the low P:O ratios that are generally observed for membrane preparations of bacteria and in whole cells (13, 28, 126). In such determinations only the actual ATP produced is included in the calculation of the P:O ratio. Because the main task of respiration is membrane energization, *E. coli* might not be competent to convert the membrane potential fully into ATP.

It has been shown that the relative growth yield in batch cultures of an ATPase-negative mutant of *B. megaterium* is approximately 67% of that of the wild type (127). This indicates that the conclusions reached above for *E. coli* also apply to *B. megaterium*.

Energy Required for Motility

It has recently been shown that the energized membrane is the driving force for motility (128). This has been included in Figure 2. For chemotaxis, in addition, ATP was supposed to be required. The influence of energy expenditure in motility and chemotaxis cannot be estimated accurately at the moment. However, a number of observations indicate that the amount of energy used for these purposes is very small (J. W. Drozd and D. Koshland, personal communications).

ENERGETIC COUPLING AND ENERGY-SPILLING MECHANISMS

Uncoupled Growth

Under a number of growth conditions the growth yield is much lower than expected on basis of the ATP yield. Senez (129) concluded that under these conditions growth and energy production are uncoupled and he introduced the term "uncoupled growth." It has been concluded that under these growth conditions there is a discrepancy between the rate of ATP production by catabolism and the rate of ATP utilization by anabolism. The capacities of bacteria to regulate cellular processes are evidently insufficient to regulate the rate of catabolism exactly to the needs of anabolism. That the rate of ATP production by catabolism is higher than the rate of ATP consumption by anabolism implies the existence of energy-spilling reactions. [Recently, Neijssel and Tempest suggested that the energy-spilling mechanisms be called "slip reactions" (130). The term "uncoupled growth" was considered by these authors to be misleading because it would imply that under other conditions growth is tightly coupled to the generation of energy in the cell. In my opinion both terms have a similar meaning, and because the term uncoupled growth has been used in the literature for a long time I will use the older term in this review.] During uncoupled growth the growth rate decreases and consequently a larger proportion of the energy source is used for maintenance purposes. However, it may be concluded from the following results that, in most cases, a change in the magnitude of m_e and/or of m_g in equation 4 must also be envisaged. In the next section the growth conditions under which uncoupled growth occurs are described. Following that, the possible mechanisms for the spilling of ATP are discussed.

Growth Conditions Giving Uncoupled Growth

The conditions giving uncoupled growth have been reviewed by Stouthamer (9, 28). The following conditions can be mentioned.

    Growth in the Presence of Excess Energy Source In tryptophan-limited cultures of *S. faecalis* in rich media, the specific rate of glucose consumption was found to be independent of μ (131). This implies that in this case Y_{glu} is a linear function of μ. Similarly, the maintenance coefficient of tryptophan-limited cultures of *A. aerogenes* growing anaerobically with glucose as carbon and energy source was very high (49). Similar results have been reported by Neijssel and Tempest (99) for aerobic cultures of *A. aerogenes*. In Figure 3 the q_{O_2} of carbon-, ammonia-, sulfate-, and phosphate-limited chemostat cultures of *A. aerogenes* growing on glucose is plotted versus μ. For the phosphate-limited culture the maintenance respiration rate is great and for the carbon-limited culture it is small. An important conclusion from the results in Figure 3 is that the q_{O_2} at the maximal specific growth rate has about the same value for the various cultures (99). This led to the opinion that in these cases the maintenance energy requirement is not independent of the specific growth rate. At the maximum specific growth rate the maintenance energy requirement is considered to be the same under the various growth conditions. This is a very logical conclusion: if we assume a constant maintenance energy requirement for all growth rates, the remaining energy would be used for growth with a greatly increased and even impossible efficiency (9, 99). In *E. coli* the maintenance energy requirement was also highly dependent on the nature

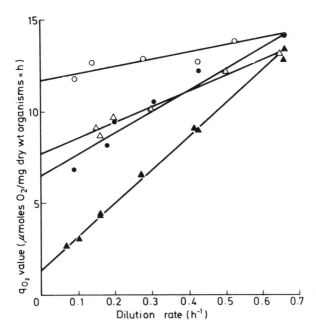

Figure 3. Effect of the specific growth rate on the rate of oxygen uptake by *Aerobacter aerogenes* during aerobic growth in the chemostat with various limiting factors. Glucose-limited, ▲———▲; ammonia-limited, ●———●; sulfate-limited, △———△; and phosphate-limited, ○———○. Reproduced from Neijssel and Tempest (99).

of the growth-limiting factor in the chemostat cultures (101). However, in *Pa. denitrificans* the maintenance energy requirement was found to be the same in chemostat cultures either limited by the carbon/energy source or limited by sulfate (116, 132). Thus in this aspect the reaction of *Pa. denitrificans* in an environment with excess carbon source is different from that of *S. faecalis, A. aerogenes,* and *E. coli.*

In *Azotobacter* spp. nitrogenase is protected against oxygen damage by a mechanism called respiratory protection (133). When batch cultures, subjected to restricted aeration, are exposed to highly aerobic conditions, there is a large increase in respiratory activity (134, 135). This is accompanied by a sharp decrease in the efficiency of energy conservation, as discussed in "Possible Mechanisms for Energy-Spilling Reactions," below. Consequently there is a very strong difference between glucose- and oxygen-limited chemostat cultures of *Az. vinelandii* growing with molecular nitrogen as nitrogen source. The results of Nagai and Aiba (98) are shown in Figure 4a and b. In the oxygen-limited cultures the rates of glucose and oxygen consumption are normal linear functions of μ. However, in glucose-

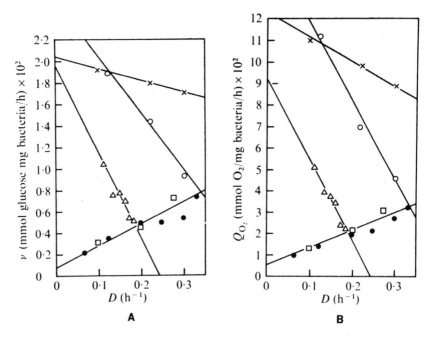

Figure 4. Specific rates of glucose consumption (A) and oxygen uptake (B) against dilution rate in glucose-limited and oxygen-limited chemostat cultures of *Azotobacter vinelandii*. (A). Rate of glucose consumption. Oxygen-limited, agitation speed in fermentor vessel at 240 rev/min, •———•; oxygen-limited at 340 rev/min, □———□; glucose-limited at 540 rev/min, △———△; glucose-limited at 780 rev/min, o———o; glucose-limited at 780 rev/min, x———x. For •, o, □, △, glucose concentration 45 mM; for x, glucose concentration 27 mM. (B). Rate of oxygen uptake: •, □, oxygen-limited; △, o, x, glucose-limited. Other details as in A. Reproduced from Nagai and Aiba (98).

limited cultures very high values for m_s (Figure 4a) and m_O (Figure 4b) are found and, in addition, negative values of Y_{glu}^{max} and $Y_{O_2}^{max}$ are obtained. Under the experimental conditions an increase in the specific growth rate led to a sharp increase in the dissolved oxygen tension. The authors proposed that the amount of energy-uncoupled growth was proportional to the dissolved oxygen tension. This means that the maintenance coefficient was supposed to be inversely related to the growth rate. The effects illustrated in Figure 4 are the most pronounced example of uncoupled growth that can be found in the literature. It is evident that, because nitrogen fixation in this case is limiting growth, the glucose-limited cultures must be regarded as growing with excess glucose.

In the chemostat cultures with excess carbon source the growth yield at low μ values is very low. Therefore the high maintenance coefficient is an indication of uncoupling between growth and energy production. As mentioned, the measured maintenance coefficient gives no indication whatsoever of the energy requirement of true maintenance processes.

The same phenomena described above for chemostat cultures have been observed in batch cultures. Some results are summarized in Table 14. In Z. mobilis growth with suboptimal concentrations of pantothenate results in a simultaneous decrease of both Y_{ATP} and μ (50). Similarly, increasing the number of vitamins present in the growth medium of *Saccharomyces carlsbergensis* and some other yeasts resulted in a simultaneous increase in Y_{ATP} and μ (136). These results are partly explained by a relative increase in the amount of energy needed for maintenance purposes at lower growth rates. However, this effect is not sufficient to explain the drastic effects on Y_{ATP} by inadequate nutrition shown in Table 14.

Growth in the Presence of Growth-Inhibitory Compounds   Growth in the presence of growth-inhibitory compounds has been mentioned as one of the conditions under which uncoupling of growth and energy production occurs (6, 28). Some data are included in Table 15. Nitrite is in general a toxic compound. In *A. aerogenes,* in which the further reduction of nitrite is not associated with ATP formation, it gives a decrease in Y_{ATP} and in μ. A similar effect has been observed in *Cl. perfringens* (137). In *Pa. denitrificans,* in which the reduction of nitrite is associated with ATP formation, the accumulation of toxic amounts of nitrite were found to lead to a loss of phosphorylation at site I (132). The effect of nitrite on *Propionibacterium pentosaceum* (138), may be due to a decrease in Y_{ATP} and/or loss of phosphorylation at site I.

Very interesting results have been obtained by studying the effect of uncoupling agents on molar growth yields. Dinitrophenol leads to a decrease in the anaerobic Y_{ATP} of *Sa. cerevisiae* (139), and *Streptococcus agalactiae* (140). The effect of dinitrophenol on aerobic growth of *S. faecalis* and *S. agalactiae* has been studied by Mickelson (77, 140). In these organisms oxidative phosphorylation occurs, although cytochromes are absent. At high dinitrophenol concentrations the molar growth yields

Table 14.   Influence on Y_{ATP} and μ of growth with suboptimal concentrations of an essential nutrient or with various vitamins

Zymomonas mobilis (50)

Pathothenate concentration ($\mu g/l$)	Synthetic medium		Minimal Medium	
	Y_{ATP}	μ	Y_{ATP}	μ
5000	6.5	0.39	4.7	0.300
4	5.2	0.353	4.2	0.275
2	4.5	0.281	3.9	0.182
1	4.2	0.261	2.9	0.151
0.1	3	0.203	2.5	0.153
0.05	2.5	0.157	1.7	0.153

Saccharomyces cerevisiae (136)

Vitamins	Minimal medium		Complex medium	
	Y_{ATP}	μ	Y_{ATP}	μ
Biotin	4.7	0.01–0.02		
Biotin + pyridoxine	5.9	0.014–0.028		
Biotin + pyridoxine	6.5	0.028–0.046		
Nine vitamins	7.7	0.105	10.7	0.200

under aerobic conditions are even lower than the anaerobic molar growth yield in the absence of inhibitor. These effects are easily explained because these compounds dissipate the energized membrane state under both aerobic and anaerobic conditions.

The studies mentioned so far were all performed with batch cultures. The influence of 0.25- and 1-mM dinitrophenol on aerobic glucose-limited chemostat cultures of *A. aerogenes* has been studied by Neijssel (141). It was observed that both concentrations caused a drastic increase in the maintenance respiration rate. However, the $Y_{O_2}^{max}$ values of the cultures did not change, indicating that the maximum growth efficiency for oxygen was not changed. These data can be interpreted with the aid of the results mentioned above in connection with Figure 2. The presence of dinitrophenol causes a certain rate of dissipation of the energized membrane state. This rate of dissipation is compensated by an increased maintenance respiration rate. After the rate of dissipation of the energized membrane state has been compensated, the remaining substrate may be used for biomass formation with the same efficiency as that during growth in the absence of the uncoupler.

It has been shown that at low pH values acetate may have an effect similar to dinitrophenol. This may be due to the proton-conducting capacity of the undissociated acid. This effect has been studied quantitatively in

Table 15. Influence of the presence of growth-inhibitory compounds on molar growth yields of a number of organisms growing anaerobically on glucose[a]

Organisms	Inhibitor	Concentration	Y_{ATP}	Reference
Aerobacter				
aerogenes			10.2	(92)
	Nitrite	0.5 mM	9.3	
	Nitrite	1 mM	7.4	
	Nitrite	1.5 mM	6.7	
	Nitrite	10 mM	no growth	
Escherichia coli			12.7	(193)
	Ferricyanide	10 mM	10.3	
	Ferricyanide	40 mM	5.9	
Proteus mirabilis			5.5	(54)
	Thiosulphate	7.5 mM	3.7	
Saccharomyces				
cerevisiae			11.1	(139)
	Dinitrophenol	0.05 mM	8.3	
	Dinitrophenol	0.1 mM	6.0	
Streptococcus				
agalactiae			10.7	(140)
	Dinitrophenol	0.1 mM	7.9	
	Dinitrophenol	0.25 mM	5.5	
	Dicoumarol	0.01 mM	8.9	
	Dicoumarol	0.1 mM	2.8	
	CCCP[b]	0.001 mM	9.6	
	CCCP	0.005 mM	8.2	
	CCCP	0.010 mM	1.35	
	PCP[c]	0.001 mM	10.5	
	PCP	0.010 mM	4.4	
	PCP	0.050 mM	no growth	

[a] After Stouthamer (192).
[b] CCCP = Carbonylcyanide-*m*-chlorophenylhydrazone.
[c] PCP = Pentachlorophenol.

acetate- and ammonium-limited chemostat cultures of *Candida utilis* (142). It was shown that the Y_{O_2} values were lower at higher extracellular acetate concentration and at lower pH values at which the relative amount of undissociated acid is higher.

Growth at Unfavorable Temperatures   Growth at unfavorable temperatures was first studied as a means of uncoupling growth and energy production by Senez (129). A more extensive study with *S. faecalis* and *Z. mobilis* was performed by Forrest (84). It was observed that at temperatures remote from the optimum the growth yields fell markedly. In *Bacillus stearothermophilus* Y_{glu} was 88.0 g/mol and 54.0 g/mol at 43 and 59°C respectively (143). In continuous culture studies it has been shown that in *Pseudomonas fluorescens* (144), *A. aerogenes* (145), *E. coli* (146, 147), and *Sa. cerevisiae* (148) the maintenance coefficient is increased by cultivation at higher temperature. However, the increase in maintenance coefficient is not the only reason for the decreased growth yields. This is clearly illustrated by the results of Farmer and Jones (147), who found Y_{ATP}^{max} values of 12.7 and 8.4 at 30 and 40°C, respectively. In *Sa. cerevisiae* grown at 39°C the situation was even more complicated, because a large number of cells in the culture were nonviable but still consumed substrate (148). It was estimated that 9 to 15% of the glucose was consumed by nonviable cells. It is evident that this is an extreme case of uncoupling between growth and energy production.

Growth in Minimal Media   The results presented earlier have shown that for any one organism the molar growth yield has the highest value for growth in complex media. Furthermore, it was concluded that very high values for Y_{ATP} are obtained for organisms growing in complex media. Neijssel and Tempest (149) have shown that the Y_{O_2} values for growth of *A. aerogenes* in minimal medium with reduced compounds like mannitol and glycerol are lower than the value for growth with glucose, suggesting that not all of the extra energy that is released in oxidizing the additional reducing equivalents (contained in mannitol and glycerol) is coupled to biomass formation. Probably the influence of the complexity of the medium and the influence of the nature of the energy source in minimal medium are related phenomena and are both due to the limited capacity for assimilation during growth of an organism in minimal medium.

Growth During Transient Periods   The effect of sudden changes in the growth conditions in chemostat cultures has been reviewed by Harrison (150). The effect of changing an anaerobic, glucose-limited chemostat culture of *A. aerogenes* to aerobic conditions was studied by Harrison and Loveless (151). Immediately after the transition the respiratory activity strongly increased, because, in addition to the supplied glucose, the products accumulated during the anaerobic period were also oxidized. However, the Y_{glu} during the transient period was lower than during the aerobic steady state. It was concluded that this represented a certain degree of uncoupling between growth and energy-conserving mechanisms. Harrison and Maitra (152) studied the effect on the respiration rate and the ATP content of pulsing small amounts of glucose and succinate into glucose-limited chemostat cultures of *A. aerogenes*. An immediate increase in respiration rate was observed. With succinate a small increase in the cellular ATP content was observed, which was quickly corrected to near the original level. When the succinate was exhausted there was a temporary fall in ATP content, but again it rapidly returned to the original value. Evidently the ATP content in the cell is carefully maintained at a constant level by a rapid adjustment of the rate of the ATP-consuming reactions. Similar pulsed additions of glucose to glucose-limited cultures of *A. aerogenes* were recently performed by Neijssel and Tempest (130). These experiments show that, in cells that have a low respiration rate due to the limited supply of oxidizable substrate, a much higher potential respiration rate is maintained.

The effect of a sudden change from a high to a low dissolved oxygen tension in glucose-limited chemostat cultures of *A. aerogenes* was also studied by Harrison and Maitra (152). Such a change was found to induce an increase in respiration rate that was not accompanied by an elevated ATP content. Furthermore, under these conditions of increased respiration the yield coefficient of the cell was diminished (153). The existence of an increased ATP turnover rate not coupled with anabolic reactions but representing a wastage of ATP was assumed to occur.

Another conclusion from these data was that the respiration rate in growing *A. aerogenes* is not controlled by the cellular content of ADP (150). A similar conclusion was obtained for resting cells of *Proteus mirabilis* by van der Beek and Stouthamer (126). These authors followed the cellular ATP, ADP, and AMP content during respiration. It was observed that respiration continued after the ATP content had risen to a maximum and the AMP content to a minimum. A rapid turnover of ATP was assumed to occur during that period. A similar observation has been made for resting cells of *Az. vinelandii* by Knowles and Smith (154).

Possible Mechanisms for Energy-Spilling Reactions

In the previous sections it has been mentioned several times that under a number of growth conditions a discrepancy exists between the rate of ATP production by catabolism and the rate of ATP consumption by anabolism. Under these conditions an adjustment of the rate of ATP production is necessary. Several possibilities exist:

1. Formation of storage compounds (27). Under these conditions growth yields are high in contrast to the experimental observations for conditions in which uncoupled growth occurs.
2. Excretion of products ["overflow metabolism" (149)]. However, this possibility does not seem to exist in all organisms.
3. Deletion of sites of oxidative phosphorylation. This has been demonstrated for sulfate-limited cultures of *E. coli* (108, 155), *A. aerogenes* (97), *Ca. utilis* (156), and *Pa. denitrificans* (116). Similarly, site I phosphorylation was shown to be lost during anaerobic growth of *Pa. denitrificans* with nitrate as hydrogen acceptor, especially when toxic concentrations of nitrite accumulated (132). In cultures of *Az. vinelandii* growing with molecular nitrogen as nitrogen source with a high dissolved oxygen tension, a sharp decrease in the efficiency of energy conservation at site I was observed by Jones et al. (135).
4. Branching of the respiratory chain. The respiratory chain of *Az. vinelandii* has a highly branched structure (157) and the cytochrome $b \rightarrow d$ pathway is nonphosphorylating (158, 159). In nitrogen-fixing cultures under high aeration the cytochrome d content is highly increased (134, 135) and therefore a larger part of the electrons follow the nonphosphorylating branch.
5. Energy-spilling mechanisms. Under a number of conditions such mecha-

nisms are the only possibility for the adjustment of the ATP production to the needs for anabolism. When energy-spilling mechanisms are involved in this adjustment the molar growth yields are generally very low.

Possible energy-spilling reactions are most easily discussed on the basis of Figure 2. From this figure, three ways may be envisaged by which energy could be wasted: 1) dissipation of the energized membrane state; 2) ATP wastage; and 3) wastage of ATP in the formation of biomass. These possibilities are discussed below.

Dissipation of the Energized Membrane State Dissipation of the energized membrane state may occur by an increase in the proton-conducting capacity of the membrane and by futile ion cycles. There is indeed evidence that bacteria have the ability to regulate the proton-conducting capacity of the membrane. In ATPase-negative mutants the permeability of the membrane to protons is increased (for a review see ref. 123). However, the $Y_{O_2}^{max}$ value of ATPase-negative mutants of *E. coli* is lower than that for wild-type *E. coli* (104). Consequently, in the mutants the oxygen uptake, and therefore also the number of protons extruded per mol of glucose consumed, are higher than in the wild type. Because in the mutants the energized membrane state is only used for a number of energy-requiring processes (see "Experimental Determination of the Amount of ATP Required for Membrane Energization," above), it is evident that in these mutants the permeability of the membrane to protons must be higher than in the wild type. After growth of these mutants under anaerobic conditions with nitrate, the permeability of the membrane to protons is much lower than after aerobic growth and is approximately the same as that in the wild-type cells (160, 161). The H^+:O and H^+:nitrate ratios for *E. coli* are both 4 (111). However, the number of oxygen atoms taken up per mol of glucose is much higher than the number of nitrate molecules consumed, because under anaerobic conditions the citric acid cycle does not function (for a review on nitrate reduction see ref. 162). Consequently the number of protons extruded per mol of glucose is smaller under anaerobic conditions with nitrate than under aerobic conditions. These considerations give an explanation for the observation that, after anaerobic growth of an ATPase-negative mutant with nitrate, the permeability of the membrane to protons is lower than after aerobic growth, and they also indicate that the permeability of the membrane to protons can be adjusted to maintain the proton-motive force of the cell at the right level. This adjustment is probably accomplished by changes in the membrane part of the ATPase, which is generally called the BF_O component (123).

It is not known whether or not in bacterial membranes proton-conducting channels also occur that can be switched off by certain nucleotides (e.g., GDP), as has been demonstrated for the inner membrane of mitochondria in brown fat adipose tissue (163).

As an example of a futile ion cycle, the uptake of ammonium ions driven by the energized membrane state, followed by dissociation of the ammonium ion into $NH_3 + H^+$ inside the cell and movement of the NH_3 through the membrane back to the medium, may be mentioned. Such a cycle has been proposed to explain the uncoupling effect of ammonium on mitochondria and chloroplasts. This cycle might be the explanation for the large increase in maintenance coefficient that is observed in anaerobic glucose-limited chemostat cultures of *A. aerogenes* when the concentration of NH_4Cl in the medium is increased (49). Whether or not futile cycles for other ions are possible in bacteria is not yet known.

Wastage of ATP   ATP degradation by an ATPase as a way for the adjustment of a too-high rate of ATP production has been suggested already by Gunsalus and Shuster (1) and by Senez (129). Some evidence for the degradation of ATP by an ATPase during growth of *Z. mobilis* in a medium with suboptimal amounts of pantothenate has been presented (164) and strong evidence for the involvement of ATPase in the spilling of energy has been obtained from a study of the effect of high hydrostatic pressure on the efficiency of growth of *S. faecalis* (165). It was found that Y_{ATP} for growth at a hydrostatic pressure of 408 atm was considerably lower than for growth at 1 atm. Pressure was found to stimulate the ATPase activity. Furthermore, the intracellular ATP level for cells growing under high pressure was lower than that for cells growing at normal pressure. N,N'-dicyclohexycarbodiimide (DCCD), an inhibitor of ATPase activity, improved the growth efficiency under pressure. Therefore it was concluded that the effects of pressure may be interpreted largely in terms of ATP supply and demand. In the author's work with the ATPase-negative mutant of *E. coli* (see "Anaerobic Batch Cultures," above), the same maintenance coefficient was observed for glucose- and ammonium-limited chemostat cultures (104), whereas a large difference was observed for similar cultures of wild-type *E. coli* (101). These observations indicate that the presence of ATPase is essential to obtain the increase in maintenance respiration rate under growth conditions in which factors other than the carbon/energy source are the growth-limiting factor.

It has been concluded that there is a fundamental difference between the membrane-bound ATPase of facultative and aerobic bacteria (166). Generally the ATPase activity in a cell-free extract is stimulated by treatment with trypsin, which is due to the degradation of an inhibitor protein. This inhibitor protein is identical with the smallest subunit (ϵ) of the ATPase molecule (167, 168). This inhibitor is a natural regulator of the ATPase complex in the hydrolytic direction (169). In aerobic organisms like *Az. vinelandii* (170), *Micrococcus lysodeikticus* (171), and *Myobacterium phlei* (172) the trypsin stimulation is very large. In *E. coli* the stimulation is small, (173, 174), and no stimulation is seen in *S. faecalis* (175). Obligate aerobes probably do not use the ATPase in the hydrolytic direction, whereas facultative and strictly anaerobic organisms use ATPase hydrolytically for the generation of the energized membrane state (see Figure 2). Whether or not in strictly aerobic organisms ATP degradation by an ATPase can be one of the possible energy-spilling mechanisms under conditions of un-coupled growth is an interesting question. In this aspect it is important to mention that Bhattacharyya and Barnes (176) have shown that a twentyfold activation of ATPase in membrane vesicles of *Az. vinelandii* can be induced by substrate oxidation. This indicates that in this organism the inhibition by the inhibitor can be changed by energization of the membrane. In *Az. vinelandii* a large uncoupling between growth and energy production can occur (98), and possibly the mechanism outlined above can play a role in the induction of a change in the ATPase to make ATP degradation possi-ble. Whether or not such changes also occur in other aerobic organisms is unknown at the moment.

When ATP degradation by a membrane-bound ATPase takes place during uncoupled growth it is evident that, in addition, dissipation of the energized membrane state must take place. In most aerobic organisms dissipation of the energized membrane state alone seems to offer sufficient possibilities for the adjustment of the rate of energy production to the needs of anabolism.

A very interesting possibility for wastage of ATP has been described for photosynthetic bacteria. Purple bacteria show a vigorous production of H_2 during photosynthetic growth on organic compounds and with an amino acid as nitrogen source. This photoproduction of H_2 is catalyzed by nitro-genase (177). Thus the ATP-dependent H_2-producing activity of nitrogenase can serve as an ATP-spilling reaction under conditions where ATP is gen-erated by catabolism at rates that exceed biosynthetic demands (178, 179).

Wastage of ATP in the Formation of Biomass   In the formation of biomass two possibilities for ATP wastage may be considered: the formation of guanosine tetra- and pentaphosphates and futile cycles. It is a well known fact that bacteria under nutritional stress (deprivation of an essential amino acid or exhaustion of the carbon source) accumulate guanosine 5'-diphosphate 3'-diphosphate (ppGpp) and guanosine 5'-triphosphate 3'-diphosphate (pppGpp) (for a review see ref. 180). The formation of these compounds requires ribosomes, messenger RNA, and uncharged transfer RNA. They are formed by a ribosomal idling reaction due to limited availability of aminoacyl-tRNA. It has recently become clear that pppGpp is a precursor of ppGpp (181, 182, 183). The concentration of ppGpp in the cell can become nearly as high as that of ATP, and furthermore both ppGpp and pppGpp show a rapid turnover, especially when the energy source is in excess. These compounds may be regarded as having a signal function, and they slow down many anabolic processes, among which is the synthesis of ribosomal RNA (for a review see ref. 184). It is evident that ppGpp and pppGpp may be of great importance for the degradation of ATP, especially because growth in chemostat cultures is in all cases limited by the availability of an essential nutrient. The quantitative contribution of the formation and degradation of these compounds to the degradation of

ATP cannot, however, be evaluated from data in the literature at this moment.

The possibility that futile cycles play a role in the wastage of ATP under conditions of uncoupled growth has been put forward several times. However, the experimental evidence for the occurrence of futile cycles in bacteria is very scarce. Lagunas (102) has considered the formation of trehalose from sucrose followed by degradation of the trehalose in *Sa. cerevisiae* as a possible ATP-consuming futile cycle. ATP degradation by a cycle of glutamine formation by glutamine synthetase and glutamine degradation by glutaminase has been considered by Neijssel and Tempest (130). The importance of such cycles as energy-spilling reactions under conditions of uncoupled growth is not known at the moment.

SUMMARY AND DISCUSSION

From the previous sections it is clear that many factors influence the molar growth yield of a microorganism on a given growth substrate. These factors are:

1. The pathway of substrate breakdown, which determines the ATP yield per mol of substrate.
2. The nature of the carbon source (see "Influence of the Carbon Source and Complexity of the Medium").
3. The complexity of the medium, which determines the monomers that must be synthesized by the organism (see "Influence of the Carbon Source and Complexity of the Medium" and "Anaerobic Batch Cultures").
4. The nature of the assimilation pathway of the carbon source (see "Influence of the Carbon Assimilation Pathway of the Growth Substrate").
5. The nature of the nitrogen source (see "Influence of the Nitrogen Source").
6. The macromolecular composition of the microbial cells.
7. The maintenance coefficient (m_e, see "Anaerobic Chemostat Cultures" and "Aerobic Cultures").
8. The specific growth rate, which determines (together with the maintenance coefficient) the relative amount of the energy source that is used for maintenance.
9. The occurrence of energy-requiring processes other than the formation of new cell material that are related to the growth rate (M_g, see "Anaerobic Chemostat Cultures").
10. The occurrence of anaerobic electron transport during fermentation.
11. The availability of and possibility of using external hydrogen acceptors.
12. The P:$2e^-$ ratio for phosphorylation coupled to electron transfer.

The interplay of these factors can sometimes result in profound differences between molar growth yields for the same organism growing under different environmental conditions or for different organisms growing under the same conditions.

In general it can be stated that the effects predicted by the theoretical calculations on the ATP requirement for the formation of biomass under various cultivation conditions are all confirmed by experimental observations. However, large differences between theoretical and experimental molar growth yields are generally observed in both anaerobic and aerobic experiments. Only in experiments in which growth is studied in complex media are the experimental molar growth yields close to the theoretical ones. Addition of preformed monomers to glucose minimal medium should not, on a theoretical basis, lead to large differences in the ATP requirement for growth in minimal, as compared to complex, media. It might be considered that the discrepancy between theoretical and experimental growth yields is due to the fact that in the theoretical calculations the ATP requirement for some processes has been underestimated. It may be mentioned, however, that this discrepancy is not found with growing maize embryos (185) or with the green alga *Scenedesmus protuberans* (186), where the same or very similar calculations were used. Therefore the above mentioned possibility seems less likely. Two factors are considered for the explanation of the discrepancy between theoretical and experimental molar growth yields: 1) the existence of a large energy requirement to maintain the membrane in the energized state, which may be due to leakiness of the membrane to protons (this factor partially explains the difference between theoretical and experimental growth yields); and 2) the occurrence of uncoupling between growth and energy production because, under a large number of conditions, there is a discrepancy between the rate of ATP production by catabolism and the rate at which ATP can be consumed by anabolism. It is clear that in this aspect bacteria fall short in regulating catabolic processes exactly to the needs of anabolism.

It is worthwhile to mention again that microorganisms growing at low specific growth rates in the chemostat with the carbon/energy source as growth-limiting factors possess a much higher potential rate of substrate consumption and oxygen uptake than is actually expressed in the culture. Similarly, the potential rate of protein synthesis is much higher than the actual rate in such cultures (187). It might be that this is of importance to the organism because in natural environments growth is mostly limited by lack of some nutrients and the rate at which those nutrients become available is not constant. The possession of a high potential rate of catabolism could make possible a more rapid adaptation to any sudden availability of nutrients, and this could be an advantage in the competition between organisms for the limited and sporadic supply of nutrients. In this aspect it is worth mentioning that in chemostat cultures of *A. aerogenes* growing with histidine as energy source, cells growing at $\mu = 0.15$ h^{-1} can start growing immediately at $\mu = 0.42$ h^{-1} when the dilution rate is increased

(188). Similar results have been reported by Mateles, Riju, and Yasadu (189) and by Koch and Deppe (190). Uncoupled growth takes place during:

1. Growth in the presence of excess energy source
2. Growth in the presence of growth-inhibitory compounds
3. Growth at unfavorable temperatures
4. Growth on minimal media
5. Growth during transient periods

Under all these conditions the rate of ATP production seems to be larger than can be used in anabolism under the prevailing growth conditions. Several responses of the organism to such circumstances are mentioned:

1. Formation of storage compounds
2. Formation of excretion products (overflow metabolism)
3. Deletion of sites of oxidative phosphorylation
4. Utilization of nonphosphorylating branches of the respiratory chain
5. Energy-spilling mechanisms

Under a number of conditions energy-spilling mechanisms seem the only possibility for adapting the rate of ATP production by catabolism to the rate of ATP consumption by anabolism. An increase in the permeability of the membrane to protons seems the most important energy-spilling mechanism during aerobic growth with substrates that do not give a substantial yield of ATP by substrate-level phosphorylation. During anaerobic growth when the ATP production is completely accomplished by substrate-level phosphorylation, ATP degradation by a membrane-bound ATPase, followed by dissipation of the energized membrane state, seems the most important energy-spilling mechanism. During aerobic growth with substrates that give a substantial ATP production by substrate-level phosphorylation, both mechanisms may be involved. In this aspect it must be stressed again that the membrane-bound ATPase is involved in the permeability of the membrane to protons (123). Elucidation of the regulatory properties of the ATPase seems very important for our understanding of its role in energy-spilling mechanisms.

A difference between the ATPases in aerobic and facultative bacteria has been described (166). Only in the latter organisms does ATPase function in the hydrolytic direction. In general, in aerobic organisms—with the exception of *Az. vinelandii*—the metabolism seems more tightly controlled than in facultative organisms. In *Pa. denitrificans* no increase in maintenance respiration rate is observed when growth is limited by factors other than the carbon/energy source (116); overflow metabolism does not occur either. The only way for adjustment of the rate of ATP production in *Pa. denitrificans* to the needs of anabolism is the regulation of the number of phosphorylation sites in the respiratory chain (118, 132). In *A. aerogenes* ADP does not control respiration (150); however, in *Pa. denitrificans* respiratory control can easily be demonstrated (191). It seems likely that ATPase structure, the occurrence of respiratory control, and the tightly controlled metabolism are related properties. Until now yield studies on microorganisms have been performed mostly with anaerobic and facultative organisms. The number of aerobic organisms that have been studied in detail is very restricted; therefore, it is not yet known whether or not the behavior of *Pa. denitrificans* in this aspect is characteristic for most aerobic organisms. More detailed studies on aerobic organisms are therefore urgently needed.

REFERENCES

1. Gunsalus, I. C., and Shuster, C. W. (1961). *In* I. C. Gunsalus and R. Y. Stanier (eds.), The Bacteria, Vol. 2, p. 1. Academic Press, New York.
2. Forrest, W. W., and Walker, D. J. (1971). Adv. Microb. Physiol. 5:213.
3. Stouthamer, A. H. (1973). Antonie van Leeuwenhoek 39:545.
4. Morowitz, H. J. (1968). Energy Flow in Biology, p. 179. Academic Press, London and New York.
5. Harder, W., and van Dijken, J. P. (1976). *In* H. G. Schlegel, N. Pfennig, and G. Gottschalk (eds.), Microbial Production and Utilization of Gases (H_2, CH_4, CO), p. 403. Erich Goltze Verlag, Göttingen.
6. Stouthamer, A. H. (1969). *In* J. R. Norris and D. W. Ribbons (eds.), Methods in Microbiology, Vol. 1, p. 629. Academic Press, London and New York.
7. Hadjipetrou, L. P., Gerrits, J. P., Teulings, F. A. G., and Stouthamer, A. H. (1964). J. Gen. Microbiol. 36:139.
8. Whitaker, A. M., and Elsden, S. R. (1963). J. Gen. Microbiol. 31:xxii.
9. Stouthamer, A. H. (1977). Symp. Soc. Gen. Microbiol. 28:285.
10. Hernandez, E., and Johnson, M. J. (1976). J. Bacteriol. 94:996.
11. Postma, P. W., and Roseman, S. (1976). Biochim. Biophys. Acta 457:213.
12. Kaback, H. R. (1972). Biochim. Biophys. Acta 265:367.
13. Harold, F. M. (1972). Bacteriol. Rev. 36:172.
14. Mitchell, P. (1970). Symp. Soc. Gen. Microbiol. 20:121.
15. Hamilton, W. A. (1975). Adv. Microb. Physiol. 12:1.
16. Hamilton, W. A. (1977). Symp. Soc. Gen. Microbiol. 27:185.
17. West, I. C. (1970). Biochem. Biophys. Res. Commun. 41:655.
18. West, I. C., and Mitchell, P. (1972). J. Bioenerg. 3:445.
19. West, I. C., and Mitchell, P. (1973). Biochem. J. 132:587.
20. Collins, S. H., Jarvis, A. W., Lindsay, R. J., and Hamilton, W. A. (1976). J. Bacteriol. 126:1232.
21. Ramos, S., and Kaback, H. R. (1977). Biochemistry 16:848.
22. Papa, S. (1976). Biochim. Biophys. Acta 456:39.
23. Dijkhuizen, L., and Harder, W. (1977). Proc. Soc. Gen. Microbiol. 4:69.
24. Tempest, D. W., Hunter, J. R., and Sykes, J. (1965). J. Gen. Microbiol. 39:355.
25. Herbert, D. (1961). Symp. Soc. Gen. Microbiol. 11:391.
26. Neidhardt, F. C. (1963). Annu. Rev. Microbiol. 17:61.
27. Dawes, E. A., and Senior, P. J. (1973). Adv. Microb. Physiol. 10:135.
28. Stouthamer, A. H. (1976). *In* Yield Studies in Micro-organisms. Meadowfield Press Ltd., Durham, England.
29. Holme, T. (1957). Acta Chem. Scand. 11:763.
30. Senior, P. J., Beech, G. A., Ritchie, G. A. F., and Dawes, E. A. (1972). J. Biochem. 128:1193.
31. Lees, H., and Postgate, J. R. (1973). J. Gen. Microbiol. 75:161.

32. Zumft, W. G., and Mortenson, L. E. (1975). Biochim. Biophys. Acta 416:1.
33. Brown, C. M., Macdonald-Brown, D. S., and Meers, J. L. (1974). Adv. Microb. Physiol. 11:1.
34. Hill, S., Drozd, J. W., and Postgate, J. R. (1972). J. Appl. Chem. Biotechnol. 22:541.
35. Hill, S. (1976). J. Gen. Microbiol. 95:297.
36. Daesch, G., and Mortenson, L. E. (1967). J. Bacteriol. 96:346.
37. Shanmugam, K. T., and Valentine, R. C. (1975). Proc. Natl. Acad. Sci. USA 72:136.
38. Ström, T., Ferenci, T., and Quayle, J. R. (1974). Biochem. J. 144:465.
39. Quayle, J. R. (1976). In H. G. Schlegel, N. Pfennig, and G. Gottschalk (eds.), Microbial Production and Utilization of Gases (H$_2$, CH$_4$, CO), p. 353. Erich Goltze Verlag, Göttingen.
40. Goldberg, I., Rock, J. S., Ben-Bassat, A., and Mateles, R. I. (1976). Biotechnol. Bioeng. 18:1657.
41. Herbert, D. (1976). In A. C. R. Dean, D. C. Ellwood, C. G. T. Evans, and J. Melling (eds.), Continuous Culture 6, Applications and New Fields, p. 1. Society of Chemical Industry, London.
42. Smith, L. A., Hill, S., and Yates, M. G. (1976). Nature 262:209.
43. de Bont, J. A. M. (1976). Antonie van Leeuwenhoek 42:255.
44. Schuberth, K. R., and Evans, H. J. (1976). Proc. Natl. Acad. Sci. USA 68:757.
45. Aleem, M. I. H. (1977). Symp. Soc. Gen. Microbiol. 27:351.
46. Bauchop, T., and Elsden, S. R. (1960). J. Gen. Microbiol. 23:457.
47. Payne, W. J. (1970). Annu. Rev. Microbiol. 24:17.
48. Decker, K., Jungermann, K., and Thauer, R. K. (1970). Angew. Chem. [Engl] 9:138.
49. Stouthamer, A. H., and Bettenhaussen, C. W. (1973). Biochim. Biophys. Acta 301:53.
50. Bélaich, J. P., Bélaich, A., and Simonpiétri, P. (1972). J. Gen. Microbiol. 70:179.
51. de Vries, W., Kapteijn, W. M. C., van der Beek, E. G., and Stouthamer, A. H. (1970). J. Gen. Microbiol. 63:333.
52. Stouthamer, A. H., and Bettenhaussen, C. W. (1976). Arch. Microbiol. 111:21.
53. Kapralek, F. (1972). J. Gen. Microbiol. 71:135.
54. Stouthamer, A. H., and Bettenhaussen, C. W. (1972). Antonie van Leeuwenhoek 38:81.
55. Payne, W. J., and Williams, M. L. (1976). Biotechnol. Bioeng. 18:1653.
56. Hasan, S. M., and Hall, J. B. (1975). J. Gen. Microbiol. 87:120.
57. Hardman, J. K., and Stadtman, T. C. (1963). J. Bacteriol. 85:1326.
58. Gaston, L. W., and Stadtman, E. R. (1963). J. Bacteriol. 85:356.
59. Brown, W. V., and Collins, E. B., (1977). Appl. Env. Microbiol. 33:38.
60. de Vries, W., Rietveld-Struyk, T. R. M., and Stouthamer, A. H. (1977). Antonie van Leeuwenhoek 43:153.
61. Rittenberg, S. C., and Hespell, R. B. (1975). J. Bacteriol. 121:1158.
62. Pritchard, M. A., Langley, D., and Rittenberg, S. C. (1975). J. Bacteriol. 121:1131.
63. Kröger, A., and Innerhöfer, A. (1976). Eur. J. Biochem. 69:497.
64. Kröger, A. (1974). Biochim. Biophys. Acta 347:273.
65. van der Beek, E. G., Oltmann, L. F., and Stouthamer, A. H. (1976). Arch. Microbiol. 110:195.
66. de Vries, W., van Wijck-Kapteijn, W. M. C., and Stouthamer, A. H. (1973). J. Gen. Microbiol. 76:31.
67. Sone, M. (1972). J. Biochem. 71:931.

68. de Vries, W., Aleem, M. I. H., Hemrika-Wagner, A., and Stouthamer, A. H. (1977). Arch. Microbiol. 112:271.
69. White, D. C., Bryant, M. P., and Caldwell, D. R. (1962). J. Bacteriol. 84:822.
70. Rizza, V., Sinclair, P. R., White, D. C., and Cuorant, P. R. (1968). J. Bacteriol. 96:665.
71. Macy, I., Probst, I., and Gottschalk, G. (1975). J. Bacteriol. 123:436.
72. de Vries, W., van Wijck-Kapteijn, W. M. C., and Oosterhuis, S. K. H. (1974). J. Gen. Microbiol. 81:69.
73. Hatchikian, E. C., and Le Gall, J. (1972). Biochim. Biophys. Acta. 267:479.
74. Kröger, A. (1977). Symp. Soc. Gen. Microbiol. 27:61.
75. Sokatch, J. T., and Gunsalus, I. C. (1957). J. Bacteriol. 73:452.
76. Beck, R. W., and Shugart, L. R. (1966). J. Bacteriol. 92:802.
77. Mickelson, M. N. (1972). J. Bacteriol. 109:96.
78. Davies, H. C., Karush, F., and Rudd, J. H. (1968). J. Bacteriol. 95:162.
79. Oxenburgh, M. S., and Snoswell, A. M. (1965). J. Bacteriol. 89:913.
80. de Vries, W., and Stouthamer, A. H. (1968). J. Bacteriol. 96:472.
81. Bulder, C. J. E. A. (1966). Arch. Mikrobiol. 53:189.
82. Bélaich, J. P., and Senez, J. C. (1965). J. Bacteriol. 89:1195.
83. Dawes, E. A., Ribbons, D. W., and Rees, D. A. (1966). Biochem. J. 98:804.
84. Forrest, W. W. (1967). J. Bacteriol. 94:1459.
85. McGill, D. J., and Dawes, E. A. (1971). Biochem. J. 125:1059.
86. Stephenson, M. P., and Dawes, E. A. (1971). J. Gen. Microbiol. 69:331.
87. Hernandez, E., and Johnson, M. J. (1967). J. Bacteriol. 94:991.
88. Hopgood, M. F., and Walker, D. J. (1967). Aust. J. Biol. Sci. 20:165.
89. Buchanan, B. B., and Pine, L. (1967). J. Gen. Microbiol. 46:225.
90. Gunsalus, I. C., and Sokatch. J. T. (1964). J. Bacteriol. 87:844.
91. Forrest, W. W. (1965). J. Bacteriol. 90:1013.
92. Hadjipetrou, L. P., and Stouthamer, A. H. (1965). J. Gen. Microbiol. 38:29.
93. Twarog, R., and Wolfe, R. S. (1963). J. Bacteriol. 86:112.
94. Thauer, R. K., Jungermann, K., Wenning, J., and Decker, K. (1968). Arch. Microbiol. 64:125.
95. Pirt, S. J. (1965). Proc. R. Soc. Lond. [Biol.] 163:224.
96. Watson, T. G. (1970). J. Gen. Microbiol. 64:91.
97. Stouthamer, A. H., and Bettenhaussen, C. W. (1975). Arch. Microbiol. 102:187.
98. Nagai, S., and Aiba, S. (1972). J. Gen. Microbiol. 73:531.
99. Neijssel, O. M., and Tempest, D. W. (1976). Arch. Microbiol. 107:215.
100. Hempfling, W. P., and Mainzer, S. E. (1975). J. Bacteriol. 123:1076.
101. Farmer, I. S., and Jones, C. W. (1976). Eur. J. Biochem. 67:115.
102. Lagunas, R. (1976). Biochim. Biophys. Acta 400:661.
103. de Kwaadsteniet, J. W., Jager, J. C., and Stouthamer, A. H. (1976). J. Theor. Biol. 57:103.
104. Stouthamer, A. H., and Bettenhaussen, C. W. (1977). Arch. Microbiol. 113:185.
105. Rogers, P. J., and Stewart, P. R. (1974). Arch. Microbiol. 99:25.
106. Jones, C. W. (1977). Symp. Soc. Gen. Microbiol. 27:23.
107. Lawford, H. G., Cox, J. C., Garland, P. B., and Haddock, B. A. (1976). FEBS Lett. 64:269.
108. Poole, R. K., and Haddock, B. A. (1975). Biochem. J. 152:537.
109. Lawford, H. G., and Haddock, B. A. (1973). Biochem. J. 136:217.
110. Ashcroft, J. R., and Haddock, B. A. (1975). Biochem. J. 148:349.

111. Brice, J. M., Law, J. F., Meyer, D. J., and Jones, C. W. (1974). Biochem. Soc. Trans. 2:523.
112. Downs, A. J., and Jones, C. W. (1975). Arch. Microbiol. 105:159.
113. Imai, K., Asano, A., and Sato, R. (1967). Biochim. Biophys. Acta 143:462.
114. Knobloch, K., Ishaque, M., and Aleem, M. I. H. (1971). Arch. Microbiol. 76:114.
115. van Verseveld, H. W., and Stouthamer, A. H. (1976). Arch. Microbiol. 107:241.
116. Meijer, E. M., van Verseveld, H. W., van der Beek, E. G., and Stouthamer, A. H. (1977). Arch. Microbiol. 112:25.
117. Cox, R. B., and Quayle, J. R. (1975). Biochem. J. 150:569.
118. van Verseveld, H. W., and Stouthamer, A. H. (1978). Arch. Microbiol. 118:21.
119. Edwards, C., Spode, J. A., and Jones, C. W. (1977). FEMS Microbiol. Lett. 1:67.
120. Lawford, H. G. (1977). Proc. Soc. Gen. Microbiol. 4:71.
121. Mitchell, P. (1966). In Chemiosmotic Coupling and Energy Transduction. Glynn Research Ltd., Bodmin, England.
122. Cox, G. B., and Gibson, F. (1974). Biochim. Biophys. Acta 346:1.
123. Simoni, R. D., and Postma, P. W. (1975). Ann. Rev. Biochem. 44:523.
124. Yamamoto, T. H., Mevel-Ninio, M., and Valentine, R. C. (1973). Biochim. Biophys. Acta 314:267.
125. Daniel, J., Roisin, M. P., Burstein, C., and Kepes, A. (1975). Biochim. Biophys. Acta 376:195.
126. van der Beek, E. G., and Stouthamer, A. H. (1973). Arch. Mikrobiol. 89:327.
127. Decker, S., and Lang, D. R. (1976). In D. Schlessinger (ed.), Microbiology 1976, p. 214. American Society for Microbiology, Washington, D.C.
128. Larsen, S. H., Adler, J., Gargus, J. J., and Hogg, R. W. (1974). Proc. Natl. Acad. Sci. USA 71:1239.
129. Senez, J. C. (1962). Bacteriol. Rev. 26:95.
130. Neijssel, O. M., and Tempest, D. W. (1976). Arch. Microbiol. 110:305.
131. Rosenberger, R. F., and Elsden, S. R. (1960). J. Gen. Microbiol. 22:727.
132. van Verseveld, H. W., Meijer, E. M., and Stouthamer, A. H. (1977). Arch. Microbiol. 112:17.
133. Dalton, H., and Postgate, J. R. (1969). J. Gen. Microbiol. 54:463.
134. Drozd, J., and Postgate, J. R. (1970). J. Gen. Microbiol. 63:63.
135. Jones, C. W., Brice, J. M., Wright, V., and Ackrell, V. A. C. (1973). FEBS Lett. 29:77.
136. Haukeli, A. D., and Lie, S. (1971). J. Gen. Microbiol. 69:135.
137. Ishimoto, M., Umeyama, M., and Chiba, S. (1974). Z. Allg. Mikrobiol. 14:115.
138. van Gent-Ruyters, M. L. W., de Vries, W., and Stouthamer, A. H. (1975). J. Gen. Microbiol. 88:36.
139. Kormancikova, V., Kovac, L., and Vidova, M. (1969). Biochim. Biophys. Acta 180:9.
140. Mickelson, M. N. (1974). J. Bacteriol. 120:733.
141. Neijssel, O. M. (1977). FEMS Microbiol. Lett. 1:47.
142. Hueting, S., and Tempest, D. W. (1977). Proc. Soc. Gen. Microbiol. 4:70.
143. Coulgate, T. P., and Sundaram, T. K. (1975). J. Bacteriol. 121:55.
144. Palumbo, S. A., and Witter, L. D. (1969). Appl. Microbiol. 18:137.
145. Topiwala, H., and Sinclair, C. G. (1971). Biotechnol. Bioeng. 13:795.
146. Mainzer, S. E., and Hempfling, W. P. (1976). J. Bacteriol. 126:251.

147. Farmer, I. S., and Jones, C. W. (1976). FEBS Lett. 67:359.
148. van Uden, N., and Madeira-Lopes, A. (1976). Biotechnol. Bioeng. 18:791.
149. Neijssel, O. M., and Tempest, D. W. (1975). Arch. Microbiol. 106:251.
150. Harrison, D. E. F. (1976). Adv. Microb. Physiol. 14:243.
151. Harrison, D. E. F., and Loveless, J. E. (1971). J. Gen. Microbiol. 68:45.
152. Harrison, D. E. F., and Maitra, P. K. (1969). Biochem. J. 112:647.
153. Harrison, D. E. F., and Pirt, S. J. (1967). J. Gen. Microbiol. 46:193.
154. Knowles, C. J., and Smith, L. (1970). Biochim. Biophys. Acta 197:152.
155. Poole, R. K., and Haddock, B. A. (1974). Biochem. Soc. Trans. 2:941.
156. Haddock, B. A., and Garland., P. B. (1971). Biochem. J. 124:155.
157. Yates, M. G., and Jones, C. W. (1974). Adv. Microb. Physiol. 11:97.
158. Ackrell, B. A. C., and Jones, C. W. (1971). Eur. J. Biochem. 20:22.
159. Downs, A. J., and Jones, C. W. (1975). FEBS Lett. 60:42.
160. Boonstra, J., Gutnick, D. L., and Kaback, H. R. (1975). J. Bacteriol. 124:1248.
161. Hasan, S. M., and Rosen, B. P. (1977). Biochim. Biophys. Acta 459:225.
162. Stouthamer, A. H. (1976). Adv. Microb. Physiol. 14:315.
163. Nicholls, D. G. (1976). TIBS 1:128.
164. Lazdunski, A., and Bélaich, J. P. (1972). J. Gen. Microbiol. 70:187.
165. Matsumura, P., and Marquis, R. E. (1977). Appl. Env. Microbiol. 33:885.
166. West, I. C. (1974). Biochem. Soc. Spec. Publ. 4:27.
167. Carreira, J., Munoz, E., Andreu, J. M., and Nieto, M. (1976). Biochim. Biophys. Acta 436:183.
168. Nieuwenhuis, F. J. R. M., and Bakkenist, A. R. J. (1977). Biochim. Biophys. Acta 459:596.
169. Asami, A., Juntti, K., and Ernster, L. (1970). Biochim. Biophys. Acta 205:307.
170. Eilermann, L. J. M., Pandit-Hovenkamp, H. G., van der Meer-van Buren, M., Kolk, A. H. J., and Feenstra, M. (1971). Biochim. Biophys. Acta 245:305.
171. Munoz, E., Salton, M. R. J., Ng, M. H., and Schorr, M. T. (1969). Eur. J. Biochem. 7:490.
172. Higashi, T., Kalra, V. K., Lees, S. H., Bogin, E., and Brodie, A. F. (1975). J. Biol. Chem. 250:6541.
173. Carreira, J., Leal, J. A., Rojas, M., and Munoz, E. (1973). Biochim. Biophys. Acta 307:541.
174. Nieuwenhuis, F. J. R. M., van der Drift, J. A. M., Voet, A. B., and van Dam, K. (1974). Biochim. Biophys. Acta 368:461.
175. Harold, F. M., and Baarda, J. R. (1968). J. Bacteriol. 96:2025.
176. Bhattacharyya, P., and Barnes, E. M., Jr. (1976). J. Biol. Chem. 251:5614.
177. Wall, J. D., Weaver, P. F., and Gest, H. (1975). Nature 258:630.
178. Hillmer, P., and Gest, H. (1977). J. Bacteriol. 129:724.
179. Hillmer, P., and Gest, H. (1977). J. Bacteriol. 129:732.
180. Cashel, M. (1975). Annu. Rev. Microbiol. 29:301.
181. Weijer, W. J., de Boer, H. A., de Boer, J. G., and Bruber, M. (1976). Biochim. Biophys. Acta 442:123.
182. Chaloner-Larsson, G., and Yamazaki, H. (1976). Can. J. Biochem. 54:291.
183. Fiil, N. P., Willumsen, B. M., Friesen, J. D., and von Meyenburg, K. (1977). Mol. Gen. Genet. 150:87.
184. Travers, A. (1976). Nature 263:641.
185. Penning de Vries, F. W. T. (1974). Neth. J. Agric. Sci. 22:40.
186. Gons, H. J. (1977). Ph.D. thesis, University of Amsterdam.
187. Koch, A. L. (1971). Adv. Microb. Physiol. 6:147.
188. Jensen, D. E., and Neidhardt, F. C. (1969). J. Bacteriol. 98:131.
189. Mateles, R. I., Riju, D. Y., and Yasuda, T. (1965). Nature 208:263.

190. Koch, A. L., and Deppe, C. S. (1971). J. Mol. Biol. 55:549.
191. John, P., and Whatley, F. R. (1975). Nature 254:495.
192. Stouthamer, A. H. (1978). *In* L. N. Ornston and T. R. Sokatch (ed.), The Bacteria, Vol. 6. Academic Press, New York.
193. Hadjipetrou, L. P., Gray-Young, T., and Lilly, M. D. (1966). J. Gen. Microbiol. 45:479.

International Review of Biochemistry
Microbial Biochemistry, Volume 21
Edited by J. R. Quayle
Copyright 1979   University Park Press   Baltimore

2
Energy Metabolism in Aerobes

C. W. JONES
Department of Biochemistry,
University of Leicester, Leicester, England

Work from the author's laboratory was supported by the United Kingdom Science Reseach
Council (Grants B/SR/59102 and GR/A 26400).

During the growth of microorganisms under anaerobic conditions in the absence of exogenous electron acceptors, energy transduction occurs solely at the expense of fermentation reactions. In the majority of these the oxidation of a particular catabolite eventually yields a relatively unstable acyl-phosphate with a high free energy of hydrolysis, which subsequently transfers its phosphoryl group to ADP with the formation of ATP. The reducing power that is released in the initial stages (usually as NADH or reduced ferredoxin) is ultimately transferred to a suitable endogenous oxidant (e.g., pyruvate, acetaldehyde, or acetone) or is released as molecular hydrogen. This method of energy conservation (substrate-level phosphorylation) is characterized by its essentially soluble nature, and by its macroscopic and scalar properties. Because fermentative pathways generally contain few substrate-level phosphorylation steps, energy yields and growth efficiencies of anaerobic cultures are low (1-4, also Gottschalk and Andreesen, this volume).

In contrast, the growth of heterotrophic microorganisms under aerobic conditions is characterized by considerably higher cell yields (2-5), which reflect the presence of an alternative major method of energy transduction (oxidative or respiratory-chain phosphorylation). In the latter process the low redox potential products of catabolism (NADH and reduced flavin) are oxidized via the respiratory chain using an exogenous high redox potential oxidant (molecular oxygen). Since the total yield of reducing power from catabolism and the total redox potential span over which reducing equivalents are transferred through the respiratory chain are both relatively large, the energy yield from oxidative phosphorylation is considerably higher than from substrate-level phosphorylation. It should be noted, however, that oxidative phosphorylation in some species of aerobic chemolithotrophic bacteria is associated with rather low energy yields, principally because these particular organisms oxidize inorganic compounds with relatively high redox potentials, e.g., NH_3, NO_2^-, or $S_2O_3^-$ (6).

Unlike fermentation and substrate-level phosphorylation, respiration and oxidative phosphorylation are vectorial membrane-bound processes that are best described by Mitchell's elegant and satisfying chemiosmotic hypothesis (7-10). Briefly, he states that respiration and ATP synthesis (or ATP hydrolysis) are anisotropic reactions that are linked by a proton current (proticity). Thus, respiration by intact bacteria or mitochondria leads to the stoichiometric ejection of protons, the electrochemical potential of which is responsible for the transmembrane protonmotive force ($\Delta\bar{\mu}_{H^+}$ or Δp) that serves as the driving power for ATP synthesis (and for a variety of other membrane-associated reactions, including reversed electron transfer, some forms of nutrient transport, and possibly bacterial movement). A multicomponent ATP synthetase (ATPase) complex spans the coupling membrane and effects the formal inward retranslocation of the ejected protons (probably as an outwardly directed transfer of O^{2-} from ADP and inorganic phosphate) with the concomitant synthesis of ATP on the inner surface and the ejection of water into the external compartment. The major tenets and possible limitations of the chemiosmotic hypothesis of oxidative phosphorylation are lucidly and critically reviewed in more detail elsewhere (10-15).

This chapter concentrates on the diversty and functional organization of the respiratory chains of heterotrophic bacteria cultured under aerobic conditions, and on the nature, mechanism, and efficiency of their associated energy conservation systems.

RESPIRATION

Respiratory Chain Composition

Species Differences   The membrane-bound respiratory systems of bacteria grown under aerobic conditions contain the same basic types of redox carriers as those present in the mitochondria of eukaryotic organisms, viz., iron-sulfur proteins and flavoproteins (organized into nicotinamide nucleotide transhydrogenases and a variety of primary dehydrogenases), quinones, cytochromes, and cytochrome oxidases. However, a closer inspection of these systems reveals that this apparent unity is largely superficial and that only a very few species of bacteria contain respiratory chains that are truly similar to those of mitochondria, e.g., *Paracoccus* (*Micrococcus*) *denitrificans* and *Alcaligenes eutrophus* (*Hydrogenomonas eutropha*). Indeed, there are convincing arguments based partly on similarities in redox carrier composition, sensitivity to electron transfer inhibitors, and various other membrane properties, that the inner membrane of the present-day mitochondrion may have evolved from the plasma membrane of an ancestral relative of *Pa. denitrificans* via endosymbiosis with a primitive host cell (16, 17).

The majority of bacterial respiratory systems exhibit redox carrier compositions that are significantly different from those of mitochondria. Although some differences in transhydrogenase or quinone content are apparent, the most striking differences occur in the terminal oxidase system, where cytochrome c may be absent and where the single cytochrome oxidase aa_3 of mitochondria may be replaced by up to three carbon monoxide-binding bacterial cytochromes—aa_3, o, d, a_1, and c_{co} (18-21). As yet, rapid kinetic analyses of these cytochromes have confirmed oxidase roles only for cytochromes aa_3, o, and d (22-24).

Large variations in redox carrier composition also occur between different species of bacteria; indeed, one of the major characteristics of bacterial respiratory systems is the immense variety of their redox carrier patterns (19-21, 25, 26). A simplified, and by no means comprehensive, example of this is shown in Table 1. It can be seen that these variations usually fall into one of the following categories:

1. The replacement of one redox carrier by another with basically similar

Table 1. The simplified respiratory chain compositions of selected species of bacteria following heterotrophic growth under conditions of excess and limiting oxygen concentrations

Oxygen concentration	Organism	Respiratory chain composition[a]								
Excess	*Paracoccus denitrificans*	Th	Ndh	Q		b	c	aa_3	(o)	
	Alcaligenes eutrophus	Th	Ndh	Q		b	c	aa_3	(o)	
	Pseudomonas fluorescens	Th*	Ndh	Q		b	c		(o)	
	Arthrobacter globiformis		Ndh	Q		b	c	aa_3	(o)	
	Micrococcus lysodeikticus	Th	Ndh		MK	b	c	aa_3	(o)	
	Micrococcus luteus		Ndh		MK	b	c	aa_3	(o)	
	Escherichia coli	(Th)	Ndh	Q	(MK)	b			o	
	Klebsiella pneumoniae	(Th)	Ndh	Q		b			o	
	Chromobacterium violaceum	(Th)	Ndh	Q		b	c		o	c_{co}
	Bacillus megaterium		Ndh		MK	b	c	aa_3	(o)	
	Azotobacter vinelandii		Ndh	Q		b	c		o	
	Aeromonas punctata	(Th*)	Ndh	Q	(MK)	b	c		o	$a_1 d$
Limiting	*Paracoccus denitrificans*	Th	Ndh	Q		b	c	aa_3	o	
	Alcaligenes eutrophus	Th	Ndh	Q		b	c	aa_3	o	
	Arthrobacter globiformis		Ndh	Q		b	c	(aa_3)	(o)	
	Escherichia coli	(Th)	Ndh	Q	MK	b	c		o	d
	Klebsiella pneumoniae	(Th)	Ndh	Q		b			(o)	$(a_1)d$
	Aeromonas punctata	(Th*)	Ndh	Q	(MK)	b	c		(o)	$(a_1)d$

[a]Abbreviations: Th, nicotinamide nucleotide transhydrogenase; Ndh, NADH dehydrogenase complex; Q, ubiquinone; MK, menaquinone; cytochromes are referred to by letter. Note that more than one b- or c-type cytochrome is often present within a single species, but these are not shown in this simplified table. Brackets indicate redox carriers of low concentration or activity; the asterisk denotes energy-independent transhydrogenases. (For further details and primary references see refs. 14, 18, 20, 21, 25, 26, 33.)

properties, e.g., the replacement of ubiquinone by menaquinone (both are lipophilic hydrogen carriers), cytochrome oxidase aa_3 by cytochrome oxidases o or d, and cytochrome oxidase o by cytochrome oxidase d (all are autoxidizable, 1-electron carriers)

2. The replacement of one redox carrier by another with significantly different properties, e.g., the replacement of energy-dependent transhydrogenase by energy-independent transhydrogenase

3. The addition/deletion of a limited number of redox carriers, e.g., transhydrogenases of both types and cytochrome c

These variations generally have little effect on the respiratory capacities of isolated membrane preparations, and in only a few instances do they influence the efficiency of oxidative phosphorylation (see "Energy Coupling Sites"). They are, however, often accompanied by altered sensitivities to classic chain inhibitors. Thus, most of the respiratory systems in the lower half of the excess oxygen section of Table 1 are resistant to both rotenone and antimycin A, and those systems that contain cytochrome oxidase d are relatively insensitive to cyanide (19, 26-29).

The Effect of Growth Conditions   Variations in the composition of aerobic respiratory chains, both qualitative and quantitative, can also occur within a single species of bacterium (14, 19, 26, 30-32). These usually reflect changes in the growth environment, particularly the decreased availability of molecular oxygen or other essential nutrients, the presence of respiratory chain inhibitors, or alterations in the nature of the carbon/energy source. Thus, growth under oxygen-limited conditions (Table 1) is often accompanied by the increased synthesis of alternative cytochrome oxidases, e.g., d (plus a_1) relative to either o (*Escherichia coli, Klebsiella pneumoniae, Aeromonas punctata,* and *Achromobacter* spp.) or aa_3 (*Arthrobacter globiformis*), or o relative to aa_3 (*Pa. denitrificans, A. eutrophus*); such changes undoubtedly reflect attempts by these organisms to combat an insufficiency of oxygen through the differential synthesis of alternative cytochrome oxidases that have higher affinities for this electron acceptor (19, 25, 26). This conclusion is supported by the results of continuous culture experiments in which pairs of obligately aerobic bacteria with different cytochrome oxidase components are made to compete for limiting concentrations of molecular oxygen. Because those organisms that contain cytochrome oxidase d invariably outgrow competitors containing cytochrome oxidases o or aa_3 (33), it would appear that the affinities of bacterial cytochrome oxidases for molecular oxygen are in the order $d > o$ or aa_3; direct estimations yield conflicting results (34-36). Changes in cytochrome oxidase content similar to those induced by oxygen limitation can also occur during growth in the presence of low concentrations of cyanide (37, 38), thus indicating that the alternative oxidases have lower sensitivities to this inhibitor (e.g., the K_i for cyanide of cytochrome d can be as high as 10 mM as compared with 0.5 mM or less for o and aa_3).

The replacement of molecular oxygen by other electron acceptors (e.g., nitrate or fumarate) leads to the differential synthesis of specific reductases at the expense of cytochrome oxidase(s) (14, 39, 40). In contrast, when glycerol is replaced as the carbon source by a powerful catabolite repressor like glucose, changes in the redox carrier pattern are solely quantitative, i.e., there is a general decrease in the concentrations of existing redox components that can be reversed by the addition of cAMP to the growth medium (41, 42).

Respiratory chain phenotypes can also be altered by growing organisms in continuous culture such that a single nutrient that is essential for the biosynthesis of a particular redox carrier becomes rate-limiting for growth; this technique is reviewed in detail elsewhere (43). In *E. coli* and the few other species of bacteria that have so far been examined in this manner, iron limitation leads to a decrease in the concentrations of membrane-bound iron-sulfur proteins and cytochromes. Sulfate limitation also decreases the content of iron-sulfur proteins (although the extent of this loss varies considerably between species) and quinones, but appears to stimulate the synthesis of alternative cytochrome components; the reason for its effect on quinone and cytochrome levels is unclear (31).

There is some evidence that bacterial cytochrome oxidase aa_3, like its counterpart in mammalian and yeast mitochondria, is associated with copper ions (44). Perturbations in cytochrome oxidase levels have been observed as a result of growing the yeast *Candida utilis* under copper-limited conditions (45), but this potentially powerful tool for investigating cytochrome oxidase functions has not yet been applied to bacterial systems.

Genetic Manipulation   The composition of bacterial respiratory chains can also be manipulated using a variety of genetic procedures, with the formation of mutant strains that are deficient in one or more redox carriers or that contain defective subunits in their ATPase complex (14, 15, 31, 46–48). Several aerobic electron transfer mutants of this type have been prepared from *Staphylococcus aureus, Bacillus subtilis,* and *Salmonella typhimurium,* but the most popular organisms for this type of work are undoubtedly *E. coli* K12 and related strains. The major advantages of using *E. coli* K12 for the isolation of respiratory chain mutants (and ATPase mutants; see "Mutants Defective in Energy Transduction") are that: 1) it is very amenable to mutagenesis and to sophisticated genetic analysis; 2) it readily allows the transfer of mutant alleles between different strains using phage or episome vectors; and 3) it is a facultative anaerobe and therefore can obtain energy for growth using either substrate-level or oxidative phosphorylation. This latter property greatly facilitates the selection and screening of putative electron transfer mutants in that these will grow readily on fermentable carbon sources (e.g., glucose) but not on a mixture of nonfermentable compounds (e.g., succinate + malate + acetate).

Most of the respiratory chain mutants that have been selected in this way are either defective in their ability to synthesize particular redox com-

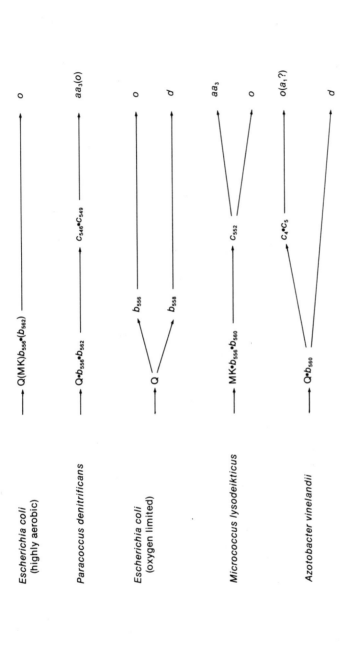

Figure 1. Bacterial respiratory chains with linear or branched terminal pathways. Redox components that are kinetically inactive or are present in low concentrations are enclosed in brackets. Abbreviations: Q, ubiquinone; MK, menaquinone: cytochromes are referred to by letter plus, for the *b*- and *c*-type cytochromes, their wavelength maxima in low temperature, reduced *minus* oxidized difference spectra. [After Jones (26)].

ponents (e.g., ubiquinone, menaquinone, and heme) or lack effective transport systems for the uptake of essential precursors (e.g., iron). Functional electron transfer activity can often be restored to these mutants by incubating whole cells or depleted membranes with an easily assimilated form of the missing redox carrier or a precursor thereof. Thus, respiratory activity can be restored to membranes from $ubiB^-$, $menA^-$, and $hemA^-$ mutants of *E. coli* by the addition of Q-1, Mk-1, and hematin + ATP, respectively (49–52).

Respiratory Pathways

The respiratory pathways of bacteria cultured under aerobic conditions are generally investigated using a mixture of classic techniques pioneered with mammalian or plant mitochondria (e.g., high resolution split-beam, dual-wavelength, and photochemical action spectrophotometry, electron paramagnetic resonance specroscopy, steady-state and stopped flow redox kinetics, redox potentiometry, and extraction-reactivation or fractionation-reconstitution experiments), together with various methods specifically tailored to microbial systems (e.g., the isolation and analysis of genotypic and phenotypic variants).

The results of experiments using these various approaches indicate that bacterial respiratory systems exhibit considerable variation in the complexity and organization of their electron transfer pathways (14, 19, 26, 53, 54). For example, although all bacterial systems exhibit extensive branching at the level of the primary dehydrogenases (thus allowing reducing equivalents from a variety of substrates to be channeled into a common respiratory chain), the terminal quinone-cytochrome system may be either branched or linear (Figure 1). Linear terminal pathways are relatively rare and are limited to the few systems that contain only one functional cytochrome oxidase, e.g., *E. coli* cultured under highly aerobic conditions, and *Pa. denitrificans* (the small amount of cytochrome *o* that is present in the latter in addition to cytochrome oxidase aa_3 appears not to exhibit significant oxidase activity) (23, 24). Branching is usually associated with the presence of more than one functional species of cytochrome oxidase and, in its simplest form, consists of electron transfer from penultimate nonautoxidizable *b*- or *c*-type cytochromes to molecular oxygen via two or more oxidases, as in *E. coli* grown under oxygen-limited conditions (23), and possibly also *Micrococcus lysodeikticus* (55). In a few systems branching is more complex and the separate terminal pathways appear to contain *b*- and/or *c*-type cytochromes as well as different cytochrome oxidases, as in *Azotobacter vinelandii* (56), *Benekea natriegens* (57), and possibly also *Ar. globiformis* and *A. punctata* cultured under oxygen-limited conditions (33). There is increasing evidence that in these complex systems one branch contains cytochrome *c* and is terminated by cytochrome oxidases aa_3 or *o*, whereas in the other branch a *b*-type cytochrome donates electrons directly to cytochrome oxidase *d*. It should be noted, however, that the presence of a branched respiratory system in respiratory membranes examined in vitro does not necessarily reflect the presence of such pathways in vivo (36, 58).

The Control of Respiration

Bacterial respiration is subject to both coarse and fine regulation; the former is effected via the repression and induction of redox carrier biosynthesis (32, 59), and the latter is determined by the inherent kinetic and thermodynamic properties of the respiratory chain and its attendant energy conservation apparatus (32). Therefore, fine control of dehydrogenase activity reflects the availability of the various reducing substrates (or, very occasionally, either the overall energy charge or the concentration of one or more adenine nucleotides), whereas cytochrome oxidase activities are dependent on the ambient concentration of molecular oxygen. In addition, and most importantly, bacterial respiratory systems exhibit classic respiratory control, as evidenced by the ability of uncoupling agents or ADP (plus inorganic phosphate) to stimulate the respiration of appropriate whole cell or membrane vesicle preparations (60-64). Indeed, the latter is probably the major control process for linear or simple branched respiratory systems under highly aerobic conditions, but kinetic regulation of cytochrome oxidase activity is probably more important at low oxygen concentrations (32).

It is more likely that the major function of a branched respiratory system is to allow some flexibility in the exact pathway of terminal electron transfer, thus enabling the organism to minimize the potentially deleterious effects of certain growth environments and to take maximum advantage of others. Therefore, in the presence of cyanide or at low concentrations of molecular oxygen, electron transfer would tend to be routed via the terminal branch that is most capable of maintaining a high potential rate of respiration. This rerouting can obviously take place immediately if multiple cytochrome oxidases are present (e.g., in *E. coli, A. eutrophus,* and *K. pneumoniae*), with the result that each oxidase carries a fraction of the total electron flux that reflects its concentration and its kinetic properties (with respect to cyanide or molecular oxygen) relative to those of the other oxidase(s).

In the complex branched systems (e.g., *Az. vinelandii*) the two terminal pathways may exhibit unequal energy conservation efficiencies as well as different affinities for molecular oxygen (33, 54, 65). Differential rates of electron flow along the available terminal pathways would therefore be determined not only by the kinetic properties of the respective oxidases and of the redox carriers at the point of branching, but also by the energy status of the cell.

It should also be noted, however, that intact bacteria can waste significant amounts of energy. This conclusion is based on repeated observations that washed suspensions can rapidly and extensively oxidize carbon substrates in the absence of growth, and that cultures growing under excess carbon conditions (and possibly also under certain carbon-limited conditions) respire very much more rapidly than is necessary simply to satisfy their energy requirements for growth (66–69). Although the nature of this energy spillage is not currently understood, there is some evidence that it does not entail a diminution in the efficiency of respiration-linked proton transloca-tion (see below), and that it is more likely to reflect either an enhanced utilization of the transmembrane protonmotive force for nonproductive pur-poses or an increased hydrolysis of ATP by ATPases associated with several types of futile cycles.

Current views on the regulation of bacterial energy metabolism and respiration are excellently reviewed in more detail elsewhere (32, 53).

The Spatial Organization of the Respiratory Chain

There is increasing evidence that the energy-conserving respiratory systems of bacteria, like those of mitochondria, are organized asymmetrically across the coupling membrane. This conclusion is based predominantly on the results of fairly recent investigations using whole cells, protoplasts, spheroplasts, or various types of membrane vesicles from several species of bacteria, including *B. subtilis, M. lysodeikticus,* and *E. coli.* Protoplasts (from gram-positive organisms) and spheroplasts (from gram-negative organisms) are usually prepared by exposing whole cells to lysozyme and lysozyme-EDTA respectively under hypertonic conditions; membrane vesicles are subsequently formed by osmotically shocking intact protoplasts, or by shearing whole cells in a French pressure cell or Ribi cell disintegrator (70, 71).

Two basic approaches have so far been employed to study the organiza-tion of the various redox and energy-transduction components within cou-pling membranes: 1) observing the location of these components in the electron microscope, using negative staining or freeze-cleave techniques; and 2) measuring the accessibility of these components to proteases, to lactoperoxidase-catalyzed iodinations, to macromolecules with specific bind-ing sites (e.g., antibodies), to nonpenetrating substrates and effectors of en-zyme activity, and to nonpenetrating functional group reagents (72). Thus, electron microscopy reveals the presence of ATPase headpieces (BF_1) on the surface of membrane vesicles prepared by shear-breakage of intact bacteria, but not on the surface of vesicles formed by osmotic shock procedures. Because these latter vesicle preparations (like intact cells, spheroplasts, and protoplasts) are generally unable either to hydrolyze ATP or to form microscopically detectable conjugates with ferritin-labeled antibodies to BF_1 (whereas the shear-breakage vesicles exhibit a rapid ATPase activity that is sensitive to these antibodies), it is clear that the ATPase is located asym-metrically in the coupling membrane and that the latter is oriented differ-ently in the two types of vesicles (73–78). These vesicles are therefore termed right-side-out vesicles (where the orientation of the membrane is the same as that of the intact cell, protoplast, or spheroplast, such that the ATPase is exposed on the inner surface and is therefore inaccessible to exogenous adenine nucleotides) and inside-out vesicles (where the orientation of the membrane is inverted, such that the ATPase is fully exposed on the outer surface).

Right-side-out vesicles are used extensively for investigating solute transport, whereas inside-out vesicles, like sonic submitochondrial particles, are more suited to the study of oxidative phosphorylation and various ATP-dependent membrane reactions. It should be noted, however, that each type of vesicle is usually contaminated by the presence of variable quantities of the other type, and also by scrambled vesicles (48, 79, 80). In addition, there is increasing evidence that some membrane proteins change their location during the preparation procedures (48, 81). The use of membrane vesicles to investigate the spatial distribution of respiratory chain components within the coupling membrane is therefore potentially hazardous and the results must be interpreted with caution; intact cells, protoplasts, and spheroplasts generally yield more reliable data.

The failure of intact cells, protoplasts, and spheroplasts from many species of bacteria to oxidize exogenous NADH or to catalyze trans-hydrogenation strongly supports the concept that the active sites of the NADH dehydrogenase and transhydrogenase complexes are located on the cytoplasmic surface of the respiratory membrane. In addition, experiments with the membrane-impermeable electron acceptors 5-N-methylphena-zonium-3-sulfonate (MPS) and ferricyanide indicate that the NADH, L-malate, L-lactate, and L-α-glycerophosphate dehydrogenases of B. $subtilis$, E. $coli$, or M. $lysodeikticus$ are located on the cytoplasmic surface, but that the succinate dehydrogenase of B. $subtilis$ is at least partly exposed on the opposite surface (55, 70, 82, 83). Furthermore, since antibodies against D-lactate dehydrogenase fail to inhibit the activity of this enzyme complex in spheroplasts and right-side-out vesicles of E. $coli$ unless these are physically disrupted alone or in the presence of the antibodies, respectively, it is likely that this dehydrogenase is also located on the cytoplasmic surface (70).

No experiments using antibodies against cytochrome c or cytochrome oxidase, or involving nonpenetrating reagents like p-diazonium benzene sulfonate (DABS) have yet been reported with bacterial respiratory systems, and the examination of membrane proteins using lactoperoxidase-catalyzed iodination has been restricted to the nitrate reductase of anaerobically grown E. $coli$ (84). Thus, except for some evidence from extraction-reactivation and ferricyanide studies that cytochrome c is located on the side of the coupling membrane that faces the cell wall (or periplasm) in the intact organism, virtually nothing is known about the intramembrane location of the various bacterial cytochromes and cytochrome oxidases. Our current knowledge of the organization and spatial distribution of redox carriers within bacterial coupling membranes is therefore extremely fragmentary. Because the ability to translocate protons across these membranes concomitant with respiration appears to be an essential prerequisite to oxidative phosphorylation in bacteria (as well as in the mitochondria of higher organisms), the completion of a detailed topographical analysis of bacterial respiratory membranes is clearly a matter of some importance.

Respiration-Linked Proton Translocation

One of the major tenets of the chemiosmotic hypothesis of oxidative phosphorylation is that the respiratory chain acts as an electrogenic proton pump, translocating protons across the osmotic barrier of the coupling membrane (the so-called M phase) and thus generating an electrochemical potential (the protonmotive force, Δp or $\Delta\bar{\mu}_H^+$) that subsequently drives a variety of energy-dependent membrane reactions, including ATP synthesis (7-10, 85). The translocation of protons by the respiratory chain is clearly dependent upon the latter being organized into an alternative sequence of hydrogen and electron carriers, such that during respiration protons are transported to one side of the coupling membrane and taken up from the other side. Therefore, in vectorial terms, we can envisage the coupling membrane as containing either a looped respiratory system plus an essentially linear osmotic barrier, or an essentially linear respiratory system plus a looped osmotic barrier (86). The known asymmetric location of several redox carriers in bacterial respiratory membranes supports the former concept, although some looping of the osmotic barrier cannot be entirely ruled out.

Following the pioneering work of Scholes and Mitchell with *Pa. denitrificans* (87), there is now ample evidence from several laboratories that respiration by whole cells, protoplasts, spheroplasts, and membrane vesicles (right-side-out and inside-out) prepared from a wide variety of bacterial species is accompanied by electrogenic proton translocation (55, 88-96). The direction of proton movement across the respiratory membrane is invariably away from the surface on which the ATPase headpiece is located, i.e., outward with whole cells, protoplasts, spheroplasts, and right-side-out vesicles, but inward with inside-out vesicles. The primary movement of protons in response to an oxygen pulse is extremely rapid and matches the rate of electron transfer. On the other hand, the decay of the proton gradient is generally much slower ($t_{1/2} < 45$ sec), is usually first order with respect to proton concentration, and reflects the inherently low permeability of the bacterial coupling membrane to protons and other ions.

Stoichiometries of respiration-linked proton translocation are routinely measured in the presence of either valinomycin plus K^+ or a permeant anion like CNS^- in order to allow the full expression of proton translocation without interference from an opposing membrane potential. Under these conditions $\rightarrow H^+/O$ quotients ranging from approximately 4 to approximately 8 g ion of H^+ per g atom of O have been obtained with a wide range of isocitrate ($NADP^+$), malate (NAD^+), succinate (flavin), QH_2 (ubiquinone), and ascorbate-TMPD (cytochrome c). This technique can be improved by adding inhibitors to block activity in unwanted segments of the respiratory chain (e.g., rotenone or piericidin A to inhibit electron transfer from endogenous NADH during the oxidation of exogenous flavin-linked substrates), and can be extended by varying the nature of the electron acceptor (e.g., Q or ferricyanide instead of molecular oxygen) (26, 90, 93, 95). Using this approach, it has become clear that bacterial respiratory chains are organized into between two and four proton-translocating respiratory segments, and most results (26, 90, 91, 93, but see also 95) indicate that each of these segments catalyzes the translocation of approximately two protons per electron pair transferred (i.e., the $\rightarrow H^+$:site ratio is 2). Recent reports suggest that the previously accepted $\rightarrow H^+$:site ratio of 2 for mammalian mitochondria may be an underestimate caused by the concomitant inflow of weak acids (carrying H^+) or phosphate (in exchange for OH^-) during respiration; in the absence of this type of counterflow, the minimum $\rightarrow H^+$:site ratio is 3 (97, 98). This value is commensurate with the use of $2H^+$ for the synthesis of each molecule of ATP (see "$\rightarrow H^+/P$ Quotients") and with the use of a third proton to compensate for the electrogenic exchange of ADP^{3-} for ATP^{4-} through the inner membrane. However, since bacterial $\rightarrow H^+$:site ratios are unaffected by the elimination of phosphate transport, and bacterial coupling membranes do not contain an adenine nucleotide translocase, there is no reason to suspect that the most frequently observed value of approximately 2 is an underestimate.

Intensive studies of wild-type strains and cytochrome-deficient mutants of *E. coli*, and of other species of heterotrophic bacteria that exhibit different respiratory chain compositions, indicate that all aerobic respiratory systems are organized into two basic proton-translocating segments during aerobic growth—segments 1 (NADH dehydrogenase) and 2 (the central quinone-cytochrome *b* region). In contrast, segments 0 (nicotinamide nucleotide transhydrogenase) and 3 (cytochrome *c* oxidase) are not exhibited by all species of bacteria because they are predicated upon the ability of the individual organism to synthesize significant amounts of an energy-linked transhydrogenase and of a high-potential cytochrome *c* (linked to cytochrome oxidases aa_3 or *o*), respectively (14, 26, 93, 99).

OXIDATIVE PHOSPHORYLATION

Energy Coupling Sites

The organization of bacterial respiratory chains into proton-translocating segments is, of course, compatible with the classic concept of separate energy coupling sites located within well-defined regions of the redox system. The location of these coupling sites by direct assay of ATP synthesis during respiration is considerably more difficult in bacteria than in mitochrondria, because the absence of an adenine nucleotide translocase from the bacterial coupling membrane renders whole cells incapable either of utilizing exogenous ADP as a phosphoryl acceptor for oxidative phosphorylation or of catalyzing the hydrolysis of exogenous ATP. Unfortunately, attempts to measure P/O (P/2e$^-$) quotients in whole cells by monitoring changes in the intracellular concentrations of adenine nucleotides and nicotinamide nucleotides following the initiation of respiration by the addition of small aliquots of oxygen-saturated buffer to anaerobic cell suspensions have yielded variable results (100-103), and some uses of this potentially powerful method have been severely criticized on a variety of technical grounds.

The detection and location of energy coupling sites in bacteria has therefore traditionally been carried out on subcellular fractions via the direct assays of P/O (P/2e$^-$) quotients (20) or, to a lesser extent, ADP/O quotients from respiratory control cycles (104, 105). Both of these methods optimally require the use of inside-out membrane vesicles but, because populations of the latter are rarely either topologically homogenous or completely closed, energy transduction efficiences are generally low, e.g., P/O quotients are rarely greater than 1.0–1.5 for the oxidation of NADH. Absolute efficiencies are therefore of limited use in determining the number of energy coupling sites. Instead, a more qualitative approach has been adopted—detecting those segments of the respiratory chain that: 1) can drive ATP synthesis or membrane energization [the latter determined by measuring the energy-dependent fluorescence changes of probe dyes (106) or the movement of synthetic ions (55, 82) during forward electron transfer]; or 2) can exhibit reversed electron transfer at the expense either of ATP hydrolysis or of forward electron transfer through a different segment (107, 108).

Using these approaches (14, 16, 26, 54, 109) classic energy coupling sites 1 and 2 have been detected in respiratory membrane vesicles from a wide range of bacteria (e.g., *Mycobacterium phlei, Acinetobacter lwoffi, A. eutrophus, E. coli, My. lysodeikticus, Az. vinelandii, Pa. denitrificans*). Energy coupling at site 3 is considerably more difficult to detect because of the poor specificities and high membrane permeabilities of potential electron transfer mediators in this region of the chain, but has been reported to occur in *My. phlei* (109) and *Az. vinelandii* (65). Interestingly, both of these organisms contain terminal respiratory chains that are comprised of cytochrome *c*, linked to cytochrome oxidases *o* (*Az. vinelandii*) or *aa₃, o* (*My. phlei*). The alternative *b → d* terminal branch in *Az. vinelandii*, which lacks cytochrome *c*, shows no evidence of energy coupling at site 3. Energy transduction at site 0 has been detected in several species of bacteria, principally by measuring energy-dependent reversed electron transfer from NADH to NADP$^+$ (e.g., *E. coli, K. pneumoniae, Pa. denitrificans*). ATP synthesis concomitant with the oxidation of NADPH by NAD$^+$ has also been detected, but only in the presence of initially very high [NADPH] [NAD$^+$]:[NADP$^+$] [NADH] ratios, in accordance with the small $\Delta E_0'$ of these two redox couples (approximately 4 mV).

These direct in vitro measurements of respiratory chain energy conservation have been supplemented by indirect in vivo determinations based upon measurements of the growth efficiencies of carbon/energy-limited continuous cultures (2, 5, 66, 68, 69, 99). During the aerobic growth of heterotrophic bacteria on nonfermentable carbon sources, oxidative phosphorylation provides at least 85% of the energy that the cell uses for growth. Thus, true molar growth yields with respect to the utilization of molecular oxygen ($Y_{O_2}^{max}$; g cells per mol O_2) or to the oxidation of carbon substrate ($Y_{substrate\ ox}^{max}$; g cells per mol substrate oxidized) closely reflect the efficiency of oxidative phosphorylation in the growing cell. Comparative studies based on measuring $Y_{O_2}^{max}$ and $Y_{substrate\ ox}^{max}$ values of bacteria with different respiratory chain compositions (99) indicate that all of the species examined contain two basic energy coupling sites. A third site, reflected in approximately 50% higher growth yields, parallels the presence of significant amounts of a membrane-bound high potential cytochrome c in association with cytochrome oxidases aa_3 or o (e.g., Pa. denitrificans, A. eutrophus, Ar. globiformis, Ae. punctata). Since the presence of an energy-linked nicotinamide nucleotide transhydrogenase has no significant stimulatory effect on bacterial growth efficiencies, it must be concluded that site 0 does not contribute significantly to energy conservation under normal growth conditions. Indeed, the thermodynamic properties of the transhydrogenase suggest that in vivo it may operate in reverse, i.e., to produce NADPH for biosynthetic purposes at the expense of the protonmotive force generated by forward electron transfer through the other energy coupling sites (protontranslocating segments) (110, 111).

The results of these studies on the proton translocation and energy conservation efficiencies of different respiratory chains, and of preliminary studies on the spatial organization of redox carriers in bacterial coupling membranes (see above), suggest that bacterial NADH oxidase complexes (by analogy with those of mitochondria) may be organized as shown in Figure 2. Therefore, in those organisms that lack significant amounts of cytochrome c (e.g., Ac. lwoffi, E. coli, K. pneumoniae, some strains of Bacillus megaterium, and B. subtilis) segment 3 is missing and electron transfer is terminated by cytochrome oxidases aa_3, o, or d immediately after the b-type cytochromes. In contrast, when cytochrome c is present (as in Pa. denitrificans, A. eutrophus, Ar. globiformis, My. lysodeikticus, and Ae. punctata) the ability to translocate an extra two protons per electron pair transferred may involve the quinone moiety in a complex protonmotive cycle (112). This ingenious reaction sequence circumvents repeated failures to detect a hydrogen carrier for segment 3 in both mitochondrial and bacterial respiratory systems by combining segments 2 and 3 into a complex redox cycle that serves to translocate 4 protons concomitant with the net oxidation of 1 molecule of QH_2 (or MKH_2) by molecular oxygen. Cytochrome c thus completes the vectorial organization of terminal limb of the respiratory chain such that it catalyzes the inward transfer of electrons to the terminal oxidase. There is some evidence that the latter is either cytochrome oxidase aa_3 or o, but not cytochrome oxidase d, which continues to terminate respiration after segment 2 even in the presence of cytochrome c (33, 65, 113).

(a)   Out   Cell   In
                                 Membrane

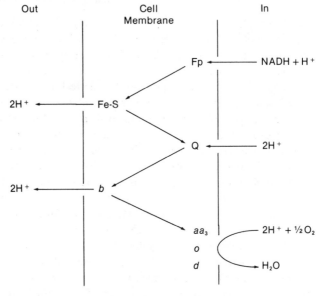

(b)   Out   Cell   In
                                 Membrane

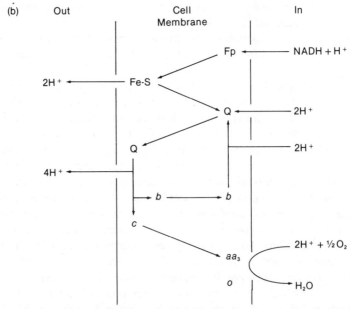

Figure 2.   The possible vectorial organization of bacterial NADH oxidase complexes: (a) in the absence of cytochrome *c* (e.g., *E. coli, K. pneumoniae, Ac. lwoffi,* and some strains of *B. megaterium* and *B. subtilis*); (b) in the presence of a high-potential cytochrome *c* (e.g., *Pa. denitrificans, A. eutrophus, Ar. globiformis, My. lysodeikticus,* and *Ae. punctata*). Abbreviations: Fp, flavoprotein; Fe-S, iron-sulfur protein; other abbreviations as for Figure 1. Note that in some organisms ubiquinone is replaced by menaquinone, and that not all of the cytochrome oxidases are necessarily present at the same time. [After Jones (26)].

The ATPase (ATP Synthetase) Complex

The properties of membrane-bound bacterial ATPase complexes have been investigated using preparations from several species of aerobically grown organisms, particularly *E. coli* and the thermophilic bacterium PS3, but also to varying extents *S. typhimurium, B. megaterium, My. phlei, My. lysodeikticus,* and *Alcaligenes faecalis* (11, 14, 15, 114–118). In each of these organisms, as in mitochondria and chloroplasts prepared from eukaryotic organisms, the ATPase complex appears to consist of two multiprotein complexes, BF_1 and BF_0. The corresponding complexes from the thermophilic bacterium PS3 are often termed TF_1 and TF_0 respectively, but in this review all bacterial ATPase complexes will be abbreviated as BF_0 and BF_1.

BF_1 is an asymmetrically located hydrophilic protein complex that is responsible for the binding and activation of adenine nucleotides and inorganic phosphate. It is easily detached from the membrane and in its soluble form catalyzes the essentially irreversible hydrolysis of ATP. In contrast, BF_0 is a hydrophobic lipoprotein complex that forms an intrinsic part of the coupling membrane. The major function of BF_0 is probably to facilitate the transmembrane passage of H^+ and water to and from the active site of BF_1. Apart from their completely different physical and biochemical properties, BF_0 and BF_1 also exhibit different sensitivities to classic inhibitors of ATPase activity. Thus, 4-chloro-7-nitrobenzofurazan (Nbf-Cl), azide, and quercetin react with BF_1, whereas N,N'-dicyclohexylcarbodiimide (DCCD) exerts its action on BF_0. Unlike the energy-transducing ATPase complexes of higher organisms, the bacterial enzymes are generally insensitive to oligomycin. The activity of the entire BF_0-BF_1 complex is, of course, reversible in that it can act, under the appropriate conditions, either as an ATPase or as an ATP synthetase. Both of these reactions are sensitive to Nbf-Cl, azide, quercetin, and DCCD.

The following discussion of the properties and functions of bacterial ATPase complexes concentrates predominantly on the ATPases from *E. coli* and thermophilic bacterium PS3. ATPases from other aerobically grown bacteria are considered only where they offer novel or extraordinary aspects that render them significantly different from the *E. coli* and PS3 enzymes.

The Isolation and General Properties of BF_1 BF_1 is readily released from bacterial coupling membranes either by washing the latter several times in Mg^{2+}-free, low ionic strength buffers (e.g., 2mM Tris-HCl), by exposing them to selected organic solvents and detergents (e.g., *n*-butanol, Triton X-100, or sodium dodecyl sulfate), or by treatment with chaotropic agents (e.g., guanidine-HCl or urea). The solubilized enzyme can then be purified to homogeneity using relatively standard procedures involving combinations of ammonium sulfate fractionation, ion-exchange or exclusion chromatography, and polyacrylamide gel electrophoresis (117, 119–122).

BF_1 prepared in this manner is a relatively large multisubunit protein with a molecular weight of approximately 340,000 (280,000–404,000) and is therefore similar in size to the corresponding enzyme from mitochondria (F_1) and chloroplasts (CF_1). Some BF_1 preparations contain tightly bound inorganic phosphate and adenine nucleotides, often in amounts commensurate with that of the enzyme itself on a molar basis. Distinct binding sites for ADP and ATP (the latter detected using a nonhydrolyzable analog of ATP) appear to be present in *My. phlei* BF_1, thus indicating that the synthesis and hydrolysis of ATP may occur at separate loci. Both reactions invariably require the presence of divalent metal ions (optimally Mg^{2+} or Ca^{2+}).

All of the BF_1 preparations examined so far exhibit certain properties that are quite different from those shown by their membrane-bound $BF_0\text{-}BF_1$ complex counterparts. Thus, in addition to their insensitivities to certain $BF_0\text{-}BF_1$ complex inhibitors (e.g., DCCD), BF_1 preparations generally lose ATPase activity on storage at $0\text{-}4\,^\circ C$ and are therefore said to be cold labile (the BF_1 from *My. phlei* and the thermophile PS3 appear to be exceptional in this respect and, like all $BF_0\text{-}BF_1$ complexes, are cold stable). Since cold lability is enhanced by high salt concentrations (e.g., 0.5 M Tris-HCl) and diminished by various protein-stabilizing agents (e.g., glycerol or methanol), the phenomenon is compatible with the dissociation of a soluble quaternary structure into its constituent subunits (see next section). This phenomenon, in which a solubilized enzyme exhibits different properties as compared with its membrane-bound form, is called allotopy and is apparently common to all energy-transducing ATPases.

The Composition and Subunit Function of BF_1 Exposure of BF_1 to low concentrations of sodium dodecyl sulfate readily causes the enzyme to dissociate into its constituent subunits, which can then be separated by polyacrylamide gel electrophoresis. The relative concentrations of the individual subunits can be determined either by colorimetric analysis of the gels after staining with an appropriate dye (e.g., Coomassie Brilliant Blue) or, more accurately, by radioactive analysis of ^{14}C-labeled samples.

This type of approach generally indicates the presence of five nonidentical subunits in BF_1; these are termed α, β, γ, δ, and ϵ, and have approximate molecular weights of 56,000, 53,000, 32,000, 15,500–21,000, and 11,000–13,000, respectively (see refs. 14 and 15 for more detailed information). In contrast, BF_1 from *A. faecalis, My. lysodeikticus,* and *B. megaterium* appear to lack either the δ or ϵ subunit. However, because the δ subunit content of *E. coli* BF_1 is known to vary with the precise nature of the isolation-purification procedure employed, it is possible that these two minor peptides are present in *A. faecalis, My. lysodeikticus,* and *B. megaterium* BF_1 in vivo but are lost during the preparation procedures.

Analysis of relative subunit concentrations indicates that BF_1 probably has the composition $\alpha_3\beta_3\gamma\delta\epsilon$, but since the δ and ϵ peptides comprise only approximately 10% by weight of the total enzyme this stoichiometry should be treated with some caution; slightly different stoichiometries have also been proposed (123).

The extremely difficult problem of determining the functions of the individual subunits of BF_1 has recently been facilitated by the advent of various methods for effecting: (1) the selective removal of individual subunits from BF_1; (2) the larger scale purification of the subunits; (3) the partial or complete reconstitution of BF_1 from its constituent subunits; and (4) the preparation of artificial inside-out membrane vesicles containing intact or reconstituted BF_1 plus BF_0 and phospholipids (proteoliposomes) (118, 121, 124–129).

Following the extensive application of several of these techniques to BF_1 from *E. coli* and PS3, there appears to be little doubt that the β subunit has a mainly catalytic function and that, possibly in association with the α subunit, it forms the active site of the ATPase (ATP synthetase). This conclusion is based on the observations that: 1) antibodies prepared against a combination of α and β subunits inhibit both ATPase activity and a variety of ATP-dependent membrane functions (but have no significant effect on the corresponding respiration-linked reactions); 2) trypsin stimulates the ATPase activity of intact BF_1 (from *E. coli,* but not PS3) while at the same time digesting significant amounts of the smaller subunits but leaving the α and β subunits largely unimpaired; 3) mixtures of $\beta\gamma$ and $\alpha\beta\delta$ subunits exhibit substantial ATPase activity, as do mixtures of $\alpha\beta$ subunits from *E. coli* (but not PS3); and 4) combinations of subunits that do not include the β subunit exhibit no significant ATPase activity (121, 122, 124, 128). It should be noted, however, that the isolated β subunit has no catalytic activity per se, and that association of this subunit with the γ and α subunits (from PS3) or the α subunit (from *E. coli*) is apparently an essential prerequisite for the expression of catalytic activity (121, 128). Studies with ^{3}H-labeled Nbf-Cl indicate that the β subunit is the site of action of this inhibitor, and that the latter probably reacts with a tyrosine residue close to the active site (124, 130).

The attractive concept, drawn mainly from studies with the ATPases of mitochondria and chloroplasts, of a trypsin-sensitive subunit of BF_1 that functions as an ATPase inhibitor and thus effects the gross regulation of ATPase/ATP synthetase activity in vivo is currently undergoing a certain amount of reappraisal. In *E. coli* this ATPase-inhibitor polypeptide is a tightly bound component of BF_1 (in contrast to its functional analog in F_1 and CF_1, which can be removed by Sephadex chromatography), has a molecular weight of 12,000–16,000, and is probably identical with the ϵ subunit; furthermore, small amounts of purified ϵ strongly inhibit the ATPase activities of intact BF_1 and of BF_1 that is deficient in the δ subunit (121, 124, 131). In contrast, trypsin inhibits rather than potentiates the ATPase activity of thermophile BF_1, and neither the intact BF_1 nor the active $\alpha\beta\gamma$ reconstituted complex is inhibited by the purified ϵ subunit (128).

The γ subunit appears to be required for the expression of inhibition by the ϵ subunit of *E. coli* BF_1 and to protect the thermolabile β subunit of the PS3 enzyme from heat denaturation. More importantly, recent studies using artificial proteoliposomes containing reconstituted ATPase complexes from PS3 suggest that the prime function of the γ subunit may be to control the passage of protons to and from the active site of BF_1, a role that is compatible with the observation that the absence of this subunit from reconstituted complexes and selected unc^- mutants (see "Mutants Defective in Energy Transduction") eliminates energy transduction (ATP synthesis, ATP-dependent energization) but not ATP hydrolysis (118, 128, 129, 132). There is some evidence from the observed sensitivities to azide of variously reconstituted BF_1 complexes that the γ subunit may contain the azide binding site (128).

There is now reasonably good evidence that the function of the δ subunit is structural, rather than catalytic or regulatory, i.e., it facilitates the binding of BF_1 to an unidentified membrane component of BF_0. This conclusion is based on the observation that only BF_1 preparations that contain the δ subunit can bind to BF_1-depleted membranes and thus reconstitute ATP-dependent membrane energization. Those preparations that lack this subunit are ineffective at this type of reconstitution unless supplemented with either a combined $\delta\epsilon$ fraction or with the δ subunit alone (121, 124, 132). It therefore appears likely that this subunit comprises at least part of the stalk that connects BF_1 to BF_0, a view that is supported by recent claims that the active form of the subunit may be elongated in shape or even dimeric. The location of this subunit between the active site and/or γ subunit of BF_1 and the cytoplasmic surface of the coupling membrane would therefore imply a proton-translocating function for the δ subunit similar to that proposed for BF_0. In this respect it is interesting to note that there is some evidence that the δ subunit of chloroplast CF_1, which is similarly located, has the necessary protonophoric properties (133). Alternatively, since the ϵ subunit of the ATPase complex from the thermophilic bacterium PS3 has some ability to bind BF_1 to BF_0, it is possible that the δ and ϵ subunits interact to form the required proton-translocating channel (129).

The General Properties and Subunit Composition of BF_0 Because BF_0 alone has no readily assayable enzyme activity, considerable efforts are currently being directed toward the isolation of BF_0-BF_1 complexes that can be assayed via the sensitivity of their ATPase activities to DCCD (134–138). Complexes of this type have recently been isolated from *E. coli* and from the thermophilic bacterium PS3 following exposure of coupling membranes to selected detergents (e.g., cholate, deoxycholate, Triton X-100). Both preparations exhibit DCCD-sensitive ATPase activity, but the full expression of these properties requires the addition of exogenous phospholipids, presumably to replace endogenous components lost during the extraction procedures.

Sodium lauryl sulfate–polyacrylamide gel electrophoresis of a crude BF_0-BF_1 complex isolated from *E. coli* yields 12 different polypeptides, five of which ($\alpha\beta\gamma\delta\epsilon$) are the hydrophilic subunits of BF_1; two of the remaining seven components have been partially purified (ζ, M_r 29,000; η, M_r 8,000) and both are strongly hydrophobic (134, 135, 137, 138). In contrast, the BF_0-BF_1 complex isolated from the thermophilic bacterium PS3 appears to contain only three polypeptides that are not attributable to BF_1; all have been partially purified, are hydrophobic, and have molecular weights of 19,000, 13,500, and 5,400 (136). Since this complex can catalyze a two-way DCCD-sensitive translocation of protons across the coupling membrane of reconstituted proteoliposomes concomitant with the hydrolysis or synthesis of ATP (136, 139), it would appear that the number of polypeptides required for BF_0 activity is surprisingly small. It is therefore possible that up to four of the polypeptides tentatively assigned to *E. coli* BF_0 are in fact artifacts of isolation.

The Interaction of BF_0 *with* DCCD   A great deal of attention has recently been focused on the DCCD-binding component of BF_0. The purification of this component was initially hindered by its hydrophobicity and by its lack of obvious enzyme activity, but these problems have been overcome by employing organic solvents in the solubilization-separation procedures and by using radioactive DCCD as an easily detectable marker. Current purification procedures involve the extraction of [^{14}C]DCCD-labeled membranes with chloroform-methanol, followed by precipitation of the extracted material with diethyl-ether and subsequent purification by ion-exchange and molecular sieve chromatography (138). The pure, DCCD-reactive component is a proteolipid with a molecular weight of 8–9,000 and is thus probably identical to the η subunit of *E. coli* BF_0 (the corresponding subunit of the thermophile BF_0 has not yet been identified) (135, 137, 138).

The exact role of the DCCD-binding proteolipid in energy transduction is currently unclear, although there is increasing evidence that it is associated with the formation of a transmembrane channel through BF_0 that normally allows protons and water to pass to and from the active site of BF_1, but that is blocked in the presence of DCCD. This evidence stems mainly from the ability of DCCD to: 1) inhibit oxidative phosphorylation and ATP-dependent membrane energization; and 2) repair the loss of respiration-linked energy transduction (caused by an increased permeability to protons) that characterizes membrane vesicles prepared from certain mutants of *E. coli* (see "Mutants Defective in Energy Transduction") or that results from the exposure of vesicles from wild-type cells to selected chaotropic agents (140–142). Additional supporting evidence is provided by the observation that incorporation of either BF_0 or the DCCD-reactive proteolipid into artificial proton-impermeable liposomes renders the latter specifically permeable to protons, i.e., they can passively translocate protons in the presence of a suitable membrane potential (129, 138). This permeability is essentially abolished following the addition of DCCD or intact BF_1, but is unaffected by the addition of the δ and/or ϵ subunits of BF_1 unless these are supplemented by the presence of the γ subunit. In view of the possible protonophoric properties of the DCCD-binding proteolipid, it is interesting to note that its molecular weight is considerably greater than that of gramicidin, valinomycin, or other membrane-active ionophores, thus making a transmembrane orientation perfectly feasible. Furthermore, the pH profile for proton translocation through thermophilic BF_0-liposomes confirms that H^+ rather than OH^- is the translocated species and indicates the presence of a monoprotic binding site in the proton-translocating channel of BF_0 that has properties commensurate with those of a histidine "gate" (118).

The interaction of the DCCD with BF_0 is also being investigated using DCCD-resistant mutants (*E. coli* RF-7, DC1; *Streptococcus faecalis dcc*-8). Such mutants are selected via their ability to grow on solid medium containing an oxidizable carbon source (e.g., succinate or malate) supplemented with normally lethal concentrations of DCCD. Phenotypically, they exhibit energy transduction properties comparable with those of the wild-type organisms, but their ATPase and ATP-dependent membrane energization reactions are up to 100-fold less sensitive to DCCD (135, 143, 144). Such mutants are useful in several ways. Thus, mixtures of BF_1 and BF_1-depleted membranes prepared from wild-type and mutant organisms can be used to reconstitute hybrid ATPase complexes, the properties of which show unambiguously that the locus of DCCD inhibition is BF_0 rather than BF_1. Furthermore, sodium dodecyl sulfate–polyacrylamide gel electrophoresis of membranes prepared from these mutants generally indicates the continued presence of the DCCD-binding proteolipid, although it is presumably modified in such a way that many of its potentially reactive protein functional groups (carboxyl, hydroxyl, thiol, amino) are no longer accessible to DCCD in vivo. In this respect it is interesting to note that the binding of DCCD to the DCCD-reactive proteolipid purified from wild-type cells is accompanied by a significant diminution of overall negative charge, as would occur if the reaction involved condensation of DCCD with a protein carboxyl group to yield the N-acyl urea derivative (137, 138).

Subunit Organization in the BF_0-BF_1 *Complex*   These various types of elegant reconstitution experiments throw considerable light not only on the functions of the individual subunits of the BF_0-BF_1 complex, but also on the organization of these subunits within this complex. Information about the latter has also been obtained via electron microscopy (145–147) and via the use of protein cross-linking reagents like dithiobis-succinimidyl propionate (DSP) (122). DSP cross-links adjacent polypeptides through the ϵ-amino groups of lysine residues (in a reaction that can be reversed by the addition of suitable reducing agents) and thus affords a powerful means of investigating the location and proximity of the constituent subunits of BF_1.

The results of these various approaches suggest that the α and β subunits are arranged alternately in the form of a planar hexagon, the central hole of which contains the γ subunit. In view of the fact that some BF_1 preparations are deficient in the δ or ϵ subunit, but still manage to form an active complex from the other subunits, it would appear that the δ and ϵ subunits are not involved in binding the γ subunit to the $\alpha_3\beta_3$ hexagon and therefore presumably lie on the periphery of the molecule. This conclusion is supported by convincing evidence that the principal functions of the δ subunit (possibly in association with the ϵ subunit) are to bind BF_0 and to facilitate the reversible transfer of protons between the DCCD-binding channel in BF_0 and the active site of BF_1. The functional, chemical, and morphological properties of BF_0-BF_1 are compatible with the general structural model shown in Figure 3, assuming that the $\alpha_3\beta_3\gamma\delta\epsilon$ composition of BF_1 is essentially correct and that the function of the ϵ subunit is as described for the thermophile ATPase.

Mutants Defective in Energy Transduction

Investigations into the functions of the BF_0-BF_1 complex have recently received considerable impetus from the isolation of mutant strains of *E. coli* that are defective in energy transduction (*unc⁻*). Phenotypically these mutants are capable of aerobic growth on glucose but not on nonfermentable carbon sources (e.g., succinate, malate, or acetate), and none of them is

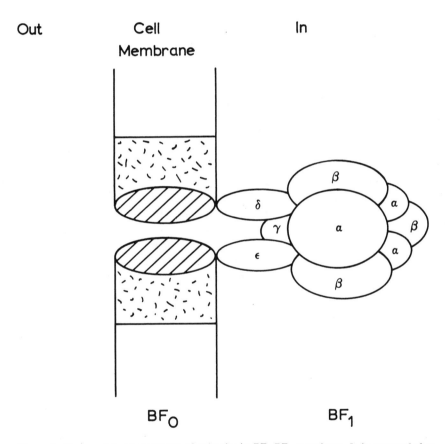

Figure 3. The possible arrangement of subunits in BF_0-BF_1 complexes. It is assumed that BF_1 has the subunit composition $\alpha_3\beta_3\gamma\delta\epsilon$, and that the δ and ϵ subunits form a proton-translocating cleft in BF_1 as described for bacterium PS3. BF_0 is composed of the DCCD-binding subunit (cross-hatches) plus phospholipids and at least two other subunits (dots). [After Yoshida et al. (129).]

able to grow anaerobically under conditions where anaerobic respiration is completely eliminated (i.e., where substrate-level phosphorylation is the sole mechanism of energy transduction). Furthermore, the aerobic molar growth yields ($Y_{glucose}$) of these mutants are substantially lower than those of the wild-type strains (46, 47). These growth patterns are compatible with the presence of defective ATPase complexes such that the mutants are energetically un-coupled (i.e., they are incapable of oxidative phosphorylation and ATP-dependent membrane energization, but their capacity for respiration is unimpaired). Because some of these mutants exhibit ATPase activity while others do not, this criterion can be used to divide them into two distinct classes—unc^- ATPase$^-$ and unc^- ATPase$^+$ (48).

A large number of mutants of both classes have now been isolated, but it is clearly beyond the scope of this chapter to discuss the detailed properties of all of them; indeed, many of these mutants are as yet incompletely characterized. The following discussion is therefore restricted to five mutants (uncA, uncB, uncC, DL54, and B_{V4}), all of which have been studied exten-sively, are fairly well characterized, and exhibit significantly different le-sions. For a more comprehensive and detailed analysis of unc^- mutants, the reader is referred to several recent reviews (14, 46–48, 148).

unc^- $ATPase^-$ (uncA, DL54)   Generally speaking, membrane vesicles prepared from these mutants catalyze neither ATP hydrolysis, oxidative phosphorylation, nor any ATP-dependent membrane reactions. On the other hand, their respiratory activities are undiminished and they exhibit a variable capacity for respiration-dependent membrane energization. It is clear, therefore, that these mutants can neither synthesize nor utilize ATP, although they have some capacity to utilize the protonmotive force generated by electron transfer. These properties thus explain why both a fermentable carbon source (e.g., glucose) and a terminal electron acceptor (e.g., oxygen, nitrate, fumarate) are obligatory for growth, since the former provides via substrate-level phosphorylation the ATP that is necessary for the biosynthesis of cell components, while the latter furnishes via respiration the proton-motive force that is required for active transport or reversed electron transfer (transhydrogenation).

The ATPase$^-$ phenotype can obviously reflect a variety of possible biochemical defects, most of which are associated with BF_1, e.g., the failure to synthesize BF_1, or the synthesis of BF_1 with a defective catalytic ability and/or an impaired ability to bind to the coupling membrane. The uncA mutant appears to fall into the second of these categories, because its BF_1 is membrane-bound but exhibits no ATPase activity. Furthermore, the unique ability of a hybrid ATPase complex constructed from wild-type BF_1 plus BF_1-depleted membranes from the uncA mutant to exhibit oxidative phosphorylation and various ATP-dependent membrane reactions confirms that the uncA lesion is located in BF_1 rather than in BF_0 (47, 149). The nor-mal proton conductance and respiratory chain energization properties of this mutant support this conclusion, and lend further credence to the current view that the α and/or β subunits of the mutant BF_1 are defective.

In contrast, BF_1 from mutant DL54 is only poorly bound to the coupling membrane. The latter exhibits relatively poor respiration-dependent energization properties and a high permeability to protons, but both defects can be repaired by the addition of DCCD or wild-type BF_1 (150, 151). These properties suggest that the BF_1 of this mutant is physically defective in such a manner that it binds improperly to the coupling membrane and fails to cover all of the internal exits of the proton translocating channels through BF_0, thus impairing the ability of the membrane to maintain the protonmotive force. Since there is good evidence that in wild-type *E. coli* the δ subunit of BF_1 is normally responsible for binding the latter to BF_0, thus preventing the useless dissipation of the protonmotive force, it is likely that this subunit is structurally defective in mutant DL54.

unc⁻ ATPase⁺ (uncB, uncC, B_V4) Generally speaking, membrane vesicles prepared from these mutants exhibit significant ATP hydrolysis, but do not catalyze either oxidative phosphorylation or ATP-dependent membrane energization. Their capacity for respiration and respiration-linked proton translocation is unimpaired, but their ability to maintain and utilize a protonmotive force at the expense of respiration is rather variable. Apart from their capacity for ATP hydrolysis, the membrane properties of *unc⁻ ATPase⁺* mutants are not grossly dissimilar to those of the *unc⁻ ATPase⁻* mutants. Furthermore, because the ability to hydrolyze ATP without concomitant energy conservation is of little general benefit to an organism, the growth characteristics of these mutants are also similar to those of the *unc⁻ ATPase⁻* mutants.

The energy transduction properties of these *unc⁻ ATPase⁺* mutants clearly indicate a diminished capacity to maintain a transmembrane protonmotive force and thus point to defects in those components of the ATPase complex that are responsible for directing the rate and pathway of proton translocation through the coupling membrane. Analyses of hybrid ATPase complexes reconstituted from BF_1 and BF_1-depleted membranes from mutant and wild-type strains indicate that the *unc⁻ ATPase⁺* phenotype can reflect alterations in either BF_0 or BF_1. Therefore these lesions may appear as defects in those components of the membrane that form the proton-translocating channel (*uncB*) (47, 149), or that interact with the stalk subunit of BF_1 (*uncC*) (152); alternatively, they may reflect structural alterations in the γ and/or δ subunits of BF_1 itself (B_{V4}) (153, 154). Since these various lesions lead to a complete loss of energy transduction at the expense of ATP hydrolysis, but not at the expense of respiration (otherwise growth would not be possible), it is clear that only the latter can translocate protons fast enough to compete effectively with the increased leakiness of the coupling membrane; however, the protonmotive force so generated is apparently sufficient only to drive active transport or transhydrogenation, not the net synthesis of ATP.

All of the *unc⁻ ATPase⁻, unc⁻ ATPase⁺*, and DCCD-resistant mutants of *E. coli* that have so far been examined genetically, usually via phage transduction techniques, map at approximately 73.5 min on the chromosome and are largely co-transducible with the *ilv* and *asn* loci. It would appear, therefore, that the structural genes that code for the various polypeptide components of BF_0 and BF_1 form a relatively tight cluster on the chromosome (47, 48, 155). There is some evidence, from work with a possible regulatory mutant of *E. coli* (AN 295), that a regulator gene for this putative operon may be located at approximately 77 min (156).

$\rightarrow H^+/P$ Quotients

In order to comply with the P/O ($P/2e^-$) and $\rightarrow H^+/O(\rightarrow^+/2e^-)$ quotients that have been determined experimentally with a variety of bacterial respiratory systems, a minimum of two protons must be retranslocated through the ATP synthetase for each molecule of ATP synthesized; the same stoichiometry (but of opposite sign) should therefore characterize ATP hydrolysis. Unfortunately, the direct determination of $\rightarrow H^+/P$ ($\rightarrow H^+/ATP$) quotients is experimentally very difficult, because the ATPase (ATP synthetase) reaction involves not only the translocation of protons but also the liberation (consumption) of protons as a result of ionization changes, i.e.:

$$x H^+_{(L)} + ADP^{3-} + HPO_4^{(1+y)-} + y H^+ \rightleftharpoons ATP^{4-} + H_2O + x H^+_{(R)}$$

It is therefore necessary to distinguish experimentally between that part of the total pH change that reflects ionization imbalances ($y H^+$, where y has a value of up to 0.8/ATP depending on the pH of the reaction) and that part that reflects transmembrane proton movements resulting from the anisotropic action of the BF_0-BF_1 complex ($x H^+$, when x is equal to the $\rightarrow H^+/P$ quotient). The problem is compounded by the absence of an adenine nucleotide translocase from the plasma membranes of bacteria, which thus rules out the use of whole cells, spheroplasts, protoplasts, and right-side-out vesicles, and restricts these measurements to inside-out vesicles and reconstituted proteoliposomes.

It is not surprising, therefore, that few attempts have been made to directly measure the $\rightarrow H^+/P$ quotients of bacterial systems. Nevertheless, it has been reported that ATP hydrolysis by inside-out vesicles from *E. coli* is accompanied by a net inflow of protons with a maximum $\rightarrow H^+/P$ quotient of approximately 0.6 (157); because this value is not corrected for the undoubted presence of leaky vesicles, the true $\rightarrow H^+/P$ quotient is likely to be considerably higher. Indeed, Kagawa and his colleagues claim an $\rightarrow H^+/P$ quotient of approximately 2 for ATP hydrolysis by reconstituted proteoliposomes from the thermophile PS3 (127), and predictions from the true molar growth yields of bacteria growing under aerobic and anaerobic conditions (see "Energy Coupling Sites") indicate $\rightarrow H^+/P$ quotients that are close to the $\rightarrow H^+/2e^-$ quotients for each proton-translocating respiratory segment (i.e., $\lessdot 2$). The influx of protons observed during ATP hydrolysis by inside-out vesicles or reconstituted proteoliposomes is of course commensurate with an inflow of protons during ATP synthesis by whole cells, spheroplasts, protoplasts, and right-side-out vesicles, exactly as predicted by the chemiosmotic hypothesis and in line with the observed ejection of protons during respiration by these latter preparations.

The Protonmotive Force

Although there is now little doubt that both respiration and ATP hydrolysis lead to the generation of an electrochemical potential difference of protons or protonmotive force (Δp or $\Delta \bar{\mu}_H{}^+$) of similar sign across the bacterial coupling membrane, only very recently have attempts been made to determine the magnitude and composition of Δp. The latter is composed of a chemical potential difference (ΔpH) and an electrical potential difference or membrane potential ($\Delta \psi$), which are related by the equation:

$$\Delta p = \Delta \psi - Z \cdot \Delta pH$$

where $Z = 2.303\, RT / F$ and has a value of approximately 60 mV at 25 °C. Experimental techniques have therefore been developed that allow the separate assay of these components following the energization of the coupling membrane. However, since microelectrode techniques are not yet sufficiently developed for these purposes, both ΔpH and $\Delta \psi$ are usually assayed indirectly via their ability to cause the unequal transmembrane distribution of suitable indicator molecules.

Many weak acids and weak bases are able to penetrate membranes much more readily in their un-ionized forms (HA, B) than in their ionized states (A^-, BH^+). The concentrations of the radioactive or fluorescent species $HA + A^-$ or $B + BH^+$ on either side of the coupling membrane following energization by respiration or ATP hydrolysis will therefore reflect the transmembrane pH differential, because the weak acid will accumulate in the alkaline compartment and the weak base in the acidic compartment. The ΔpH ($pH_{out} - pH_{in}$) of energized whole cells or right-side-out vesicles (inside alkaline) is therefore routinely determined from the distribution of the weak acids [1-^{14}C]acetate, [1-^{14}C]propionate and 5,5'-[2-^{14}C]dimethyloxazolidine-2,4-dione (DMO), whereas the weak bases [^{14}C]methylamine and 9-aminoacridine (fluorescent) are more effective indicators of the ΔpH of inside-out vesicles or reconstituted proteoliposomes (inside acidic) (127, 158).

For the measurement of $\Delta \psi$, selected organic and inorganic ions (most frequently cations) are employed that are capable of freely penetrating the coupling membrane and that therefore distribute themselves according to the prevailing electrical potential difference. The transmembrane distributions of these ions following energization are measured using appropriate radioactive, spectroscopic, or fluorimetric techniques and $\Delta \psi$ is calculated via the Nernst equation:

$$\Delta \psi = \frac{2.303\, RT}{n\, F} \log \frac{[\text{cation}]_{in}}{[\text{cation}]_{out}}$$

where n is the valency of the chosen ion. For this method of determining $\Delta\psi$ to be valid, it is important that the penetrating species diffuses passively through the membrane, that it is not chemically altered as a result of transport, and that it does not interfere with the metabolism or integrity of the whole cells or vesicles. The most commonly used ions that satisfy these criteria are: the lipophilic cations [^{3}H]dibenzyl-dimethyl-ammonium (DDA$^+$), [^{3}H]triphenylmethyl-phosphonium (TPMP$^+$) and 3,3'-dipropylthio-dicarbocyanine (DiS·C$_3^+$); the lipophilic anions [^{3}H]tetraphenylboron (TPB$^-$), [^{14}C]thiocyanate (SCN$^-$) and 8-anilinonaphthalene-1-sulfonate (ANS$^-$; fluorescent); and either ^{36}Rb$^+$ or K$^+$ in the presence of a suitable ionophore, such as valinomycin (127, 158, 159).

The determination of [cation]$_{in}$, like that of pH$_{in}$, requires an accurate knowledge of the internal volume of the cells or vesicles. In both cases this can be achieved by measuring, in parallel experiments, the external and total (external plus internal) concentrations of permeant ([^{3}H]water) and impermeant ([^{14}C]inulin or [^{3}H]sorbitol) marker molecules; [cation]$_{out}$ and pH$_{out}$ are determined by direct assay. These techniques for measuring the components of the protonmotive force are excellently reviewed in more detail elsewhere (160).

ΔpH, $\Delta\psi$, and Δp have recently been determined for several organisms, including whole cells and membrane vesicles of *Staphylococcus aureus, E. coli,* and *Pa. denitrificans* energized by respiration, and reconstituted proteoliposomes from the thermophile PS3 (BF$_0$-BF$_1$ complex plus PS3 phospholipids) energized by ATP hydrolysis (127, 158, 159, 161, 162). Maximum values of Δp are in the range of 130–308 mV, of which $\Delta\psi$ comprises 75–134 mV and $Z\Delta$pH comprises 0–195 mV (0–3.10 pH units); in whole cells and right-side-out vesicles the internal compartment is invariably alkaline and electrically negative, whereas the reverse is true for inside-out vesicles and proteoliposomes (Table 2). In each case the generation of the protonmotive force is abolished both by uncoupling agents and by inhibitors of respiration or ATP hydrolysis, according to the nature of the energy source.

The Phosphorylation Potential

If the proton-translocating ATPase of intact cells or membrane vesicles is allowed to catalyze ATP hydrolysis to equilibrium, the force generated by proton translocation (Δp · →H$^+$:P) is poised against that exerted by the free energy of hydrolysis ($\Delta G_0'$ + 2.303 RT log [ATP]/[ADP][P$_i$]); the latter is called the phosphorylation potential (ΔGp), and hence:

$$\Delta p \cdot \rightarrow \frac{H^+}{P} = \Delta Gp = \Delta G_0' + 2.303\,RT\log\frac{[ATP]}{[ADP][P_i]}$$

ΔGp is conventionally expressed in terms of kJ per mol, but in order to facilitate comparison with proton translocation forces it is routinely converted to electrical units (V) by dividing by the Faraday constant (96.5 kJ per volt per mol). This relationship thus provides a potentially powerful method for determining the $\rightarrow H^+/P$ quotient indirectly from the measured values of Δp and ΔGp. Taking 230 mV as a typical value for Δp (Table 2), the $\rightarrow H^+/P$ quotient of 2 that has been obtained by direct measurements with mitochondria (163, 164), submitochondrial particles (165), and reconstituted bacterial proteoliposomes (127) would necessitate a ΔGp of 460 mV (44.4 kJ per mol). Values of this order or slightly higher have recently been reported for bacterial membrane vesicles (159), thus indicating $\rightarrow H^+/P$ quotients of not less than 2 g ion H^+ per mol phosphate esterified and possibly higher.

Oxidative Phosphorylation Under Artificial Conditions

It is now clear that bacterial respiration and ATP hydrolysis both lead to the formation of a transmembrane protonmotive force. Therefore, since both of these processes are fully reversible, it should be possible experimentally to drive ATP synthesis, reversed electron transfer, and other energy-dependent membrane functions at the expense of artificially imposed pH gradients and/or membrane potentials of the correct size and polarity. Indeed, relatively simple techniques for the transient generation of these forces across closed coupling membranes are now available (166, 167). Thus, with suspensions of intact cells or right-side-out vesicles, the necessary Δ pH (inside alkaline) can be established by rapidly acidifying the external environment following an initial incubation at a slightly alkaline pH (e.g., pH_{out} 3.0 as compared with pH_{in} 8.0), and the required $\Delta\psi$ (inside negative) can be produced by placing K^+-loaded preparations in a K^+-free medium and then adding valinomycin (the outward movement through the ionophore of K^+ down its own concentration gradient generates a K^+-diffusion potential, negative inside). In contrast, the ΔpH and $\Delta\psi$ that are required by inside-out vesicles and reconstituted proteoliposomes (inside acidic and positive) can be generated by essentially converse procedures.

Protonmotive forces induced by these techniques have recently been shown to be capable of driving the transient synthesis of ATP (detected by standard luminescence or spectrophotometric techniques) in whole cells, membrane vesicles, or reconstituted proteoliposomes of E. coli, Streptococcus lactis, or the thermophilic bacterium PS3 (139, 158, 167–170). In each of these organisms it is the magnitude, rather than the composition, of Δp that appears to be of major importance. Thus, although oxidative phosphorylation can be effected at the expense of either ΔpH or $\Delta\psi$ alone, the extent of ATP synthesis is dependent on the overall value of Δp such that net synthesis occurs only when Δp exceeds a threshold value of approximately 200–215 mV. ATP synthesis by these imposed protonmotive forces is abolished by uncoupling agents (e.g., carbonyl cyanide p-trifluorophenyl hydrazone (FCCP), nigericin plus valinomycin) and by inhibitors of the BF_0-BF_1 ATPase complex (e.g., DCCD, azide), but not by inhibitors of respiration [e.g., 2-n-heptyl 4-hydroxyquinoline-N-oxide (HQNO), cyanide]. Fur-

Table 2. Quantitative aspects of membrane energization by respiration or ATP hydrolysis[a]

Organism	Structure	Energy source	$\Delta\psi$ (mV)	$Z\Delta pH$ (mV)	Δp (mV)	Internal compartment
Escherichia coli	Whole cells	Respiration	132	98	230	Alkaline, negative
	Right-side-out vesicles	Respiration	75	120	195	Alkaline, negative
Staphylococcus aureus	Whole cells	Respiration	134	77	211	Alkaline, negative
Paracoccus denitrificans	Inside-out vesicles	Respiration	130	0	130	Acidic, positive
Thermophilic bacterium PS3	Reconstituted proteoliposomes	ATP hydrolysis	113	195	308	Acidic, positive

[a]These data are taken from refs. 127, 158, 159, 161, and 162.

thermore, although mutants of *E. coli* that are defective in oxidative phosphorylation (e.g., *unc* A or *unc* B) are no longer able to catalyze the artificial synthesis of ATP, this capacity is fully retained by cytochrome-deficient mutants of this organism (166, 170).

These results thus strongly support a chemiosmotic mechanism for bacterial oxidative phosphorylation in which an appropriately directed protonmotive force drives the synthesis of ATP via a reversible proton-translocating ATPase complex (without any interaction with redox components). Furthermore, the threshold values of Δp (200–215 mV) are similar to the Δp values that are generated following the energization of bacterial coupling membranes via respiration or ATP hydrolysis (see "The Protonmotive Force") and, very importantly, are in good agreement with the values predicted by the chemiosmotic hypothesis on the assumption that the $\rightarrow H^+/P$ quotient is 2 (7).

CONCLUDING REMARKS

In spite of the increased pace of research into bacterial respiration and oxidative phosphorylation over the last five years, and of the almost general acceptance of the major tenets of chemiosmosis to explain these phenomena, a number of problems remain. Therefore, several respiratory chain components still await purification (i.e., transhydrogenase, NADH dehydrogenase, and the various cytochrome oxidases) prior to carrying out detailed analysis of their composition and electron transfer properties. In addition, the control of respiratory chain synthesis and the spatial organization of the respiratory chain within the coupling membrane both merit further investigation, as do the complex patterns of terminal electron flow that are exhibited by certain organisms. Ultimately, of course, it should be possible to reconstitute proton-translocating bacterial respiratory chains into vesicular form using purified redox components and phospholipids.

On the energy conservation side, it is clear that the complete, functional reconstitution of the BF_0-BF_1 complex from its constituent subunits is now close at hand for at least two organisms, but the exact roles of several of the subunits are currently unclear and the events that occur at the active site of the enzyme during ATP synthesis are still largely a matter of conjecture. In addition, many of the unc^- mutants that have been isolated have yet to be properly characterized, and this rather confusing area urgently demands clarification. Finally, some experimental time might also be profitably devoted to the accurate parallel determination of ΔGp and Δp, thus allowing unambiguous $\rightarrow H^+/P$ quotients to be calculated.

Since most of the biochemical and genetic techniques that are required for these studies are currently available, one might risk a modicum of future embarrassment by predicting that within another five years the detailed nature of respiration and oxidative phosphorylation in heterotrophic bacteria will be largely resolved.

REFERENCES

1. Bauchop, T., and Elsden, S. R. (1960). J. Gen. Microbiol. 23:457.
2. Stouthamer, A. H., and Bettenhaussen, C. W. (1973). Biochim. Biophys. Acta 301:53.
3. Stouthamer, A. H., and Bettenhaussen, C. W. (1975). Arch. Microbiol. 102:187.
4. Thauer, R. K., Jungermann, K., and Decker, K. (1977). Bacteriol. Rev. 41:100.
5. Stouthamer, A. H. (1977). *In* B. A. Haddock and W. A. Hamilton (eds.), Microbial Energetics, p. 285. Cambridge University Press, Cambridge, Eng.
6. Aleem, M. I. H. (1977). *In* B. A. Haddock and W. A. Hamilton (eds.), Microbial Energetics, p. 351. Cambridge University Press, Cambridge, Eng.
7. Mitchell, P. (1966). Biol. Rev. 41:445.
8. Mitchell, P., and Moyle, J. (1974). Biochem. Soc. Special Publ. 4:91.
9. Mitchell, P. (1976). Biochem. Soc. Trans. 4:399.
10. Boyer, P. D., Chance, B., Ernster, L., Mitchell, P., Racker, E., and Slater, E. C. (1977). Annu. Rev. Biochem. 46:955.
11. Harold, F. M. (1972). Bacteriol. Rev. 36:172.
12. Williams, R. J. P. (1974). Ann. N.Y. Acad. Sci. 227:98.
13. Garland, P. B. (1977). *In* B. A. Haddock and W. A. Hamilton (eds.), Microbial Energetics, p. 1. Cambridge University Press, Cambridge, Eng.
14. Haddock, B. A., and Jones, C. W. (1977). Bacteriol. Rev. 41:47.
15. Harold, F. M. (1977). Curr. Top. Bioenergetics 6:83.
16. John, P., and Whatley, F. R. (1975). Nature 254:495.
17. John, P., and Whatley, F. R. (1977). Biochim. Biophys. Acta 463:129.
18. Lemberg, R., and Barrett, J. (1973). Cytochromes. Academic Press, New York and London.
19. Jurtshuk, P., Mueller, T. J., and Acord, W. C. (1975). C. R. C. Crit. Rev. Microbiol. 3:399.
20. Gel'man, N. S., Lukoyanova, M. A., and Ostrovskii, D. N. (1975). *In* Bacterial Membranes and the Respiratory Chain: Biomembranes, Vol. 6, p. 129. Plenum Press, New York and London.
21. Jones, C. W., and Meyer, D. J. (1978). *In* H. LeChevalier and A. I. Laskin (eds.), Handbook of Microbiology, Vol. III. C. R. C. Press, Cleveland, Oh. (in press)
22. Smith, L., White, D. C., Sinclair, P., and Chance, B. (1970). J. Biol. Chem. 245:5096.
23. Haddock, B. A., Downie, J. A., and Garland, P. B. (1976). Biochem. J. 154:285.
24. Lawford, H. G., Cox, J. C., Garland, P. B., and Haddock, B. A. (1976). FEBS Lett. 64:369.
25. Smith, L. (1968). *In* T. P. Singer (ed.), Biological Oxidations, p. 55. Interscience Publications Inc., New York.
26. Jones, C. W. (1977). *In* B. A. Haddock and W. A. Hamilton (eds.), Microbial Energetics, p. 23. Cambridge University Press, Cambridge, Eng.
27. Jones, C. W. (1973). FEBS Lett. 36:347.
28. Kauffman, H. F., and Van Gelder, B. F. (1973). Biochim. Biophys. Acta 314:276.
29. Pudek, M. R., and Bragg, P. D. (1974). Arch. Biochem. Biophys. 164:682.
30. Shipp, W. S. (1972). Arch. Biochem. Biophys. 150:459.
31. Haddock, B. A. (1977). *In* B. A. Haddock and W. A. Hamilton (eds.), Microbial Energetics, p. 95. Cambridge University Press, Cambridge Eng.

32. Harrison, D. E. F. (1976). Adv. Microb. Physiol. 14:243.
33. Jones, C. W., Brice, J. M., and Edwards, C., unpublished results.
34. White, D. C. (1962). J. Biol. Chem. 238:3757.
35. Meyer, D. J., and Jones, C. W. (1973). FEBS Lett. 33:101.
36. Linton, D. J., Bull, A. T., and Harrison, D. E. F. (1977). Arch. Microbiol. 114:111.
37. Arima, K., and Oka, T. (1965). J. Bacteriol. 90:734.
38. Ashcroft, J. R., and Haddock, B. A. (1975). Biochem. J. 148:349.
39. Stouthamer, A. H. (1976). Adv. Microbial Physiol. 14:315.
40. Kröger, A. (1977). In B. A. Haddock and W. A. Hamilton (eds.), Microbial Energetics, p. 61. Cambridge University Press, Cambridge, Eng.
41. Broman, R. L., Dobrogosz, W. J., and White, D. C. (1974). Arch. Biochem. Biophys. 162:595.
42. Wright, L. F., and Knowles, C. J. (1977). FEMS Microbiol. Lett. 1:259.
43. Garland, P. B. (1970). Biochem. J. 118:329.
44. Lund, T., and Raynor, J. B. (1975). J. Bioenergetics 7:161.
45. Downie, J. A., and Garland, P. B. (1973). Biochem. J. 134:1051.
46. Gibson, F., and Cox, G. B. (1973). Essays Biochem. 9:1.
47. Cox, G. B., and Gibson, F. (1974). Biochim. Biophys. Acta 346:1.
48. Simoni, R. D., and Postma, P. (1975). Annu. Rev. Biochem. 44:523.
49. Cox, G. B., Newton, N. A., Gibson, F., Snoswell, A. M., and Hamilton, J. A. (1970). Biochem. J. 117:551.
50. Newton, N. A., Cox, G. B., and Gibson, F. (1971). Biochim. Biophys. Acta 244:155.
51. Haddock, B. A., and Downie, J. A. (1974). Biochem. J. 142:703.
52. Rockey, A. E., and Haddock, B. A. (1974). Biochem. Soc. Trans. 2:957.
53. White, D. C., and Sinclair, P. R. (1971). Adv. Microb. Physiol. 5:173.
54. Yates, M. G., and Jones, C. W. (1974). Adv. Microb. Physiol. 11:97.
55. Tikhonova, G. V. (1974). Biochem. Soc. Special Publ. 4:131.
56. Jones, C. W., and Redfearn, E. R. (1967). Biochim. Biophys. Acta 143:340.
57. Weston, J. A., Collins, P. A., and Knowles, C. J. (1974). Biochim. Biophys. Acta 368:148.
58. Linton, J. D., Harrison, D. E. F., and Bull, A. T. (1976). FEBS Lett. 64:358.
59. Cole, J. (1976). Adv. Microb. Physiol. 14:1.
60. Eilermann, L. J. M., Pandit-Hovenkamp, H. G., and Kolk, A. H. J. (1970). Biochim. Biophys. Acta 197:25.
61. John, P., and Hamilton, W. A. (1970). FEBS Lett. 10:246.
62. Jones, C. W., Ackrell, B. A. C., and Erickson, S. K. (1971). Biochim. Biophys. Acta 245:54.
63. Beatrice, M., and Chappell, J. B. (1974). Biochem. Soc. Trans. 2:151.
64. Neijssel, O. M. (1977). FEMS Microbiol. Lett. 1:47.
65. Ackrell, B. A. C., and Jones, C. W. (1971). Eur. J. Biochem. 20:22.
66. Neijssel, O. M., and Tempest, D. W. (1976). Arch. Microbiol. 107:215.
67. Neijssel, O. M., and Tempest, D. W. (1976). Arch. Microbiol. 106:251.
68. Downs, A. J., and Jones, C. W. (1975). Arch. Microbiol. 105:159.
69. Farmer, I. S., and Jones, C. W. (1976). Eur. J. Biochem. 67:115.
70. Konings, W. N. (1977). Adv. Microb. Physiol. 15:175.
71. Kaback, H. R. (1971). Methods Enzymol. 22:99.
72. DePierre, J. W., and Ernster, L. (1977). Annu. Rev. Biochem. 46:201.
73. Kaback, H. R. (1972). Biochim. Biophys. Acta 265:367.
74. Klein, W. L., and Boyer, P. D. (1972). J. Biol. Chem. 247:7257.
75. Van Thienen, G., and Postma, P. W. (1973). Biochim. Biophys. Acta 323:429.
76. Altendorf, K. H., and Staehelin, L. A. (1974). J. Bacteriol. 117:888.

77. Salton, M. J. R. (1974). Adv. Microb. Physiol. 11:213.
78. Houghton, R. L., Fisher, R. J., and Sanadi, D. R. (1975). Biochim. Biophys. Acta 396:17.
79. Hare, J. F., Olden, K., and Kennedy, E. P. (1974). Proc. Natl. Acad. Sci. USA 71:4843.
80. Hinds, T. R., and Brodie, A. F. (1974). Proc. Natl. Acad. Sci. USA 71:1202.
81. Adler, L. W., and Rosen, B. P. (1977). J. Bacteriol. 129:959.
82. Grinius, L. L., Il'ina, M. D., Mileykovskaya, E. I., Skulachev, V. P., and Tikhonova, G. V. (1972). Biochim. Biophys. Acta 283:442.
83. Hampton, M. L., and Freese, E. (1974). J. Bacteriol. 118:497.
84. Boxer, D. H., and Clegg, R. A. (1975). FEBS Lett. 60:54.
85. Mitchell, P. (1976). J. Theor. Biol. 62:327.
86. Mitchell, P. (1977). FEBS Lett. 78:1.
87. Scholes, P., and Mitchell, P. (1970). J. Bioenergetics 1:309.
88. Reeves, J. P. (1971). Biochem. Biophys. Res. Commun. 45:931.
89. West, I. C., and Mitchell, P. (1972). J. Bioenergetics 3:445.
90. Lawford, H. G., and Haddock, B. A. (1973). Biochem. J. 136:217.
91. Beatrice, M., and Chappell, J. B. (1974). Biochem. Soc. Trans. 2:151.
92. Hertzberg, E. L., and Hinckel, P. C. (1974). Biochem. Biophys. Res. Commun. 58:178.
93. Jones, C. W., Brice, J. M., Downs, A. J., and Drozd, J. W. (1975). Eur. J. Biochem. 52:265.
94. Tsuchiya, T. (1976). J. Biol. Chem. 251:5315.
95. Lawford, H. G. (1977). Can. J. Biochem. (in press)
96. Navon, G., Ogawa, S., Schulman, R. G., and Yamane, T. (1977). Proc. Natl. Acad. Sci. USA 74:888.
97. Brand, M. D., Reynafarje, B., and Lehninger, A. L. (1976). J. Biol. Chem. 251:5670.
98. Brand, M. D., Reynafarje, B., and Lehninger, A. L. (1976). Proc. Natl. Acad. Sci. USA 73:437.
99. Jones, C. W., Brice, J. M., and Edwards, C. (1977). Arch. Microbiol. 115:85.
100. Hempfling, W. P. (1970). Biochim. Biophys. Acta 205:169.
101. Hempfling. W. P. (1970). Biochem. Biophys. Res. Commun. 41:9.
102. Knowles, C. J., and Smith, L. (1970). Biochim. Biophys. Acta 197:152.
103. Baak, J. M., and Postma, P. W. (1971). FEBS Lett. 19:189.
104. John, P., and Hamilton, W. A. (1970). FEBS Lett. 10:246.
105. John, P., and Hamilton, W. A. (1971). Eur. J. Biochem. 23:528.
106. Eilermann, L. J. M. (1970). Biochim. Biophys. Acta 216:231.
107. Asano, A., Imai, K., and Sato, R. (1967). Biochim. Biophys. Acta 143:477.
108. Asano, A., Imai, K., and Sato, R. (1967). J. Biochem. (Tokyo) 62:210.
109. Asano, A., and Brodie, A. F. (1965). J. Biol. Chem. 240:4002.
110. Bragg, P. D., Davies, P. L., and Hou, C. (1972). Biochem. Biophys. Res. Commun. 47:1248.
111. Csonka, L. N., and Fraenkel, D. G. (1977). J. Biol. Chem. 252:3382.
112. Mitchell, P. (1975). FEBS Lett. 59:137.
113. Downs, A. J., and Jones, C. W. (1975). FEBS Lett. 60:42.
114. Abrams, A., and Smith, J. B. (1974). In P. D. Boyer (ed.), The Enzymes, Vol. X, p. 395. Academic Press, New York and London.
115. Salton, M. J. R. (1974). Adv. Microb. Physiol. 11:213.
116. Simoni, R. D., and Postma, P. W. (1975). Annu. Rev. Biochem. 44:523.
117. Yoshida, M., Sone, N., Hirata, H., and Kagawa, Y. (1975). J. Biol. Chem. 250:7910.
118. Kagawa, Y. (1978). Proceedings of the 11th FEBS Meeting (Copenhagen), Abstract B7-2(L2).

119. Kobayashi, H., and Anraku, Y. (1972). J. Biochem. (Tokyo) 71:387.
120. Mirsky, R., and Barlow, V. (1973). Biochim. Biophys. Acta 291:480.
121. Futai, M., Sternweis, P. C., and Heppel, L. A. (1974). Proc. Natl. Acad. Sci. USA 71:2725. *
122. Bragg, P. D., and Hou, C. (1975). Arch. Biochem. Biophys. 167:311.
123. Vogel, G., and Steinhart, R. (1976). Biochemistry 15:208.
124. Nelson, N., Kanner, B. I., and Gutnik, D. L. (1974). Proc. Natl. Acad. Sci. USA 71:2720.
125. Sone, N., Yoshida, M., Hirata, H., and Kagawa, Y. (1977). J. Biochem. (Tokyo) 81:519.
126. Smith, J. B., and Sternweis, P. C. (1977). Biochemistry 16:306.
127. Sone, N., Yoshida, M., Okamoto, H., Hirata, H., and Kagawa, Y. (1977). J. Membr. Biol. 30:121.
128. Yoshida, M., Sone, N., Hirata, H., and Kagawa, Y. (1977). J. Biol. Chem. 252:3480.
129. Yoshida, M., Okamoto, H., Sone, N., Hirata, H., and Kagawa, Y. (1977). Proc. Natl. Acad. Sci. USA 74:936.
130. Ferguson, S. J., John, P., Lloyd, W. J., Radda, G. K. and Whatley, F. R. (1974). Biochim. Biophys. Acta 368:461.
131. Nieuwenhuis, F. J. R. M., van der Drift, J. A. M., Voet, A. B., and Van Dam, K. (1974). Biochim. Biophys. Acta 368:461.
132. Bragg, P. D., Davies, P. L., and Hou, C. (1973). Arch. Biochem. Biophys. 159:664.
133. Younis, H. M., Winget, G. D., and Racker, E. (1977). J. Biol. Chem. 252:1814.
134. Hare, J. E. (1975). Biochem. Biophys. Res. Commun. 66:1329.
135. Fillingame, R. H. (1975). J. Bacteriol. 124:870.
136. Sone, N., Yoshida, M., Hirata, H., and Kagawa, Y. (1975). J. Biol. Chem. 250:7917.
137. Altendorf, K., and Zitzmann, W. (1975). FEBS Lett. 59:268.
138. Altendorf, K. (1977). FEBS Lett. 73:271.
139. Sone, N., Yoshida, M., Hirata, H., and Kagawa, Y. (1977). J. Biol. Chem. 252:2956.
140. Patel, L., Schuldinger, S., and Kaback, H. R. (1975). Proc. Natl. Acad. Sci. USA 72:3387.
141. Patel, L., and Kaback, H. R. (1976). Biochemistry 15:2741.
142. Hasan, S. M., and Rosen, B. P. (1977). Biochim. Biophys. Acta 459:225.
143. Abrams, A., Smith, J. B., and Baron, C. (1972). J. Biol. Chem. 247:1484.
144. Friedl, P., Schmid, B. I., and Schairer, H. U. (1977). Eur. J. Biochem. 73:461.
145. Ishida, M., and Mizushima, S. (1969). J. Biochem. (Tokyo) 66:33.
146. Schnebli, H. P., Vatter, A. E., and Abrams, A. (1970). J. Biol. Chem. 245:1122.
147. Kagawa, Y., Sone, N., Yoshida, M., Hirata, H., and Okamoto, H. (1976). J. Biochem. (Tokyo) 80:141.
148. Gutnick, D. L., and Fragman, D. (1978). In E. Quagliariello, F. Palmieri, and T. P. Singer (eds.), Horizons in Biochemistry and Biophysics, Vol. 3, p. 192. Addison-Wesley, Reading, Mass.
149. Cox, G. B., Gibson, F., and McCann, L. (1973). Biochem. J. 134:1015.
150. Bragg, P. D., and Hou, C. (1973). Biochem. Biophys. Res. Commun. 50:729.
151. Altendorf, K., Harold, F. M., and Simoni, R. D. (1974). J. Biol. Chem. 249:4587.
152. Gibson, F., Cox, G. B., Downie, J. A., and Radik, J. (1977). Biochem. J. 164:193.
153. Kanner, B. I., and Gutnick, D. L. (1972). J. Bacteriol. 111:287.

84   Jones

154. Nieuwenhuis, F. J. R. M., Kanner, B. I., Gutnick, D. L., Postma, P. W., and Van Dam, K. (1973). Biochim. Biophys. Acta 325:62.
155. Schairer, H. U., Friedl, P., Schmid, B. I., and Vogel, G. (1976). Eur. J. Biochem. 66:257.
156. Cox, G. B., Gibson, F., McCann, L. M., Butlin, J. D., and Crane, F. L. (1973). Biochem. J. 132:689.
157. West, I. C., and Mitchell, P. (1974). FEBS Lett. 40:1.
158. Ramos, S., and Kaback, H. R. (1977). Biochemistry 16:848.
159. Kell, D., John, P., and Ferguson, S. J. (1977). Proc. Soc. Gen. Microbiol. 4:66.
160. Malony, P. C., Kashket, E. R., and Wilson, T. H. (1975). Methods Membr. Biol. 5:1.
161. Jeacocke, R. E., Niven, D. F., and Hamilton, W. A. (1972). Biochem. J. 127:57.
162. Collins, S. H., and Hamilton. W. A. (1976). J. Bacteriol. 126:1224.
163. Mitchell, P., and Moyle, J. (1968). Eur. J. Biochem. 4:530.
164. Brand, M. D., and Lehninger, A. L. (1977). Proc. Natl. Acad. Sci. USA 74:1955.
165. Thayer, W. S., and Hinkel, P. C. (1973). J. Biol. Chem. 248:5395.
166. Wilson, D. M., Alderete, J. F., Maloney, P. C., and Wilson, T. H. (1976). J. Bacteriol. 126:327.
167. Tsuchiya, T., and Rosen, B. P. (1976). Biochem. Biophys. Res. Commun. 86:497.
168. Maloney, P. C., and Wilson, T. H. (1975). J. Membr. Biol. 25:285.
169. Tsuchiya, T., and Rosen, B. P. (1976). J. Bacteriol. 127:154.
170. Tsuchiya, T. (1977). J. Bacteriol. 129:763.

International Review of Biochemistry
Microbial Biochemistry, Volume 21
Edited by J. R. Quayle
Copyright 1979   University Park Press   Baltimore

3
Energy Metabolism in Anaerobes

G. GOTTSCHALK AND J. R. ANDREESEN

Institute of Microbiology
University of Göttingen, F.R.G.

The experimental work from our laboratory was supported by the "Deutsche Forschungs-gemeinschaft," the "Stiftung Volkswagenwerk," and by "Forschungsmittel des Landes Niedersachsen."

The authors would like to dedicate this article to H. A. Barker on the occasion of his 70th birthday (29 November 1977).

WHAT DOES ENERGY METABOLISM MEAN?

All organisms require a continuous input of energy for growth and for maintenance. This energy is provided either as chemical energy or as physical energy (light) that is transformed into biologically utilizable energy by the organisms. Nature has selected adenosine 5′-triphosphate (ATP) as the carrier of this biologically utilizable energy, and all endergonic processes in living cells are directly or indirectly coupled to the conversion of adenosine 5′-triphosphate to adenosine 5′-diphosphate (ADP) and inorganic phosphate (P_i),

$$\text{ATP} + H_2O \xrightarrow{\quad\text{endergonic processes in living cells}\quad} \text{ADP} + P_i$$

or to the conversion of ATP to adenosine 5′-monophosphate (AMP) and pyrophosphate (PP_i),

$$\text{ATP} + H_2O \xrightarrow{\quad\text{endergonic processes in living cells}\quad} \text{AMP} + PP_i$$

and are thus made feasible.

The free energy of hydrolysis of ATP to ADP + P_i and AMP + PP_i is $\Delta G_{obs}^{0\,\prime} = -7.6$ kcal/mol and $\Delta G_{obs}^{0\,\prime} = -9.96$ kcal/mol, respectively [all data concerning $G_0{}'$ and $E_0{}'$ were taken from Thauer et al. (1)]. Therefore, ATP is a very useful donor of phosphoryl or adenylyl (AMP) groups in enzyme-catalyzed reactions, and phosphorylated or adenylylated compounds are common intermediates in biosyntheses. AMP can be converted to ADP and PP_i converted to P_i by the enzymes adenylate kinase and pyrophosphatase, respectively:

$$\text{AMP} + \text{ATP} \xrightarrow{\text{adenylate kinase}} 2\,\text{ADP}$$

$$PP_i + H_2O \xrightarrow{\text{pyrophosphatase}} 2\,P_i$$

The principal products formed in energy-requiring processes of organisms are therefore ADP and P_i. Hence, provision of organisms with biologically utilizable energy means that ATP is produced from ADP and P_i, and the reactions involved in ATP synthesis are usually described as the energy metabolism of the organisms.

$$\text{ADP} + P_i \xrightarrow{\quad\text{energy metabolism}\quad} \text{ATP} + H_2O$$

DIFFERENCES IN ENERGY METABOLISM AMONG ORGANISMS

Organisms differ in their energy metabolism. Plants and phototrophic bacteria use light as energy source for ATP synthesis by photophosphorylation (2, 3, 4). Aerobic organisms take advantage of oxygen as terminal electron acceptor. Organic substrates are oxidized to carbon dioxide and water and most of the ATP is produced by electron transport phosphorylation in the respiratory chain (5, 6, 7). Several mol of ATP can be produced per mol of substrate; the oxidation of glucose, for instance, may be coupled to the synthesis of as many as 38 mol of ATP:

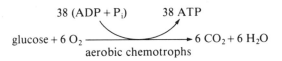

Organisms that live under anaerobic conditions and that cannot take advantage of light energy have undoubtedly the greatest difficulty in producing the necessary ATP. With the exception of some protozoa, only bacteria are adapted in such a way that they can live permanently under anaerobic conditions. A remarkable species is *Methanobacterium thermoautotrophicum*, which is not only an obligate anaerobe but also carries out a chemolitho*autotrophic* metabolism under *thermophilic* conditions (8).

Anaerobes can only form a few moles of ATP per mol of substrate; the degradation of 1 mol of glucose yields, for instance, 1-4 mol of ATP depending on the pathway used for substrate breakdown:

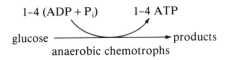

The only anaerobic process that gives higher ATP yields is the nitrate-dependent respiration carried out by a number of facultative anaerobic bacteria. If nitrate is present, these organisms use it under anaerobic conditions as terminal electron acceptor, reducing it to nitrite, N_2O, or molecular nitrogen. This electron transport is coupled to ATP formation and organic substrates can be oxidized to CO_2, as with oxygen, as terminal electron acceptor (7, 9, 10, 11). Nitrate-dependent respiration has much in common with oxygen-dependent respiration, although one difference between the two types is that oxygenase reactions are not feasible in nitrate-respiring bacteria. Nevertheless, some of them are capable of degrading aromatic substances (12, 13). Nitrate-dependent respiration is not discussed further in this chapter.

It is tempting to describe the process of sulfate-dependent respiration in analogy to nitrate-dependent respiration. However, fundamental differences are apparent. In contrast to the nitrate respirers, the sulfate-reducing bacteria are obligate anaerobes (14). Furthermore, the two processes are energetically very dissimilar. In Table 1 the free-energy changes for the reduction of oxygen, nitrate, sulfate, and carbon dioxide with H_2, NADH, or reduced flavoproteins are given. The values per two electrons transferred to the terminal acceptor differ considerably. Taking into account that an amount of approximately 10 kcal/mol is required for ATP synthesis from ADP and P_i, the transport of electrons from NADH and reduced flavoproteins to oxygen or nitrate can be coupled to ATP synthesis by electron transport phosphorylation. This is not so when sulfate and carbon dioxide are used as acceptors. In these processes the reduction of only one or two intermediates can be coupled to ATP synthesis, and the ATP yield is comparable to the ATP yield of classic fermentations like the alcohol fermentation.

The energy metabolism of the anaerobic organisms is intrinsically connected to the formation of large amounts of reduced compounds, which are excreted. Most of these are organic compounds, such as ethanol, butyrate, methane, etc. In addition, molecular hydrogen and carbon dioxide are produced, and sulfate-reducing bacteria form large amounts of hydrogen sulfide.

SCOPE OF THIS CHAPTER

Anaerobes gain ATP by substrate-level and/or electron transport phosphorylation. The reactions leading to the formation of ATP and the consequences of these reactions on the overall metabolism of these organisms are described in this chapter.

ENERGY-YIELDING REACTIONS IN ANAEROBES

ATP Synthesis by Substrate-Level Phosphorylation

A great number of anaerobes synthesize ATP exclusively by substrate-level phosphorylation. This is true for organisms fermenting carbohydrates and other substrates to the following products or mixtures thereof: lactate, butyrate, caproate, ethanol, acetone, isopropanol, n-butanol, 2,3-butanediol, carbon dioxide, and molecular hydrogen. Organisms like yeasts or Zymomonas mobilis (15) that ferment carbohydrates to ethanol and CO_2 form ATP only by substrate-level phosphorylation and so do the lactic acid bacteria, most clostridia, and many other anaerobic bacteria.

Only a few compounds can serve as substrates in ATP synthesis (Table 2). Five of these contain "energy-rich" phosphoryl bonds, and ATP is formed by transfer of the phosphoryl group to ADP. In the acetate kinase reaction, the intermediate formation of a phosphorylated enzyme has been shown (16). The sixth compound, N^{10}-formyltetrahydrofolate (N^{10}-formyl FH_4) is acted upon in such a way that presumably a phosphorylated intermediate is formed that reacts with ADP to give ATP (17).

Table 1. Free energy changes in oxygen- and nitrate-dependent respirations and in sulfide and methane fermentations

Redox reaction	$\Delta G_0'$ (kcal/mol acceptor)	$\Delta G_0'$ (kcal/2e⁻) Electron donor		
		H_2	NADH	flavoprotein[a]
$2\,H_2 + O_2 \longrightarrow 2\,H_2O$	−113.4	−56.7	−52.4	−42.3
$5/2\,H_2 + NO_3' + H^+ \longrightarrow 1/2\,N_2 + 3\,H_2O$	−133.9	−53.6	−49.3	−39.2
$H_2 + NO_3' \longrightarrow NO_2' + H_2O$	−39.0	−39.0	−34.7	−24.6
$4\,H_2 + SO_4'' + H^+ \longrightarrow 1\,HS' + 4\,H_2O$	−36.3	−9.1	−4.8	+5.3
$4\,H_2 + CO_2 \longrightarrow CH_4 + 2\,H_2O$	−31.2	−7.8	−3.5	+6.6

[a] For the flavoprotein a redox potential of $E_0' = -100$ mV has been assumed.

Of the six reactions listed in Table 2, the last three are important only to a few microorganisms. In *Clostridium cylindrosporum,* N^{10}-formyl FH_4 is an intermediate in purine fermentation (18, 19). In the course of xanthine degradation formiminoglycine is formed, which is converted to glycine and formate as shown in Figure 1. *Methanosarcina barkeri* ferments methanol in such a way that 4 mol of methanol are converted to 3 mol of methane and 1 mol of CO_2 (20). The formation of CO_2 proceeds via N^{10}-formyl FH_4 and formate. During degradation of histidine, formiminoglutamate is formed by *Clostridium tetanomorphum* (18), and transfer of the formimino-group to tetrahydrofolate and subsequent formation of formate increases the growth yield of this organism (21).

Streptococcus faecalis (22), *Clostridium botulinum* (23), and *Clostridium perfringens* (24) grow on arginine, which is first converted to citrulline. Phosphorolytic cleavage of citrulline yields ornithine and carbamyl phosphate, the latter of which is used for ATP synthesis by substrate-level phosphorylation:

$$arginine + H_2O \longrightarrow citrulline + NH_3$$

$$citrulline + P_i \longrightarrow ornithine + carbamyl\ phosphate$$

In the anaerobic degradation of allantoin by *Streptococcus allantoicus* (25) and by enterobacteria (26), carbamyl phosphate is also formed and used for ATP synthesis. Also, during anaerobic breakdown of uracil and orotic acid, N-carbamyl compounds are formed. So far, however, the generation of carbamyl phosphate from these compounds has not been reported (19).

1,3-Bisphosphoglycerate and phosphoenolpyruvate (PEP) are intermediates of sugar breakdown, and the ATP synthesis connected with the further metabolism of these compounds is important to all anaerobes fermenting carbohydrates and related compounds. It should be mentioned that PEP also plays a role as a direct energy carrier in anaerobes in that these organisms use phosphotransferase systems for the transport of sugars into the cells (27). At the expense of PEP a small protein (HPr) is phosphorylated,

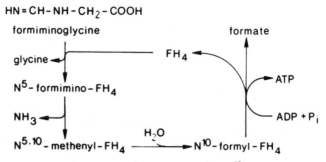

Figure 1. ATP synthesis by *Clostridium cylindrosporum* from N^{10}-formyl-tetrahydrofolate (formyl FH_4). Formiminoglycine is an intermediate of purine degradation.

Table 2. Reactions yielding ATP by substrate-level phosphorylation in anaerobes

Reaction	Enzyme	$\Delta G_{obs}^{0'}$ (kcal/mol)
1,3-bisphosphoglycerate + ADP $\rightleftharpoons$ 3-phosphoglycerate + ATP	phosphoglycerate kinase	−5.8
phosphoenolpyruvate + ADP $\rightleftharpoons$ pyruvate + ATP	pyruvate kinase	−5.7
acetyl phosphate + ADP $\rightleftharpoons$ acetate + ATP	acetate kinase	−3.1
butyryl phosphate + ADP $\rightleftharpoons$ butyrate + ATP	butyrate kinase	−3.1
carbamyl phosphate + ADP $\rightleftharpoons$ carbamate + ATP	carbamate kinase	−1.8
N^{10}-formyl FH$_4$ + ADP + P$_i$ $\rightleftharpoons$ formate + FH$_4$ + ATP	formyl FH$_4$ synthetase	+2.0

which subsequently donates its phosphate group to the sugar entering the membrane:

$$PEP + HPr \xrightarrow{\text{enzyme I}} P \sim HPr + pyruvate$$

$$P \sim HPr + sugar \xrightarrow{\text{enzyme II}} sugar\text{-}P + HPr$$

Acetyl phosphate is a very important agent for substrate-level phosphorylation in anaerobes. In most fermentations, the acetate found among the end products is formed from acetyl phosphate. Only a few enzymes are known that yield acetate in reactions not involving acetyl phosphate or acetyl-CoA. Citrate lyase—the key enzyme of the anaerobic breakdown of citrate by streptococci (28) and enterobacteria (29)—splits citrate into oxaloacetate and acetate:

$$citrate \longrightarrow oxaloacetate + acetate$$

Similarly, citramalate lyase of *C. tetanomorphum* cleaves citramalate (an intermediate in glutamate degradation) into pyruvate and acetate (30). Free acetate is formed during the CO_2 reduction to acetate as catalyzed by *Clostridium aceticum, Clostridium thermoaceticum, Clostridium formicoaceticum,* and *Acetobacterium woodii* (31). By the action of glycine reductase, which occurs in *Clostridium sticklandii, Clostridium lentoputrescens, Clostridium sporogenes,* and other clostridia carrying out the Stickland reaction, free acetate is also formed. This enzyme system is membrane-bound. Ferredoxin, a selenoprotein (protein A), a second protein (protein B), and additional factors are involved in this reaction (32):

$$glycine + NADH \xrightarrow[\substack{selenoprotein A \\ protein B \\ protein fraction C}]{\text{ferredoxin, factors}} acetate + NH_3 + NAD$$

How is acetyl phosphate generated? The reactions leading to the formation of acetyl phosphate in anaerobes are summarized in Figure 2. *Lactobacillus delbrückii* contains a flavin adenine dinucleotide (FAD)–linked pyruvate dehydrogenase that is CoA-independent and forms acetyl phosphate directly (33). Another enzyme yielding acetyl phosphate is phosphoketolase. It occurs in lactic acid bacteria, such as *Lactobacillus plantarum* (34) and *Leuconostoc mesenteroides* (35), and in *Bifidobacterium bifidum* (36). Phosphoketolases catalyze the phosphorolytic cleavage of xylulose 5-phosphate or fructose 6-phosphate (37).

The most important precursor of acetyl phosphate is acetyl-CoA, and the enzyme phosphotransacetylase, which catalyzes the reversible interconversion of these two acetyl compounds, is widespread among anaerobes

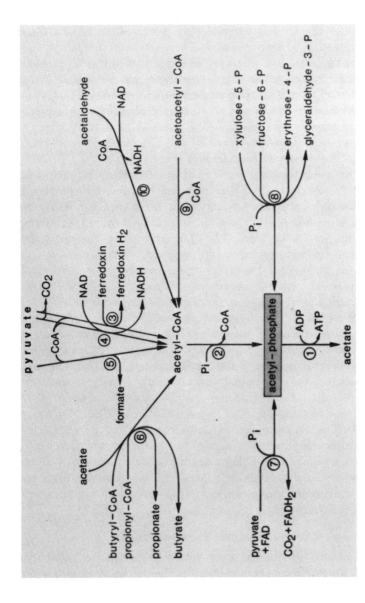

Figure 2. Reactions involved in the generation of acetyl phosphate in anaerobes. 1) acetate kinase; 2) phosphotransacetylase; 3) pyruvate:ferredoxin oxidoreductase; 4) pyruvate dehydrogenase multienzyme complex (NAD-dependent); 5) pyruvate formate lyase; 6) coenzyme A transferase; 7) pyruvate dehydrogenase of *Lactobacillus delbrückii*; 8) phosphoketolase; 9) β-ketothiolase; 10) acetaldehyde dehydrogenase (CoA-dependent).

(38). Numerous pathways lead to acetyl-CoA. As is shown in Figure 2, CoA-dependent acetaldehyde dehydrogenase makes acetyl-CoA from acetaldehyde. This reaction is important in the fermentation of ethanol and acetate to butyrate and caproate by *Clostridium kluyveri* (39, 40) and during growth of sulfate-reducing bacteria on ethanol plus sulfate (41).

Propionate (42, 43) and butyrate (44) fermentations involve the corresponding thioesters as intermediates. These esters are not simply hydrolyzed, but are subjected to CoA transferase reactions in which the thioester group is transferred to acetate. Consequently, the enzyme acetate kinase is also involved in ATP synthesis from propionyl-CoA and butyryl-CoA. A more direct ATP synthesis from butyryl-CoA via butyryl phosphate is also possible. The necessary enzymes, phosphotransbutyrylase and butyrate kinase, have been found in *Clostridium butyricum* (45) and *C. tetanomorphum* (21). The distribution of these enzymes in anaerobes is not known.

Three enzyme systems are known that convert pyruvate to acetyl-CoA. The NAD-dependent pyruvate dehydrogenase multienzyme complex is a characteristic enzyme of aerobes. Nevertheless, it has been shown to occur in *Rhodospirillum rubrum* (46), and it might be involved in pyruvate breakdown in propionibacteria (42). The two other enzyme systems, pyruvate:ferredoxin oxidoreductase and pyruvate formate lyase, are chiefly responsible for acetyl-CoA formation from pyruvate in anaerobes. The ferredoxin-dependent enzyme occurs in clostridia (47, 48, 49), sulfate-reducing bacteria (50, 51), *Eubacterium limosum* (*Butyribacterium rettgeri*) (49), *Megasphaera* (*Peptostreptococcus*) *elsdenii* (49), *Peptococcus anaerobius* (*Diplococcus glycinophilus*) (49), *Veillonella alcalescens* (*Micrococcus lactilyticus*) (49, 52), *Sarcina maxima* (53), *Sarcina ventriculi* (54), *Ruminococcus albus* (55), and some *Spirochaeta* spp. (56). The enzyme pyruvate formate lyase is present in anaerobically growing enterobacteria (57), in *S. faecalis* (58), in photosynthetic bacteria fermenting pyruvate anaerobically in the dark (59, 60, 61), and in some clostridia for anabolic reactions (62). The last two enzyme systems lose their activity when exposed to oxygen, which underlines their importance exclusively to anaerobes.

Although anaerobes are able to degrade a large number of different organic compounds, the catabolic routes lead to a comparatively small number of "energy-rich" intermediates, which are further metabolized in ATP-yielding reactions.

ATP Synthesis by Electron Transport Phosphorylation

So far one enzyme system has been found in facultative and obligate anaerobes in which electron transport is coupled to the phosphorylation of ADP—the fumarate reductase system. In *Escherichia coli* this enzyme is distinct from succinate dehydrogenase, as shown with the help of mutants (63). It is membrane-bound and is associated with redox carriers that take part in electron transfer from molecular hydrogen, formate, lactate, glycerol phosphate, or NADH to fumarate (10). The fumarate reductase systems of various organisms differ to a certain extent as to the nature of the electron donor that can be used and as to the types of redox carriers involved. Examples are given in Table 3. Menaquinone (in some organisms 2-demethyl-menaquinone) and a b-type cytochrome are the characteristic electron carriers involved in the fumarate reductase system. However, mutants of $E.$ $coli$ devoid of cytochromes are still able to carry out fumarate reduction (69), but this reduction might not be coupled to the phosphorylation of ADP (70). In $Bacteroides$ $fragilis,$ on the other hand, the presence of cytochrome b is essential, and this organism accumulates fumarate when grown in the absence of hemin, which is required for cytochrome biosynthesis (65). Menaquinone is absolutely essential for the formation of active fumarate reductase in $E.$ $coli$ (64) and in propionibacteria (68).

   $A.$ $woodii$ (80), which, like $C.$ $aceticum,$ ferments molecular hydrogen and carbon dioxide to acetate (81), and methanogenic bacteria contain fumarate reductase (78). The carriers involved are not known, except that in methanogenic bacteria factor F_{420} is reduced by molecular hydrogen (79) and cytochromes and menaquinone are not present (1). The function of fumarate reductase in these organisms is not known at present.

   Fumarate reductase is also involved in the energy metabolism of certain eukaryotes (82), some protozoa (83), and helminths. Moreover, a soluble fumarate reductase serves as an anabolic enzyme in a number of anaerobically growing organisms for the provision of succinate in tetrapyrrole biosynthesis, since most of these bacteria lack α-oxoglutarate dehydrogenase activity (84, 85, 86). Fumarate also serves as an oxidant in pyrimidine (87) and protoporphyrin (88) biosyntheses.

   The redox potential of the fumarate/succinate couple is $E_0' = +33$ mV, and the potential span between the couples $2H^+/H_2$ ($E_0' = -414$ mV), CO_2/formate ($E_0' = -432$ mV), or $NAD^+/NADH$ ($E_0' = -320$ mV) and the fumarate/succinate couple is large enough to allow the synthesis of at least 1 mol ATP per mol of fumarate reduced. But what is the evidence that ATP is indeed formed in the course of fumarate reduction? The evidence derives from three types of observations:

1. Particulate preparations of $Desulfovibrio$ $gigas$ were shown to couple the reduction of fumarate with H_2 to the phosphorylation of ADP. Approximately 0.3 mol of ATP was formed per mol of H_2 consumed (89, 90). Similar results have been obtained using spheroplasts of $Vibrio$ $succinogenes$ (10). Fumarate-dependent proton translocations have been reported for vesicles from $E.$ $coli$ (91, 92).

2. Some anaerobes are able to grow on substrate combinations that yield succinate but do not allow ATP synthesis by substrate-level phosphorylation. $Vi.$ $succinogenes$ (93), $E.$ $coli$ (75) and several other enterobacteria (94) grow with H_2 and fumarate or L-malate, the only end product being succinate. It is apparent from the scheme of succinate formation (Figure 3) that these organisms must gain their energy by electron trans-

Table 3.   Electron donors and electron carriers involved in fumarate reduction

Donors	Carriers probably involved	Organisms	References
H_2, formate	MK^a and cyt. b	*Escherichia coli*, other enterobacteria	(64, 65)
		Vibrio succinogenes	(66)
glycerol phosphate	MK and cyt. b	*E. coli*, propionibacteria	(67, 68)
glycerol phosphate	MK, no cytochrome	*E. coli hem*A$^-$ mutant	(69, 70)
L-lactate	MK and cyt. b	*Veillonella alcalescens*,	(71, 72)
		Desulfovibrio gigas	(73, 74)
NADH	MK and cyt. b	*Bacteroides fragilis*,	(71, 75)
		propionibacteria	(76, 77)
H_2	not known	*Acetobacterium woodii*	(78)
H_2	F_{420}, other carriers not known	Methanogenic bacteria	(78, 79)

^{a}MK, menaquinone.

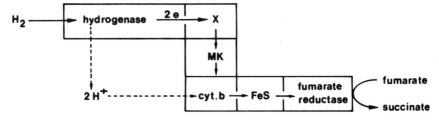

Figure 3. Succinate formation from fumarate and H_2 by *Escherichia coli* and *Vibrio succinogenes*. X, unknown carrier; MK, menaquinone; FeS, iron-sulfur protein.

port phosphorylation. *Desulfuromonas acetoxidans* (95) grows on acetate plus fumarate; the former is oxidized to CO_2, thus providing reducing power for fumarate reduction (Figure 4). Again, ATP must come from the fumarate reductase system. The same is true for *Proteus rettgeri* (96), which ferments fumarate to succinate and carbon dioxide.

3.   Organisms fermenting carbohydrates to acetate, propionate, and succinate exhibit unusually high growth yields. Per mol of glucose fermented, 62 g (dry weight) of *Selenomonas ruminantium* (97), 50 g of *B. fragilis* (65), and 65 g of *Propionibacterium freudenreichii* (98) are produced. However, lactic acid bacteria, ethanol-producing organisms, and saccharolytic clostridia form in the order of 20–35 g (dry weight) of cells per mol of glucose (99), much less than those organisms in which the fumarate reductase system is involved.

There is, therefore, now convincing evidence that ATP is produced in the course of fumarate reduction.

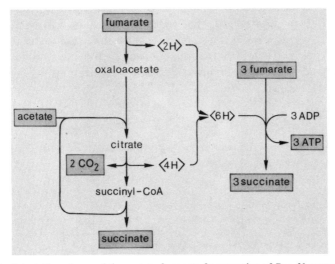

Figure 4. A tentative scheme of the acetate-fumarate fermentation of *Desulfuromonas acetoxidans*.

Occurrence of a Fumarate-Succinate Cycle?

The data summarized in Table 3 show that fumarate reductase is present in such organisms as *A. woodii* and methanogenic bacteria (e.g., *Methanobacterium ruminantium*), which do not produce succinate or propionate (78). Fumarate reductase has also been found in sulfate-reducing bacteria (51, 73). All these organisms must synthesize ATP by electron transport phosphorylation. The reduction of CO_2 to methane or acetate does not proceed via intermediates yielding ATP by substrate-level phosphorylation. During growth of sulfate-reducing bacteria with L-lactate and sulfate, ATP surplus cannot be produced by substrate-level phosphorylation (14, 100). According to the fermentation equation, two "energy-rich" phosphoryl bonds are formed in acetate production from the two acetyl phosphates:

$$2 \text{ lactate} + H_2O \longrightarrow 2 \text{ acetate} + 2 CO_2 + 8 H$$

$$\text{sulfate} + 8 H \longrightarrow \text{sulfide} + 4 H_2O$$

The reduction of sulfate to sulfide, however, is associated with the conversion of ATP to AMP, so that net ATP yield is zero.

Synthesis of ATP for growth of all these organisms can be understood if the operation of a fumarate-succinate cycle is assumed. Its principle is that the low potential electrons generated from H_2 or reduced ferredoxin are not used directly for all the reduction steps in methane or sulfide formation. Instead, some H_2 may be used to reduce fumarate to succinate, and succinate may then function as reducing agent, thereby regenerating fumarate. As indicated in Figure 5, such a cycle could be coupled to the last step in methane formation since the redox potential of the methanol/methane couple ($E_0' = +170$ mV) is more positive than that of the fumarate/succinate couple ($E_0' = +33$ mV). Similarly, the redox potential of $S_3O_6^{2-}/S_2O_3^{2-} + HSO_3^-$ ($E_0' = +225$ mV) is also more positive than that of the fumarate/succinate couple. Therefore, the operation of such a cycle would, even at standard conditions, have no thermodynamic complications.

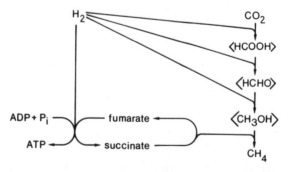

Figure 5. Tentative scheme of the operation of a fumarate-succinate cycle in methane bacteria.

It should be mentioned that Grossman and Postgate (101) had already postulated in 1955 that "a reversible succinate/fumarate system may form a link between sulfate reduction and the oxidation of organic compounds or hydrogen." The link in the form of the fumarate-succinate cycle may allow ATP synthesis by electron transport phosphorylation and may couple oxidation and reduction reactions in otherwise unrelated fermentations.

Some other reduction reactions have also been associated with ATP synthesis by electron transport phosphorylation: the crotonyl-CoA/butyryl-CoA couple ($E_0' = -15$ mV) in connection with the ethanol-acetate fermentation of C. *kluyveri* (102); the acrylyl-CoA/propionyl-CoA couple ($E_0' = -15$ mV) in connection with the formation of propionate by *Megasphaera elsdenii* (103); and the methylenetetrahydrofolate/methyltetrahydrofolate couple ($E_0' = -182$ mV) in connection with ATP formation during acetate synthesis from CO_2 by C. *thermoaceticum* (31, 104). Energy generation in the course of these reduction reactions has not been demonstrated so far.

Pyrophosphate as Energy Carrier

It has been outlined that in anaerobes, which ferment carbohydrates, PEP plays some role as energy carrier in sugar transport. In the propionibacteria (105) and the parasitic protozoan *Entamoeba histolytica* (106), pyrophosphate replaces ATP in some phosphorylation reactions. During growth of these organisms on sugars or other compounds metabolized via PEP, pyrophosphate is formed by the enzyme PEP-carboxytransphosphorylase:

$$\text{PEP} + P_i + CO_2 \rightleftharpoons \text{oxaloacetate} + PP_i$$

During growth on L-lactate the enzyme pyruvate-orthophosphate dikinase produces pyrophosphate:

$$\text{ATP} + P_i + \text{pyruvate} \rightleftharpoons \text{PEP} + \text{AMP} + PP_i$$

The free energy change of pyrophosphate hydrolysis is $\Delta G_{obs}^{0'} = -5.27$ kcal/mol and pyrophosphate is used in some phosphorylation reactions as energy donor:

$$PP_i + \text{D-fructose-6-P} \xrightarrow{\text{phosphofructokinase (PP}_i)} \text{D-fructose-1,6-P}_2 + P_i$$

$$PP_i + \text{L-serine} \xrightarrow{PP_i\text{-serinephosphotransferase}} \text{L-serine-P} + P_i$$

$$PP_i + \text{acetate} \xrightarrow{PP_i\text{-acetate kinase}} \text{acetyl} \sim P + P_i$$

Except in the cases of the propionibacteria and *Entamoeba histolytica*, not much is known about the function of a pyrophosphate as an energy-conserving agent.

THE AMBIGUOUS ROLE OF HYDROGEN IN FERMENTATIONS

The Advantage of H_2 Evolution

In most fermentations the breakdown of substrates is intrinsically connected to redox reactions. Electrons are transferred to electron carriers like NAD ($E_0' = -320$ mV) or FAD ($E_0' = -220$ mV), and the reoxidation of these carriers requires an electron acceptor, which in most cases is formed by intramolecular rearrangement of the substrate. This is apparent for the alcohol and lactate fermentations where dehydrogenating and hydrogenating reactions are coupled to a NAD/NADH cycle (Figure 6a). If in a fermentation of this type molecular hydrogen is produced, this is advantageous because the redox balance allows the formation of less-reduced compounds and thus more ATP is available for the organism. In the butyrate fermentation, as carried out by clostridia and several other anaerobes, H_2 is evolved by the enzyme system pyruvate:ferredoxin oxidoreductase + ferredoxin hydrogenase (49, 107) (Figure 6b). The oxidation of pyruvate to acetyl-CoA + CO_2 is coupled to the reduction of ferredoxin, the redox potential of which ($E_0' = -398$ mV) is low enough to allow its further use as electron donor for the reduction of protons to H_2. Consequently, acetyl-CoA instead of pyruvate becomes available for NAD regeneration; 2 acetyl-CoA molecules are condensed and can be reduced to butyrate by 4 hydrogen equivalents (H), and additional ATP is formed after CoA transferase reaction from acetyl-CoA. A similar ATP yield is achieved in certain H_2-producing organisms [e.g., *R. albus* (55) and *C. sphenoides* (108)] by the formation of ethanol + acetate from acetyl-CoA and NADH:

2 acetyl-CoA + ADP + P_i + 2 NADH $\longrightarrow$

ethanol + acetate + 2 NAD + 2 CoA + ATP

Experiments have shown that even more hydrogen gas can be evolved by some organisms than pyruvate is degraded, the source being NADH (109). The occurrence of this thermodynamically unfavorable reaction has been demonstrated in various clostridia (110, 111, 112), the S-organism (113), and *R. albus* (55). The electron transfer from NADH to ferredoxin requires the involvement of an NADH-ferredoxin oxidoreductase that is activated by acetyl-CoA and inhibited by CoA (111). NADH is a potent inhibitor of the back reaction. Therefore, the NADH-ferredoxin oxidoreductase seems to effectively catalyze the reduction of ferredoxin, from which, in turn, electrons can be liberated as hydrogen gas through ferredoxin hydrogenase. Calculations have shown that about 25% of the hydrogen gas evolved by *Clostridium pasteurianum* is derived from NADH (114), thereby increasing the energy available for the organism. In *C. kluyveri,* any growth and ATP production is dependent on H_2 evolution (115, 116, 117). Only if part of the electrons that are transferred to carriers in the course of ethanol oxidation are subsequently transferred to protons does acetyl-CoA become available to this

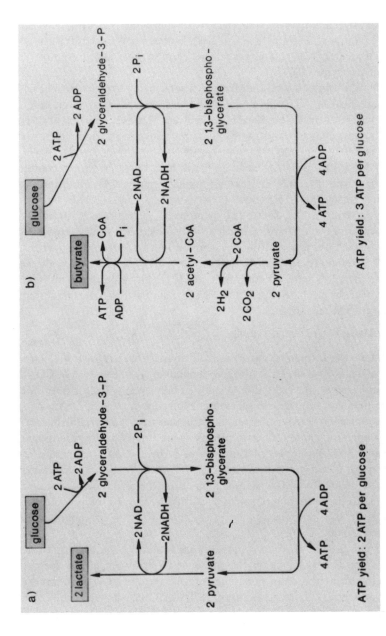

Figure 6. Scheme of the lactate fermentation (a) and the butyrate fermentation (b). In both fermentations glucose is degraded to pyruvate via the Embden-Meyerhof pathway. In (b) pyruvate is further degraded to acetyl-CoA, H₂, and CO₂. Acetyl-CoA is converted to butyrate via acetoacetyl-CoA, β-hydroxy-butyryl-CoA, crotonyl-CoA, and butyryl-CoA.

organism for ATP production. The dehydrogenases involved in ethanol oxidation by *C. kluyveri* are particulate (40), which might facilitate ethanol dehydrogenation and H_2 evolution taking place simultaneously. It is not yet understood why *C. kluyveri* is so rich in flavoproteins (118) or why this organism contains an extremely active NADP-dependent β-hydroxybutyryl-CoA dehydrogenase (119), whereas all other clostridia tested contain an NAD-specific enzyme (107).

In general it can be stated that fermentations involving dehydrogenation reactions and yielding ATP only by substrate-level phosphorylation give more ATP per mol of substrate the more H_2 is evolved. However, the extent of H_2 evolution has mechanistic and thermodynamic limitations. By mechanistic limitation is meant that the organisms must contain the necessary enzyme equipment for H_2 evolution. Yeasts or lactic acid bacteria do not contain these enzymes, and therefore they cannot take advantage of H_2 evolution as a means of increasing the ATP yield.

Instead of transferring electrons to protons, *C. perfringens* (120) can use nitrate as an electron sink, reducing it to nitrite without an accompanying electron transport phosphorylation as observed in nitrate-respiring bacteria. The nitrate fermentation will also increase the energy yield by allowing the organism to form less-reduced end products (121). Therefore, fermentations involving nitrate as electron acceptor show some similarities to fermentations proceeding with H_2 evolution.

Hydrogen-Dependent Fermentations

In contrast to the fermentations discussed above, those involving electron transport phosphorylation are hydrogen-dependent: the reduction of CO_2 to methane and acetate; the reduction of sulfate and elemental sulfur to sulfide; and the reduction of fumarate to succinate. Molecular hydrogen can function as electron donor in these fermentations, although not for all organisms. In a number of bacteria, these reductive processes are linked to the oxidation of organic substrates. Examples of this type of fermentation are: the acetate fermentation of *C. thermoaceticum* and *C. formicoaceticum*; the oxidation of organic substrates by sulfate-reducing bacteria; the oxidation of acetate to CO_2 coupled to the reduction of sulfur to sulfide by *Desulfuromonas acetoxidans*; and the succinate-propionate fermentation of the propionibacteria and *Veillonella* spp.

Anaerobic growth of *Vi. succinogenes* on fumarate, malate, aspartate, and asparagine is dependent on the presence of hydrogen gas or formate (93), as is the fermentation of malate and fumarate by *E. coli* (75). An oral *Campylobacter* sp. (122) and *Spirillum* 5175 (123) also show a requirement for H_2 or formate, whereas a free-living *Campylobacter* (124) can grow on aspartate, asparagine, L-malate, or fumarate in the presence or absence of H_2. *P. rettgeri* (96) cannot utilize molecular hydrogen, but oxidizes part of the fumarate to CO_2 and uses the other part of fumarate as electron acceptor, as has also been reported for *Proteus mirabilis* (125). In *C. formi-*

coaceticum (126) fumarate can act as both electron donor and acceptor and forms acetate, CO_2, and succinate, as do some sulfate-reducing bacteria (51, 127). *Enterobacter* (*Aerobacter*) *aerogenes* (128) produces succinate, acetate, CO_2, ethanol, formate, and sometimes H_2 from fumarate or L-malate.

In the fermentations discussed above, fumarate or a direct precursor thereof (L-malate or aspartate) is added as substrate. However, fumarate can also be generated by a number of organisms in the catabolism of other structurally unrelated substrates. Glucose is fermented by the propionibacteria according to the following equation (105):

$$1.5 \text{ glucose} \longrightarrow 2 \text{ propionate} + \text{acetate} + CO_2$$

One third of the glucose is oxidized to acetate and CO_2, and the reduced coenzymes generated in this oxidation are used to form propionate via fumarate and succinate. This fermentation involves the fumarate reductase system and yields ATP by electron transport phosphorylation (98). The fermentation pathway is rather complicated (Figure 7). Oxaloacetate is formed from pyruvate in a transcarboxylation reaction (129) with methylmalonyl-CoA as donor of the carboxyl group. If succinate is excreted as fermentation end product, net CO_2 fixation must occur. The enzyme responsible for that is PEP-carboxytransphosphorylase (130), which has been mentioned in connection with the function of pyrophosphate as energy carrier. *Bacteroides* spp. that also carry out a propionate fermentation use another energy-conserving reaction for oxaloacetate synthesis: PEP-carboxykinase (131):

$$\text{PEP} + CO_2 + \text{ADP} \rightleftharpoons \text{oxaloacetate} + \text{ATP}$$

V. alcalescens transfers electrons from lactate to oxaloacetate via a malate-lactate transhydrogenase (132). Propionate is formed from succinate via methylmalonyl-CoA by decarboxylation, not by transcarboxylation (133). *V. alcalescens*, as the propionate-forming *Selenomonas ruminantium* and *Anaerovibrio lipolytica*, contains cytochrome *b*, which is active in fumarate reduction (72). However, not all the propionate produced by anaerobes is synthesized via fumarate. *Clostridium propionicum* (134) and *Megasphaera elsdenii* (103) have been shown to reduce lactate to propionate with acrylyl-CoA and propionyl-CoA as intermediates. Whether or not this pathway is also coupled to an electron transport phosphorylation reaction is not known.

An interesting relationship exists between sulfur and fumarate reduction. *Spirillum* 5175 (123) (now tentatively classified as a *Campylobacter*) can reduce sulfite, thiosulfite, or sulfur to sulfide using hydrogen gas or formate as reductants. On the other hand, sulfide or hydrogen gas can act as electron donor for reduction of fumarate to succinate. The reduction of sulfur to sulfide with a $\Delta G_0'$ of -7.8 kcal/mol does not allow a stoichiometric coupling to ATP synthesis, as in substrate-level phosphorylation. However, stoichiometric coupling is not necessary according to the chemiosmotic

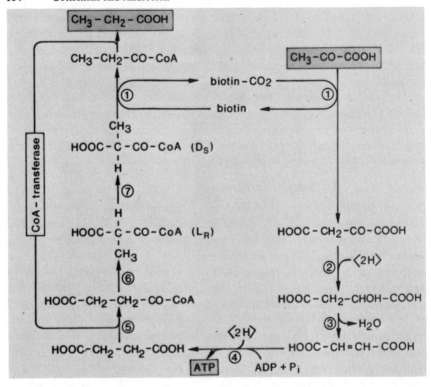

Figure 7. Formation of propionate from pyruvate by propionibacteria. Pyruvate is formed from glucose by the Embden-Meyerhof pathway. 1) (D_s)-methylmalonyl-CoA pyruvate transcarboxylase; 2) malate dehydrogenase; 3) fumarase; 4) fumarate reductase; 5) CoA transferase; 6) (L_R)-methylmalonyl-CoA mutase; 7) methylmalonyl CoA racemase.

model. *Desulfuromonas acetoxidans* (95) cannot use hydrogen gas as electron donor for sulfur reduction, but can use acetate, ethanol, propanol, and butanol. Fumarate also functions as electron acceptor instead of sulfur. Acetate can be oxidized to CO_2, which by itself is an endergonic process ($\Delta G_0' = +25$ kcal/mol), but when coupled to sulfur reduction the overall reaction becomes exergonic. *Desulfotomaculum acetoxidans* (135) can oxidize acetate with the concomitant reduction of sulfate but not of sulfur.

During the breakdown of organic substrates some species of the genera *Desulfotomaculum* and *Desulfovibrio* (51, 101, 127, 135) can use fumarate as electron acceptor instead of sulfate. This demonstrates again the relationship between sulfate/sulfur and fumarate metabolism. In addition to the unsolved problem of the location of energy generating sites, the pathway of sulfate reduction to sulfide is still a matter of some controversy. It is still uncertain whether trithionate and thiosulfate are regular intermediates of bisulfite reduction or artifacts (136) (Figure 8).

In the acetate fermentation of *C. thermoaceticum* (104) and *C. for-*

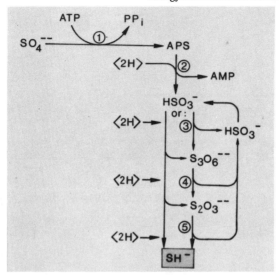

Figure 8. Pathway of dissimilatory sulfate reduction. Both alternatives of bisulfite reduction have been indicated: a) one 6-electron reduction step, b) three 2-electron reduction steps. 1) ATP sulfurylase; 2) APS (adenylyl-sulfate) reductase; 3) bisulfite reductase; 4) trithionate reductase; 5) thiosulfate reductase.

micoaceticum (137), hexoses are fermented according to the following equations:

$$\text{hexose} + 2\,H_2O \longrightarrow 2\,\text{acetate} + 2\,CO_2 + 8\,H$$

$$2\,CO_2 + 8\,H \longrightarrow \text{acetate} + 2\,H_2O$$

The oxidation of the hexose to acetate and CO_2 is linked to the exergonic process of CO_2 reduction to acetate. Both organisms have been shown to contain cytochromes and menaquinones (138), and the process of CO_2 reduction may also be connected to ATP production, as it has to be in *A. woodii* (80, 81) and *C. aceticum* (31). CO_2 reduction is an obligatory part of the fermentative metabolism of these two clostridia because they are devoid of a hydrogenase and are unable to form lactate, ethanol, or butyrate from acetyl-CoA. CO_2 reduction proceeds via formate, N^{10}-formyl-tetrahydrofolate, N^5-methyl-tetrahydrofolate, and methyl-B_{12} coenzyme, which after (trans-) carboxylation yields acetate (31) (Figure 9). The formate dehydrogenase that is involved in formate synthesis from CO_2 is a protein containing tungsten and selenium (31, 139). In contrast, corresponding bacterial enzymes involved in formate oxidation are usually molybdo-proteins (140).

Interspecies Hydrogen Transfer

Interactions between hydrogen-producing and hydrogen-consuming fermentation pathways located in different organisms are also found (141). The pro-

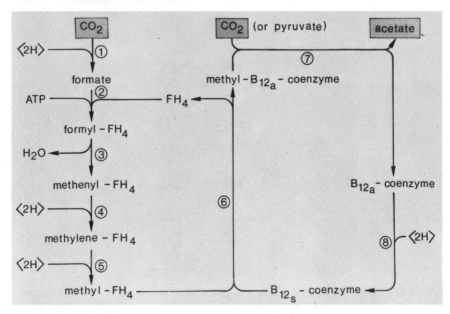

Figure 9. Pathway of acetate formation from CO_2 by *Clostridium thermoaceticum* and *Clostridium formicoaceticum*. FH_4 = tetrahydrofolate. B_{12s} contains Co^+ and is oxidized to B_{12a} (Co^{3+}) by transfer of the methyl group. 1) formate dehydrogenase; 2) formyl FH_4 synthetase; 3) methenyl FH_4 cyclohydrolase; 4) methylene FH_4 dehydrogenase; 5) methylene FH_4 reductase; 6) "methyl transferase"; 7) "methyl corrinoid (trans-) carboxylase"; 8) "corrinoid reductase."

cess of methane formation is very effective and well suited for trapping any hydrogen evolved in fermentative processes, and methanogenic bacteria may keep the partial pressure of H_2 as low as 1.5×10^{-3} atm (10^{-6} M) (142). At such low pressures of H_2 the oxidation of NADH to NAD^+ and H_2 becomes thermodynamically possible. Thus, in a mixed culture with methanogenic bacteria, a number of organisms can improve their ATP yield by evolving hydrogen gas from NADH. Good examples are *Ruminococcus* spp. that, in combination with *Vi. succinogenes* (143) or a methanogenic organism (144), ferment glucose to acetate and CO_2 (Figure 10). Other fermentations only become possible if the partial pressure of H_2 is kept low; the S-organism oxidizes ethanol to acetate if the H_2 produced is used for methane formation (145):

$$2\ C_2H_5OH + 2\ H_2O \xrightarrow{\text{S-organism}} 2\ CH_3\text{—COOH} + 4\ H_2$$

$$4\ H_2 + CO_2 \xrightarrow{\text{Methanobacterium MOH}} CH_4 + 2\ H_2O$$

Interspecies hydrogen transfer as observed with *R. albus* or the S-organism requires, of course, that the hydrogen-donating organism contains the en-

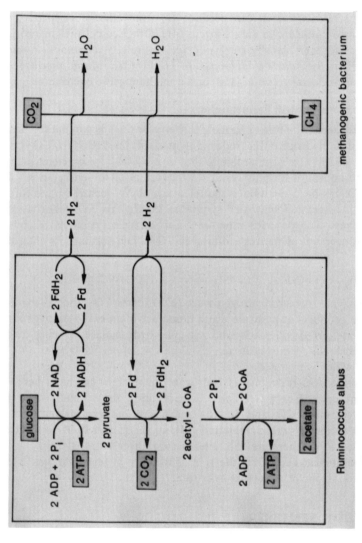

Figure 10. Hydrogen transfer from *Ruminococcus albus* to a methanogenic bacterium.

zymes necessary for H_2 evolution from NADH at low pH_2. These are NADH: ferredoxin oxidoreductase and hydrogenase.

Not only the methane bacteria can function as H_2-consuming companions (146): sulfate-reducing bacteria also utilize H_2 in the presence of sulfate and are able to keep the H_2 pressure low for the benefit of H_2-producing organisms (147, 148). On the other hand, at low sulfate concentrations, *Desulfovibrio* can produce H_2 from ethanol or lactate, thus allowing a co-culture with methanogens (41). Other co-cultures involving interspecies hydrogen transfer consist of, for example, *Selenomonas ruminantium* (149), *Clostridium cellobioparum* (150), *Clostridium thermocellum* (151), or *Citrobacter freundii* (152) and a methanogenic bacterium.

The Role of Formate in Fermentations

In some respects, the role of formate in fermentations is comparable to that of molecular hydrogen: the redox potential of the $HCOOH/CO_2$ couple ($E_0' = -432$ mV) is comparable to that of $2 H^+/H_2$; in some enzyme reactions the removal of formate results in the formation of "energy-rich" compounds. It has been mentioned that cleavage of N^{10}-formyl FH_4 to formate and FH_4 is associated with ATP synthesis (18). In this connection pyruvate formate lyase also deserves attention. This enzyme system is responsible for the conversion of pyruvate into acetyl-CoA and formate by enterobacteria (57):

$$CH_3-CO-COOH + CoASH \rightleftharpoons CH_3-CO-SCoA + HCOOH$$

As the enzyme formate hydrogen lyase (153) forms CO_2 and H_2 from formate, the products of pyruvate degradation are identical to those produced by pyruvate:ferredoxin oxidoreductase plus hydrogenase. During the late growth phase *C. formicoaceticum* (137) and *R. albus* (154) reduce CO_2 to formate, which is excreted.

Formate also functions as hydrogen donor (155) in many hydrogen-dependent fermentations (see Table 3). In addition, several methane bacteria ferment formate to CO_2 and CH_4 (8), and sulfate-reducing bacteria use it as hydrogen donor in sulfide formation (14). In the rumen fluid no formate is detectable (156); therefore, an interspecies formate transfer might have biological significance. Thus the role of formate in fermentations is comparable to the one of molecular hydrogen.

CONCLUDING REMARKS

In this chapter we have shown that there has been a significant increase in our knowledge of the energy metabolism of anaerobes in recent years. Not too long ago it was generally accepted that anaerobes gain ATP predominantly by substrate-level phosphorylation (157, 158). It is now apparent that this is not true for many anaerobes and that electron carriers, such as cytochromes and menaquinones, are common to a number of anaerobes. Also, we have attempted to emphasize the importance of the fumarate reductase system for electron transport phosphorylation in anaerobes.

There is increasing evidence that a number of compounds hitherto known to be attacked by aerobes and phototrophs only can also be fermented. In some cases this can only be achieved by interspecies hydrogen transfer to a methanogenic bacterium that acts as a final electron acceptor. The anaerobic degradation of benzoic acid (159) might be given as an example.

Not much has been said about the pathways used by anaerobes to convert the various substrates to key intermediates of fermentations, such as pyruvate or acetyl-CoA. Some reactions of these pathways are catalyzed by enzymes that are typical for anaerobic metabolism and that do not occur in aerobes. The lysine fermentation as carried out by some clostridia will be given as an example of the versatility of these organisms in rearranging functional groups of a given substrate, thereby preparing it for the final reactions of energy generation and end product formation (160, 161) (Figure 11). As in the lysine fermentation, B_{12} coenzymes play an important role in the anaerobic breakdown of glutamate, ornithine, leucine, glycerol, ethylene glycol, ethanolamine, and nicotinic acid (162).

Under standardized conditions anaerobic bacteria show characteristic patterns of end products formed, which can be used for taxonomical purposes. However, depending on the nature of the substrate and other nutritional factors these patterns are subject to some variation:

1. The redox state of the substrate has considerable influence on the products that are formed (163), e.g., the degradation of sugar acids allows the formation of more oxidized end products than the catabolism of sugar alcohols. If more reduced substrate has to be fermented, the energy gain often decreases.
2. The chain length of a substrate and the nature of its functional groups influence the pathway that can be used for degradation. Lactate has the same redox status as glucose. Whereas glucose is an excellent substrate for many anaerobes, only a few are able to degrade lactate.
3. The capacity of the enzymes involved in the degradation of a certain substrate determines the rate of carbon flow to the end products. If the activity of a particular enzyme becomes limiting, the substrate might be channeled to an energetically less favorable pathway. Omission of ferrous ions from the medium impairs the pyruvate degradation in clostridia, and the fermentation is shifted from a butyrate type to a lactate type (164). This decreases the energy yield from 3 to 2 mol of ATP per mol of glucose.
4. A change of the pH might also have consequences. At low pH, some clostridia change from a butyrate fermentation to an acetone-butanol fermentation (165), which is connected with a decrease of the ATP yield.

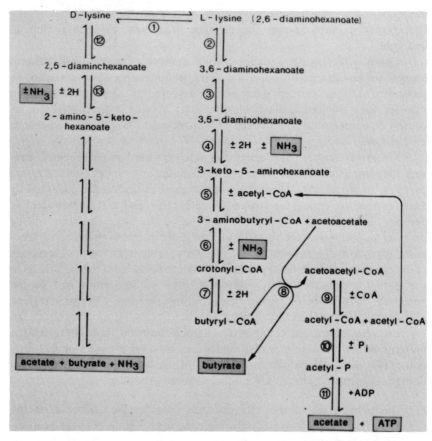

Figure 11. Fermentation of lysine to acetate and butyrate by *Clostridium sticklandii* and *Clostridium* SB4. 1) lysine racemase; 2) L-lysine-2,3-aminomutase (pyridoxal-P and Fe^{2+}-dependent; activated by S-adenosyl-methionine); 3) β-lysine mutase (B_{12} coenzyme and pyridoxal-P dependent); 4) 3,5-diaminohexanoate dehydrogenase [NAD(P)-dependent]; 5) 3-keto-5-aminohexanoate cleavage enzyme (acetyl-CoA requiring); 6) L-3-aminobutyryl coenzyme A deaminase; 7) butyryl coenzyme A dehydrogenase; 8) coenzyme A transferase; 9) β-ketothiolase; 10) phosphotransacetylase; 11) acetokinase; 12) D-α-lysine mutase (B_{12} coenzyme and pyridoxal-P dependent, activated by ATP); 13) 2,5-diaminohexanoate dehydrogenase [NAD(P)-dependent].

Whereas the energy metabolism of anaerobes is so distinct from that of aerobes, differences in biosynthetic reactions are not so pronounced. A number of anaerobes use reactions of the tricarboxylic acid cycle for glutamate synthesis (84,166); many other routes are also the same as in aerobes. One reaction, however, deserves to be mentioned in this connection: the reductive carboxylation of CoA esters to form α-oxo acids. Organisms growing on C_2 compounds (*C. kluyveri* and sulfate-reducing bacteria) use the pyruvate:ferredoxin oxidoreductase for pyruvate synthesis from acetyl-CoA and CO_2 (167,168) (the so-called pyruvate synthase reaction). Similarly,

certain ruminal bacteria also synthesize the α-oxo acids for leucine, valine, isoleucine, phenylalanine, tryptophan, and glutamate synthesis by the reductive carboxylation of the corresponding CoA esters (169,170,171).
Little is known about transport processes in obligate anaerobic bacteria (91). Recently, it has been demonstrated that membrane-bound ATPases are present in these organisms and that a transmembrane protonmotive force can be established by this enzyme (172,173,174).

ACKNOWLEDGMENTS

We are grateful to Dr. M. Dorn for helpful discussions.

REFERENCES

1. Thauer, R. K., Jungermann, K., and Decker, K. (1977). Bacteriol. Rev. 41:100.
2. Hauska, G., and Trebst, A. (1977). In D. R. Sanadi (ed.), Current Topics in Bioenergetics, Vol. 6, p. 151. Academic Press, New York.
3. Parson, W. W. (1974). Annu. Rev. Microbiol. 28:41.
4. Jones, O. T. G. (1977). In B. A. Haddock and W. A. Hamilton (eds.), Microbial Energetics, p. 151. Cambridge University Press, Cambridge, Eng.
5. Harold, F. M. (1977). In D. R. Sanadi (ed.), Current Topics in Bioenergetics, Vol. 6, p. 84. Academic Press, New York.
6. Boyer, P. D., Chance, B., Ernster, L., Mitchell, P., Racker, E., and Slater, E. C. (1977). Annu. Rev. Biochem. 46:955.
7. Haddock, B. A., and Jones, C. W. (1977). Bacteriol. Rev. 41:47.
8. Zeikus, J. G. (1977). Bacteriol. Rev. 41:514.
9. Stouthamer, A. H. (1976). Adv. Microb. Physiol. 14:315.
10. Kröger, A. (1977). In B. A. Haddock and W. A. Hamilton (eds.), Microbial Energetics, p. 61. Cambridge University Press, Cambridge, Eng.
11. Payne, W. J. (1973). Bacteriol. Rev. 37:409.
12. Bakker, G. (1977). FEMS Microbiol. Lett. 1:103.
13. Williams, R. J., and Evans, W. C. (1975). Biochem. J. 148:1.
14. Le Gall, J., and Postgate, J. R. (1973). Adv. Microb. Physiol. 10:81.
15. Swings, J., and DeLey J. (1977). Bacteriol. Rev. 41:1.
16. Todhunter, J. A., Reichel, K. B., and Purich, D. L. (1976). Arch. Biochem. Biophys. 174:120.
17. Buttlaire, D. H., Himes, R. H., and Reed, G. H. (1976). J. Biol. Chem. 251:4159.
18. Barker, H. A. (1961). In I. C. Gunsalus and R. Y. Stanier (eds.), The Bacteria, Vol. II: Metabolism, p. 151. Academic Press, New York.
19. Vogels, G. D., and van der Drift, C. (1976). Bacteriol. Rev. 40:403.
20. Stadtman, T. C. (1967). Annu. Rev. Microbiol. 21:121.
21. Twarog, R., and Wolfe, R. S. (1963). J. Bacteriol. 86:112.
22. Deibel, R. H. (1964). J. Bacteriol. 87:988.
23. Mitruka, B. M., and Costilow, R. N. (1967). J. Bacteriol. 93:295.
24. Venugopal, V., Nadkarni, G. B. (1977). J. Bacteriol. 131:693.
25. Valentine, R. C., and Wolfe, R. S. (1960). Biochim. Biophys. Acta 45:389.
26. Tigier, H., and Grisolia, S. (1965). Biochem. Biophys. Res. Commun. 19:209.
27. Postma, P. W., and Roseman, S. (1976). Biochim. Biophys. Acta 457:213.
28. Kümmel, A., Behrens, G., and Gottschalk, G. (1975). Arch. Microbiol. 102:111.

112   Gottschalk and Andreesen

29. Srere, P. A. (1975). Adv. Enzymol. 43:57.
30. Buckel, W., and Bobi, A. (1976). Eur. J. Biochem. 64:255.
31. Ljungdahl, L. G., and Andreesen, J. R. (1976). *In* H. G. Schlegel, G. Gottschalk, and N. Pfennig (eds.), Microbial Production and Utilization of Gases, p. 163. Akademie der Wissenschaften zu Göttingen, Göttingen.
32. Stadtman, T. C. (1974). Science 183:915.
33. Hager, L. P., and Lipmann, F. (1961). Proc. Natl. Acad. Sci. USA 47:1768.
34. Heath, E. C., Hurwitz, J., Horecker, B. L., and Ginsburg, A. (1958). J. Biol. Chem. 231:1009.
35. Hurwitz, J. (1958). Biochim. Biophys. Acta 28:599.
36. de Vries, W., Gerbrandy, S. J., and Stouthamer, A. H. (1967). Biochim. Biophys. Acta 136:415.
37. Sgorbati, B., Lenaz, G., and Casalicchio, F. (1976). Antonie von Leeuwenhoek 42:49.
38. Whiteley, H. R., and Pelroy, R. A. (1972). J. Biol. Chem. 247:1911.
39. Burton, R. M., and Stadtman, E. R. (1953). J. Biol. Chem. 202:873.
40 Hillmer, P., and Gottschalk, G. (1974). Biochim. Biophys. Acta 334:12.
41. Bryant, M. P., Campbell, L. L., Reddy, C. A., and Crabill, M. R. (1977). Appl. Environ. Microbiol. 33:1162.
42. Allen, S. H. G., Kellermeyer, R. W., Stjernholm, R. L., and Wood, H. G. (1964). J. Bacteriol. 87:171.
43. Tung, K. K., and Wood, W. A. (1975). J. Bacteriol. 124:1462.
44. Barker, H. A., Stadtman, E. R., and Kornberg, A. (1955). *In* S. P. Colowick and N. O. Kaplan (eds.), Methods in Enzymology, Vol. I, p. 599. Academic Press, New York.
45. Valentine, R. C., and Wolfe, R. S. (1960). J. Biol. Chem. 235:1948.
46. Lüderitz, R., and Klemme, J. H. (1977). Z. Naturforsch. 32c:351.
47. Uyeda, K., and Rabinowitz, J. C. (1971). J. Biol. Chem. 246:3120.
48. Mortenson, L. E., Valentine, R. C., and Carnahan, J. E. (1963). J. Biol. Chem. 238:794.
49. Valentine, R. C. (1964). Bacteriol. Rev. 28:497.
50. Akagi, J. M. (1967). J. Biol. Chem. 242:2478.
51. Hatchikian, E. C., and Le Gall, J. (1970). Ann. Inst. Pasteur 118:288.
52. Whiteley, H. R., and McCormick, N. G. (1963). J. Bacteriol. 85:382.
53. Kupfer, D. G., and Canale-Parola, E. (1970). J. Bacteriol. 94:984.
54. Stephenson, M. P., and Dawes, E. A. (1971). J. Gen. Microbiol. 69:331.
55. Glass, T. L., Bryant, M. P., and Wolin, M. J. (1977). J. Bacteriol. 131:463.
56. Hespell, R. B., and Canale-Parola, E. (1973). J. Bacteriol. 116:931.
57. Knappe, J., Blaschkowski, H. P., Gröbner, P., and Schmitt, T. (1974). Eur. J. Biochem. 50:253.
58. Lindmark, D. G., Paolella, P., and Wood, N. P. (1969). J. Biol. Chem. 244:3605.
59. Jungermann, K., and Schön, G. (1974). Arch. Microbiol. 99:109.
60. Gorrell, T. E., and Uffen, R. L. (1977). J. Bacteriol. 131:533.
61. Gürgün, V., Kirchner, G., and Pfennig, N. (1976). Z. Allg. Mikrobiol. 16:573.
62. Thauer, R. K., Kirchniawy, F. H., and Jungermann, K. A. (1972). Eur. J. Biochem. 27:282.
63. Hirsch, C. A., Raminsky, M., Davis, B. D., and Lin, E. C. C. (1963). J. Biol. Chem. 238:3770.
64. Guest, J. R. (1977). J. Bacteriol 130:1038.
65. Macy, J., Probst, I., and Gottschalk, G. (1975). J. Bacteriol. 123:436.
66. Kröger, A., and Innerhofer, A. (1976). Eur. J. Biochem. 69:497.
67. Miki, K., and Lin, E. C. C. (1975). J. Bacteriol. 124:1282.

68. de Vries, W., Aleem, M. I. H., Hemrika-Wagner, A., and Stouthamer, A. H. (1977). Arch. Microbiol. 112:271.
69. Singh, A. P., and Bragg, P. D. (1975). Biochim. Biophys. Acta 396:229.
70. Singh, A. P., and Bragg, P. D. (1976). Biochim. Biophys. Acta 423:450.
71. Gibbons, R. J., and Engle, L. P. (1964). Science 146:1307.
72. de Vries, W., van Wijck-Kapteyn, W. M. C., and Oosterhuis, S. K. H. (1974). J. Gen. Microbiol. 81:69.
73. Hatchikian, E. C. (1974). J. Gen. Microbiol. 81:261.
74. Hatchikian, E. C., and Le Gall, J. (1972). Biochim. Biophys. Acta 267:479.
75. Macy, J., Kulla, H., and Gottschalk, G. (1976). J. Bacteriol. 125:423.
76. Sone, N. (1974). J. Biochem. 76:137.
77. Schwartz A. C., and Krause, A. E. (1975). Z. Allg. Mikrobiol. 15:99.
78. Gottschalk, G., Schoberth, S., and Braun, K. (1977). In K. Skryabin, M. V. Ivanov, E. N. Kondratjeva, G. A. Zavarzin, Y. A. Trotsenko, and A. I. Nesterov (eds.). Microbial Growth on C_1-compounds, p. 157. Scientific Center for Biological Research, USSR Academy of Sciences, Pushchino.
79. Tzeng, S. F., Wolfe, R. S., and Bryant, M. P. (1975). J. Bacteriol. 121:184.
80. Balch, W. E., Schoberth, S., Tanner, R. S., and Wolfe, R. S. (1977). Int. J. Syst. Bacteriol. 27:355.
81. Schoberth, S. (1977). Arch. Microbiol. 114:143.
82. Köhler, P., and Saz, H. J. (1976). J. Biol. Chem. 251:2217.
83. Müller, M. (1975). Annu. Rev. Microbiol. 29:467.
84. Gottschalk, G., and Barker, H. A. (1967). Biochemistry 6:1027.
85. Gray, C. T., Wimpeny, J. W. T., and Mossman, M. R. (1966). Biochim. Biophys. Acta 117:33.
86. Lewis, A. J., and Miller, J. D. A. (1977). Can. J. Microbiol. 23:916.
87. Lascelles, J. (1974). Ann. N.Y. Acad. Sci. 236:96.
88. Jacobs, N. J., and Jacobs, J. M. (1977). Biochim. Biophys. Acta 459:141.
89. Barton, L. L., Le Gall, J., and Peck, H. D. (1970). Biochem. Biophys. Res. Commun. 41:1036.
90. Barton, L. L., Le Gall, J., and Peck, H. D. (1972). In A. San Pietro and H. G. Gest (eds.), Horizons in Bioenergetics, p. 33. Academic Press, New York.
91. Konings, W. N., and Boonstra, J. (1977). In F. Bronner and A. Kleinzeller (eds.), Current Topics in Membranes and Transport, Vol. 9, p. 177. Academic Press, New York.
92. Singh, A. P., and Bragg, P. D. (1977). Biochim. Biophys. Acta 464:562.
93. Wolin, M. J., Wolin, E. A., and Jacobs, N. J. (1961). J. Bacteriol. 81:911.
94. Didzun, D. (1976). Die Verbreitung der wasserstoffabhängigen Vergärung von L-Malat und Fumarat bei Enterobacteriaceae. Diplomarbeit, Universitat Göttingen, Göttingen.
95. Pfennig, N., and Biebl, H. (1976). Arc. Microbiol. 110:3.
96. Kröger, A. (1974). Biochim. Biophys. Acta 347:273.
97. Hobson, P. N., and Summers, R. (1972). J. Gen. Microbiol. 70:351.
98. de Vries, W., van Wyck-Kapteyn, W. M. C., and Stouthamer, A. H. (1973). J. Gen. Microbiol. 76:31.
99. Stouthamer, A. H. (1977). In B. A. Haddock and W. A. Hamilton (eds.), Microbial Energetics, p. 285. Cambridge University Press, Cambridge, Eng.
100. Peck, H. D. (1974). In M. J. Carlie and J. J. Skehel (eds.). Evolution in the Microbial World, p. 241. Cambridge University Press, Cambridge, Eng.
101. Grossman, J. P., and Postgate, J. R. (1955). J. Gen. Microbiol. 12:429.
102. Barker, H. A. (1956). Bacterial Fermentations. John Wiley and Sons, Inc., New York.
103. Brockmann, H. L., and Wood, W. A. (1975). J. Bacteriol. 124:1447.

104. Andreesen, J. R., Schaupp, A., Neurauter, C., Brown, A., and Ljungdahl, L. G. (1973). J. Bacteriol. 114:743.
105. Wood, H. G. (1977). Fed. Proc. 36:2197.
106. Reeves, R. E. (1976). Trends in Biochem. Sci. 1:53.
107. v. Hugo, H., Schoberth, S., Madan, V. K., and Gottschalk, G. (1972). Arch. Mikrobiol. 87:189.
108. Walther, R., Hippe, H., and Gottschalk, G. (1977). Appl. Environ. Microbiol. 33:955.
109. Jungermann, K., Thauer, R. K., Leimenstoll, G., and Decker, K. (1973). Biochim. Biophys. Acta 305:268.
110. Gottschalk, G., and Chowdhury, A. A. (1969). FEBS Lett. 2:342.
111. Jungermann, K., Rupprecht, E., Ohrloff, C., Thauer, R., and Decker, K. (1971). J. Biol. Chem. 246:960.
112. Petitdemange, H., Cherrier, C., Bengone, J. M., and Gay, R. (1977). Can. J. Microbiol. 23:152.
113. Reddy, C. A., Bryant, M. P., and Wolin, M. J. (1972). J. Bacteriol. 110:126.
114. Jungermann, K., Kern, M., Riebeling, V., and Thauer, R. K. (1976). In H. G. Schlegel, G. Gottschalk, and N. Pfennig (eds.), Microbial Production and Utilization of Gases, p. 85. Akademie der Wissenschaften zu Göttingen, Göttingen.
115. Stadtman, E. R., and Barker, H. A. (1949). J. Biol. Chem. 180:1085.
116. Thauer, R. K., Jungermann, K., Henninger, H., Wenning, J., and Decker, K. (1968). Eur. J. Biochem. 4:173.
117. Schoberth, S., and Gottschalk, G. (1969). Arch. Mikrobiol. 65:318.
118. Peel, J. L. (1958). Biochem. J. 69:403.
119. Madan, V. K., Hillmer, P., and Gottschalk, G. (1973). Eur. J. Biochem. 32:51.
120. Hasan, S. M., and Hall, J. B. (1975). J. Gen. Microbiol. 87:120.
121. Ishimoto, M., Umeyama, M., and Chiba, S. (1974). Z. Allg. Mikrobiol. 14:115.
122. van Palenstein Helderman, W. H., and Rosman, I. (1976). Antonie von Leeuwenhoek 42:107.
123. Wolfe, R. S., and Pfennig, N. (1977). Appl. Environ. Microbiol. 33:427.
124. Laanbroek, H. J., Kingma, W., and Veldkamp, H. (1977). FEMS Microbiol. Lett. 1:99.
125. van der Beek, E. G., Oltmann, L. F., and Stouthamer, A. H. (1976). Arch. Microbiol. 110:195.
126. Dorn, M., Andreesen, J. R., and Gottschalk, G. (1978). J. Bacteriol. 133:26.
127. Miller, J. D. A., and Wakerley, D. S. (1966). J. Gen. Microbiol. 43:101.
128. Barker, H. A. (1936). Proc. Konigl. Akad. Wet. Amst. 39:674.
129. Wood, H. G., Chiao, J. P., and Poto, E. M. (1977). J. Biol. Chem. 252:1490.
130. Wood, H. G., O'Brien, W. E., and Michaels, G. (1977). Adv. Enzymol. 45:89.
131. Scardovi, V., and Chiappini, M. G. (1966). Ann. Microbiol. Enzymol. 16:119.
132. Allen, S. H. G. (1973). Eur. J. Biochem. 35:338.
133. Galivan, J. H., and Allen, S. H. G. (1968). J. Biol. Chem. 243:1253.
134. Goldfine, H., and Stadtman, E. R. (1960). J. Biol. Chem. 235:2238.
135. Widdel, F., and Pfennig, N. (1977). Arch. Microbiol. 112:119.
136. Jones, H. E., and Skyring, G. W. (1975). Biochim. Biophys. Acta 377:52.
137. Andreesen, J. R., Gottschalk, G., and Schlegel, H. G. (1970). Arch. Mikrobiol. 72:154.
138. Gottwald, M., Andreesen, J. R., Le Gall, J., and Ljungdahl, L. G. (1975). J. Bacteriol. 122:325.
139. Ljungdahl, L. G., and Andreesen, J. R. (1975). FEBS Lett. 54.279.

140. Thauer, R. K., Fuchs, G., and Jungermann, K. (1977). *In* W. Lovenberg (ed.), Iron-Sulfur Proteins, Vol. III, p. 121. Academic Press, New York.
141. Wolin, M. J. (1976). *In* H. G. Schlegel, G. Gottschalk, and N. Pfennig (eds.), Microbial Production and Utilization of Gases, p. 141. Akademie der Wissenschaften zu Göttingen, Göttingen.
142. Hungate, R. E. (1975). Annu. Rev. Ecology System. 6:39.
143. Iannotti, E. L., Kafkewitz, D., Wolin, M. J., and Bryant, M. P. (1973). J. Bacteriol. 114:1231.
144. Wolin, M. J. (1974). Am. J. Clin. Nutr. 27:1320.
145. Reddy, C. A., Bryant, M. P., and Wolin, M. J. (1972). J. Bacteriol. 109:539.
146. Cappenberg, T. E. (1975). Microb. Ecol. 2:60.
147. Sorokin, Y. I. (1966). Microbiology 35:643.
148. Khosrovi, B., and Miller, J. D. A. (1975). Plant Soil 43:171.
149. Scheifinger, C. C., Linehan, B., and Wolin, M. J. (1975). Appl. Microbiol. 29:480.
150. Chung, K. T. (1976). Appl. Environ. Microbiol. 31:342.
151. Weimer, P. J., and Zeikus, J. G. (1977). Appl. Environ. Microbiol. 33:289.
152. Sifieriz, F., and Pirt, S. J. (1977). J. Gen. Microbiol. 101:57.
153. Chippaux, M., Pascal, M. C., and Casse, F. (1977). Eur. J. Biochem. 72:149.
154. Miller, T. L., and Wolin, M. J. (1973). J. Bacteriol. 116:836.
155. Krebs, H. A. (1937). Biochem. J. 31:2065.
156. Hungate, R. E., Smith, W., Bauchop, T., Yu, I., and Rabinowitz, J. C. (1970). J. Bacteriol. 102:389.
157. Decker, K., Jungermann, K., and Thauer, R. K. (1970). Angew. Chem. Int. Ed. 9:138.
158. Barker, H. A. (1972). *In* A. San Pietro and H. Gest (eds.), Horizons of Bioenergetics, p. 7. Academic Press, New York.
159. Ferry, J. G., and Wolfe, R. S. (1976). Arch. Microbiol. 107:33.
160. Stadtman, T. C. (1973). Adv. Enzymol. 38:413.
161. Yorifuji, T., Jeng, I. M., and Barker, H. A. (1977). J. Biol. Chem. 252:20.
162. Barker, H. A. (1972). Annu. Rev. Biochem. 41:55.
163. Johnson, M. J., Peterson, W. H., and Fred, E. B. (1931). J. Biol. Chem. 91:569.
164. Pappenheimer, A. M., and Shaskan, E. (1944). J. Biol. Chem. 155:265.
165. Davies, R., and Stephenson, M. (1941). Biochem. J. 35:1320.
166. Gottschalk, G. (1968). Eur. J. Biochem. 5:346.
167. Andrew, J. G., and Morris, J. G. (1965). Biochim. Biophys. Acta 97:176.
168. Buchanan, B. B. (1973). *In* W. Lovenberg (ed.), Iron-Sulfur Proteins, Vol. I, p. 129. Academic Press, New York.
169. Allison, M. J. (1969). J. Animal Sci. 29:797.
170. Allison, M. J., Robinson, I. M., and Baetz, A. L. (1974). J. Bacteriol. 117:175.
171. Sauer, F. D., Erfle, J. D., Mahadevan, S. (1975). Biochem. J. 150:357.
172. Riebeling, V., and Jungermann, K. (1976). Biochim. Biophys. Acta 430:434.
173. Booth, I. R., and Morris, J. G. (1975). FEBS Lett. 59:153.
174. Clarke, D. J., and Morris, J. G. (1976). Biochem. J. 154:725.

International Review of Biochemistry
Microbial Biochemistry, Volume 21
Edited by J. R. Quayle
Copyright 1979   University Park Press   Baltimore

4
Interrelation Between Modes of Carbon Assimilation and Energy Production in Phototrophic Purple and Green Bacteria

E. N. KONDRATIEVA
Department of Microbiology,
Moscow State University, Moscow, USSR

Purple and green bacteria are related by being photosynthetic organisms possessing a common pigment (bacteriochlorophyll *a*) (1-3). Until recently the name "phototrophic bacteria" was applied only to such microorganisms, but now it also includes cyanobacteria (3, 4), formerly known as blue-green algae (5, 6). Procaryotic organization of their cells allows them to be considered as bacteria. Moreover, halobacteria producing the retinal protein complex bacteriorhodopsin may be included in facultative phototrophs as being capable of light-dependent ATP synthesis. These bacteria are the only known organisms realizing photosynthesis without participation of chlorophyll (7-9).

By contrast, cyanobacteria reveal all principal peculiarities of photosynthesis proper to eucaryotic organisms. Like green plants, they contain chlorophyll *a*, use water as electron donor, and evolve oxygen in the process of photosynthesis (3, 5, 6).

Purple and green bacteria do not produce oxygen during photosynthesis, using as electron donor reduced compounds other than water. Thus in contrast to other phototrophic organisms that contain chlorophyll pigments, purple and green bacteria perform anoxygenic photosynthesis (2, 3, 10). Like other phototrophic procaryotes, these bacteria are gram-negative and predominantly aquatic microorganisms (1-3, 10-14).

The studies of purple and green bacteria provide much useful information for comprehension of photosynthesis and other biologic processes. The purpose of this chapter is to summarize recent data on the interrelation between carbon metabolism and modes of energy production in these bacteria.

PURPLE BACTERIA

General Characteristics

There are more than forty known species of purple bacteria, most of them isolated in pure culture (1, 2). They are included in the order Rhodospirillales and divided into several genera. The classification at genera level is based on morphological properties (Table 1). Purple bacteria are also divided into the families Chromatiaceae and Rhodospirillaceae (2). Most members of the Chromatiaceae grow only under anaerobic phototrophic conditions. They are autotrophs; some of them need vitamin B_{12}. All species are able to utilize hydrogen sulfide and elemental sulfur as electron donor in photosynthesis (1, 2, 11, 14, 15). Some of these bacteria also use thiosulfate and other reduced sulfur compounds and oxidize molecular hydrogen (1, 2, 11, 14). Usually they are called purple sulfur bacteria.

Most species of the family Rhodospirillaceae are facultative anaerobes. They grow well under photoorganotrophic conditions. Most species require B group vitamins, and many species can use CO_2 as the main or even only carbon source. Microorganisms of this family are known as purple nonsulfur bacteria, although some of them are capable of oxidizing sulfide, thiosulfate, and H_2 (2, 14-16).

The photosynthetic apparatus in purple bacteria is arranged into different types of intracytoplasmic membrane systems that are formed as a result of development and invagination of the cytoplasmic membrane (1, 2, 17-21). Such membranes contain pigments and electron carriers that are organized in a transport system (17, 18, 22).

The capacity of purple bacteria for photosynthesis is connected with the presence of specific chlorophylls. Most of these microorganisms synthesize bacteriochlorophyll a_p, which contains phytol. Some species also

Table 1.   The genera of purple bacteria (refs. 1, 2)

Genus	Cell shape, arrangement, and motility[a]
Sulfur bacteria	
Chromatium	rod or ovoid, motile
Thiospirillum	spiral, motile
Thiocystis	sphere, motile
Thiosarcina	sphere, packets, motile
Lamprocystis	sphere, contain gas vacuoles, motile
Thiocapsa	sphere
Thiodictyon	rod, contain gas vacuoles, may form nets
Thiopedia	sphere or ovoid, platelets, contain gas vacuoles
Amoebobacter	sphere, contain gas vacuoles
Ectothiorhodospira	vibrioid. motile
Nonsulfur bacteria	
Rhodospirillum	spiral, motile
Rhodopseudomonas	rod, ovoid or sphere, motile
Rhodomicrobium	ovoid, forming filaments, motile
Rhodocyclus	half-ring shaped

[a] Division by binary fission, except *Rhodomicrobium* and some *Rhodopseudomonas* spp. which multiply by budding. Motility by polar flagella, except *Rhodomicrobium*, which is peritrichous. All purple sulfur bacteria, except *Ectothiorhodospira*, may store sulfur globules inside the cells.

produce, in small quantity, bacteriochlorophyll a_{Gg} esterified by geranyl-geraniol. Three species have bacteriochlorophyll b (1, 2, 14, 22). Purple bacteria species containing bacteriochlorophyll a have long wavelength maxima in vivo from 800 to 890 nm, and species synthesizing bacteriochlorophyll b have maxima from 850 to 1,040 nm. Thus purple bacteria can use for photosynthesis light that contains less energy than that used by other phototrophs.

Like other phototrophs, purple bacteria produce carotenoids. These vary in different species, and most purple bacteria contain aliphatic carotenoids only. A few species of the purple sulfur bacteria have monocyclic aryl pigments and two species of purple nonsulfur bacteria synthesize γ-carotene, which belongs to the alicyclic pigments (1–3, 12, 23). Carotenoid glucosides have been found (24).

The presence of c-type cytochromes is common to all purple bacteria. As a general rule no less than two and often more of such cytochromes are present in these microorganisms (25–32). The purple nonsulfur bacteria also contain b-type cytochromes (26–29, 33). Recently the same cytochromes were discovered in *Chromatium vinosum* (34) and some other purple sulfur bacteria. In two species of purple nonsulfur bacteria the capacity for synthesis of a-type cytochromes was found (35–37).

Purple bacteria contain ubiquinones and, in some cases, menaquinones (25, 26, 31, 38–40). Among flavin-containing compounds (11, 26), flavodoxin similar to that in *Clostridium* was found in *Rhodospirillum rubrum* (41). The concentration of NAD in different species of purple bacteria is higher than that of NADP (42).

The purple bacteria also contain different iron-sulfur proteins with high and low redox potential. They include ferredoxins with 2-4Fe:4S and 1-4Fe:4S active centers (41, 43–48). Purple bacteria contain both conjugated and nonconjugated pteridines. Such compounds are localized in the photosynthetic membrane systems but their function is undefined (25, 26).

The Initial Stages of Photosynthesis

Photochemical Process   Light energy conversion in purple bacteria proceeds with participation of bacteriochlorophyll a or b. The bulk of these pigments, as with carotenoids, perform a light-harvesting function and belong to the so-called antenna pigments. Light energy absorbed by such pigments migrates to other molecules of antenna pigments and a part of it reaches reaction centers where the photochemical process takes place (30, 31, 49, 50, 51). Such protein-pigment complexes were isolated from some species of purple bacteria (30, 31, 50, 51) and they contain, according to all available evidence, bacteriochlorophyll that has absorption bands bleaching at 870, 890, or 960 nm on losing an electron. Usually such centers are called P_{870}.

The primary reaction consists of photooxidation of P_{870} and the second-

ary reaction consists of its reduction as a result of interaction with a
c-type cytochrome (30, 31, 49-53). According to recent data, the oxidation
of bacteriochlorophyll, which seems to be present in the reaction center as
a dimer (Bchl)$_2$, is connected with reduction first of all of bacteriopheophytin (Bph), which is contained in the same center. It is thought that the
primary electron acceptor is a quinone (Q) that contains iron (i.e., ferro-
quinone). Therefore the photochemical process occurring in the reaction
center is believed to be as follows (31, 51-65):

$$(Bchl)_2BphQ \xrightarrow{h\nu} (Bchl)_2^*BphQ \xrightarrow{10psec} (Bchl)_2^+ Bph^- Q \xrightarrow{\sim 170psec} (Bchl)_2^+ BphQ^-$$

According to redox titrations, the E_m of P_{870} is equal to 450-490 mV
and that of the primary electron acceptor varies between -20 and -200
mV in different species of purple bacteria and at different values of pH
(51, 55, 57, 58).

Electron Transport The photochemical process occurring in the
reaction center makes possible electron transport by a chain of carriers
according to a thermodynamic gradient. The studies of some purple bacteria (mainly *R. rubrum, Rhodopseudomonas palustris, Rhodopseudomonas capsulata, C. vinosum,* and *Ectothiorhodospira shaposhnikovii*)
have shown that a cyclic electron flow operates during photosynthesis by
these microorganisms. Such a system seems to involve quinones different
from those of the primary electron acceptor, c-type cytochromes, and, in
some species, b-type cytochromes (25-30, 49, 66-74). The possible existence
of other electron carriers should be investigated.

A similar system of electron flow operates in plants and cyanobacteria
(6, 74, 75) but in such photosynthetic organisms there also functions a
noncyclic electron flow involving two successive photochemical reactions.
As a result of the activity of such a system, electrons from water reduce
ferredoxin and NADP (Figure 1).

The photosynthesis of purple bacteria is also connected with production of a reduced nicotinamide adenine dinucleotide, but instead of
NADPH, NADH is formed (11, 25, 26, 30). According to some investigators, there is no fundamental difference between NAD reduction in
photosynthesis by purple bacteria and NADPH production by phototrophs
evolving oxygen (25, 26, 30, 49, 67, 76, 77). The only difference is that
purple bacteria use as electron donor for NAD reduction not water but
compounds with higher redox potential. In this way only one photoreaction
is sufficient for reduction of NAD and hence the possibility in purple
bacteria of a noncyclic system of electron flow directly connected with the
reaction center. However, the data of many researchers (68, 74, 78-82)
show that reduction of NAD in purple bacteria occurs as a result of energy-
dependent reversed electron transport (Figure 2). This follows from the fact
that isolated photosynthetic membrane structures of different purple bac-

122   Kondratieva

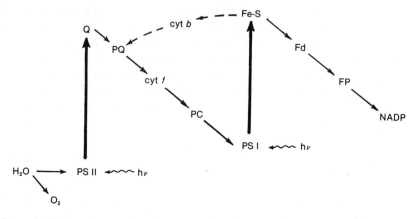

Figure 1. Photosynthetic electron transport system in green plants: PS, photosystem; Q, primary electron acceptor; PQ, plastoquinone; cyt f, cytochrome f; cyt b, cytochrome b; PC, plastocyanin; Fe-S, iron-sulfur protein; Fd, ferredoxin; FP, flavoprotein; NADP, nicotinamide adenine dinucleotide phosphate.

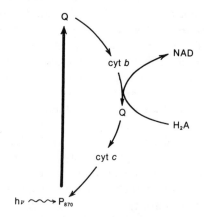

Figure 2. Photosynthetic electron transport system in purple bacteria: P_{870}, pigment (bacteriochlorophyll of reaction center; Q, quinone; cyt c, cytochrome c; cyt b, cytochrome b; NAD, nicotinamide adenine dinucleotide; H_2A, electron donor.

teria are capable of reducing NAD in the presence of succinate and some other electron donors (thiosulfate included), not only in the light but also in the dark, in the presence of ATP (68, 78, 79, 81). The reduction of NAD in the light and in the dark is inhibited by uncouplers. A similar effect of uncouplers on NAD(P) reduction is observed in the light in intact cells of purple nonsulfur and sulfur bacteria (80, 81). The value of the redox potential of the primary electron acceptor also argues against direct NAD reduction occurring as a result of the photochemical process in purple bacteria (51, 55, 57, 58, 74).

All this evidence shows the noncyclic pathway of electron transport in purple bacteria to be a purely dark process. However, its function depends on the photosynthetic cyclic system of electron flow as a source of energy. In some cases NAD reduction in purple bacteria seems not to need any expenditure of light energy. Some observations show that pigment-containing preparations from *C. vinosum* can produce NADH in the dark in the presence of molecular hydrogen (76). NAD reduction without light energy expenditure is also possible in purple bacteria using malate and other organic compounds in whose oxidation NAD(P)-linked dehydrogenases are involved. At the same time, assimilation of organic compounds by purple bacteria may considerably decrease their need of reduced NAD in comparison with growth under autotrophic conditions. Some reactions of bacterial photosynthesis, e.g., formation of poly-β-hydroxybutyrate from hydroxybutyrate, do not need reductant at all (3).

Energy Conservation The function of the photosynthetic electron transport system in purple bacteria results in the synthesis of ATP (11, 25, 26, 30, 51, 66–68, 74, 76) and pyrophosphate (68, 83–86). The synthesis of pyrophosphate by cell-free preparations of purple bacteria considerably increases when ATP formation is limited by ADP deficiency. Pyrophosphate, along with ATP, may serve as a source of energy for reversed electron transport to NAD and the transhydrogenase reaction resulting in reduction of NADP (68, 78, 79, 83–85). ATP synthesis coupled with pyrophosphate hydrolysis is also possible (87), and purple bacteria, as do other microorganisms, synthesize high molecular weight pyrophosphates. (86, 88, 89).

Several recent investigations have shown that synthesis of ATP and pyrophosphate coupled with the photosynthetic electron transport system in purple bacteria proceeds through generation of transmembrane electrochemical potential of hydrogen ions ($\Delta\bar{\mu}_H{}^+$). This question is discussed in some reviews (74, 90, 91). Studies of purple bacteria have also confirmed the generation of transmembrane electrochemical gradients of H^+ in illuminated cells and in chromatophore preparations, making possible some energy-dependent reactions without the participation of ATP and other energy-rich compounds. Such reactions include at least some transport processes (74, 90–92). At the expense of $\Delta\bar{\mu}_H{}^+$, electron flow against a thermodynamic gradient, photoinduced changes in the absorption of carotenoids as well as in cellular motility are possible (74, 90–93).

Many studies have been directed toward the investigation of properties of the coupling factor and ATPase activity of purple bacteria. Such a factor has been separated from membranes of some species of these microorganisms (17, 74, 94–97).

Inorganic Electron Donors

Oxidation of Sulfur Compounds The use of sulfide as electron donor results in accumulation of sulfur globules in cells of most species of purple sulfur bacteria. Only members of the genus *Ectothiorhodospira* accumulate sulfur in the medium (2). The sulfur is usually later oxidized to sulfate and the final result of H_2S utilization by purple sulfur bacteria during photosynthesis may correspond closely to (10, 11, 13):

$$H_2S + 2CO_2 + 2H_2O \xrightarrow{\text{light}} H_2SO_4 + 2(CH_2O)$$

where (CH_2O) represents reduced cellular carbon. Purple nonsulfur bacteria oxidize sulfide to sulfur (which may be detected in the medium) or to sulfate (2, 13, 15). According to recent data the latter process may proceed via production of elemental sulfur (16), but not in all species (2, 15). In *Thiocapsa roseopersicina*, H_2S oxidation with production of sulfur is followed by reduction of soluble cytochrome c_{550} (98).

There is more information about thiosulfate oxidation in purple bacteria (2, 11, 15, 16, 99-108). The result of this process in purple sulfur bacteria may be as follows:

$$Na_2S_2O_3 + 2CO_2 + 3H_2O \xrightarrow{\text{light}} 2NaHSO_4 + 2(CH_2O)$$

The capacity of purple bacteria to utilize external sulfite is seldom observed (101, 102), although this compound plays a principal role in sulfur metabolism (103, 104). Purple bacteria are able to act on thiosulfate by three different pathways (15, 103):

$$2S_2O_3^{2-} \xrightarrow[\text{cytochrome } c \text{ reductase (?)}]{\text{thiosulfate}} S_4O_6^{2-} + 2e^-$$

$$S_2O_3^{2-} \xrightarrow{\text{rhodanese}} SO_3^{2-} + S^0$$

$$S_2O_3^{2-} + 2e^- \xrightarrow[\text{reductase}]{\text{thiosulfate}} SO_3^{2-} + S^{2-}$$

For growing cultures of purple sulfur bacteria the most important method of thiosulfate cleavage is that catalyzed by thiosulfate reductase (15). The action of this enzyme in vivo seems to be connected with participation of dihydrolipoate as electron donor. Sulfite produced as a result of thiosulfate degradation can be further oxidized:

$$SO_3^{2-} + AMP \xrightarrow{\text{APS reductase}} \text{adenylylsulfate (APS)} + 2e^-$$

$$APS + P_i \xrightarrow{\text{ADP sulfurylase}} ADP + SO_4^{2-}$$

The presence of these enzymes has been found in several species of purple sulfur bacteria (15, 105, 106). There are also some data relating to sulfite:cytochrome c oxidoreductase participation in sulfite oxidation by *T. roseopersicina* (107).

In purple nonsulfur bacteria thiosulfate reductase has been discovered, but not APS-reductase (15). It has also been shown that thiosulfate oxidation by *R. palustris* occurs with appearance in the medium of tetrathionate, trithionate, and sulfite but not with appearance of elemental sulfur (108), and that oxidation of thiosulfate by this bacterium is connected with reduction of a *c*-type cytochrome. In the presence of ATP or pyrophosphate, cell-free preparations of *R. palustris* may reduce NAD in the dark with electrons derived from thiosulfate as follows (81, 108):

$$S_2O_3^{2-} \rightarrow \text{cyt } c \rightarrow \text{cyt } b \rightarrow \text{flavin} \rightarrow NAD$$

Tetrathionate production has also been observed in cell suspensions of *C. vinosum* but only in cases when the pH was unfavorable for growth (103). There are also no data about tetrathionate oxidation by this species or by other purple sulfur bacteria.

Many purple sulfur bacteria need reduced sulfur compounds as source of sulfur for assimilative processes because they cannot use sulfates (2, 15, 109, 110). The same behavior is shown by two species of purple nonsulfur bacteria (14, 111).

The Use of Molecular Hydrogen Several species of the purple sulfur and nonsulfur bacteria can grow under photoautotrophic conditions by assimilating CO_2 and using H_2 (2, 11, 12, 112-117):

$$2H_2 + CO_2 \xrightarrow{\text{light}} H_2O + (CH_2O)$$

The capacity for H_2 oxidation is due to the presence in purple bacteria of hydrogenase, which is an iron-sulfur protein (10, 45, 118-122).

Purified preparations of hydrogenase have been obtained from *C. vinosum* strain D (119, 120), *T. roseopersicina* (121, 122), and *R. rubrum* (389). The enzymes catalyze utilization and production of H_2 in the presence of oxidized and reduced methyl viologen, respectively. The exchange reaction ($D_2 \leftrightarrow H_2O$) is also catalyzed by hydrogenase from *T. roseopersicina* (122).

Photometabolism of Carbon

Information on carbon metabolism in purple bacteria is summarized in several reviews reflecting different stages in its study (11, 76, 99, 112, 123-128).

Assimilation of Carbon Dioxide The autotrophic carbon dioxide photoassimilation in purple bacteria, as in most other organisms, proceeds mainly via the ribulose diphosphate cycle. The functioning of such a cycle in purple sulfur and nonsulfur bacteria is evidenced by the nature of the labeled products at short-time exposure of cells with $^{14}CO_2$, the kinetics of their appearance, and the presence of ribulose 1,5-diphosphate carboxylase and associated enzymes. The studies of this aspect have been made in particular with *C. vinosum* (129), *Chromatium okenii* (130), *E. shaposhnikovii* (131), *T. roseopersicina* (132, 133), *R. rubrum* (134–137), *R. capsulata* (138), and *R. palustris* (139, 140).

Carbon dioxide fixation with participation of the ribulose diphosphate cycle often occurs in purple bacteria growing on media containing carbon compounds other than CO_2. Such is the principal way of carbon dioxide photoassimilation in *Rhodopseudomonas acidophila* (141) in the presence of methanol and in *R. palustris* (139) in the presence of formate. The formation from $^{14}CO_2$ of labeled products typical of the ribulose diphosphate cycle has been found during growth of *R. rubrum* (136, 142–144) and some other purple bacteria (145–147) in the presence of acetate, malate, or propionate.

The presence of organic compounds in the growth medium affects the amount of CO_2 assimilation via the ribulose diphosphate cycle and, accordingly, the level of ribulose diphosphate carboxylase may be much lower than under autotrophic conditions (126, 128, 148–150). However, this does not necessarily happen with all species of purple bacteria and it also depends on the nature of the organic compounds present (133, 149, 151). It is believed (3) that the operation of the ribulose diphosphate cycle in the presence of some organic compounds may be important for disposal of excess reducing power. However, not all organic compounds are capable of completely satisfying the carbon needs of purple bacteria. Part of the necessary metabolites may continue to be formed from CO_2 through its photoassimilation via the ribulose diphosphate cycle and by other pathways, even in the presence of organic substrates in the growth medium. Some organic compounds may be metabolized by purple bacteria only as a result of carboxylation. The use of propionate by purple bacteria may serve as an example (125, 146, 152, 153).

The need of purple bacteria for CO_2 for different metabolic processes is satisfied by the existence in them of several carboxylases. The content of such enzymes often varies in different species and, on the whole, is more diverse than in other phototrophs and chemoautotrophs (Table 2).

Along with 3-phosphoglycerate and sugar phosphates, different species of purple bacteria using $^{14}CO_2$ in mineral medium usually produce, rapidly and in considerable quantity, labeled organic acids such as malate, succinate, aspartate, glutamate, and some others (128, 129, 135, 140). This shows that the carboxylation of C_3 compounds derived from 3-phosphoglycerate proceeds simultaneously with operation of the ribulose diphosphate cycle. Phosphoenolpyruvate carboxylase (EC 4.1.1.31) seems to be involved often in such reactions, since this enzyme is present in many purple bacteria.

The carboxylation of organic acids and their acyl-CoA derivates may also be of great importance in purple bacteria using some organic substrates, particularly by enzymes whose action is connected with ferredoxin.

Table 2. Carboxylases of purple and green bacteria (ref. 128)

Enzyme	Reaction	Organisms in which enzymes have been reported present		
		Purple nonsulfur bacteria	Purple sulfur bacteria	Green bacteria
Ribulose 1,5-diphosphate carboxylase (EC 4.1.1.39)	D-ribulose 1,5-diphosphate + CO_2 + H_2O → 2(3-phosphoglycerate)	*R. rubrum; R. palustris; R. sphaeroides; R. vannielii; R. sulfoviridis*	*C. okenii; C. vinosum; T. roseopersicina; Thiopedia* sp.; *Thiospirillum* spp.; *E. shaposhnikovii*	*C. limicola* f. *thiosulfatophilum* (?)
Phosphoenolpyruvate carboxylase (EC 4.1.1.31)	phosphoenolpyruvate + CO_2 + H_2O → oxaloacetate + P_i	*R. rubrum; R. palustris*	*C. vinosum; C. okenii; E. shaposhnikovii; T. roseopersicina*	*C. limicola* f. *thiosulfatophilum*
Phosphoenolpyruvate carboxykinase (EC 4.1.1.32)	phosphoenolpyruvate + CO_2 + {ADP/GDP/IDP} ⇌ oxaloacetate + {ATP/GTP/ITP}	*R. rubrum; R. palustris; R. capsulata; R. gelatinosa*	*C. vinosum; C. okenii; E. shaposhnikovii; T. roseopersicina*	
Phosphoenolpyruvate carboxytransphosphorylase (EC 4.1.1.38)	phosphoenolpyruvate + CO_2 + P_i ⇌ oxaloacetate + PP_i	*R. palustris; R. rubrum*	*E. shaposhnikovii* (?)	

continued

Table 2.   (continued)

Enzyme	Reaction	Organisms in which enzymes have been reported present		
		Purple nonsulfur bacteria	Purple sulfur bacteria	Green bacteria
Pyruvate carboxylase (EC 6.4.1.1.)	pyruvate + CO_2 + H_2O + ATP → oxaloacetate + ADP + P_i	R. sphaeroides; R. rubrum	C. vinosum; C. okenii; T. roseopersicina	
Propionyl-CoA carboxylase (EC 6.4.1.3)	propionyl-CoA + CO_2 + H_2O + ATP ⇌ (S)-methylmalonyl-CoA + ADP + P_i	R. rubrum	E. shaposhnikovii	C. limicola f. thiosulfatophilum (?)
Malate dehydrogenase (decarboxylating) (NAD(P)) "malic enzyme" (EC 1.1.1.40)	pyruvate + CO_2 + NAD(P)H ⇌ malate + NAD(P)	C. vinosum		
Pyruvate synthase (EC 1.2.7.1.)	acetyl-CoA + CO_2 + ferredoxin (red) → pyruvate + CoA + ferredoxin (ox)	C. vinosum; E. shaposhnikovii; T. roseopersicina	R. rubrum	C. limicola f. thiosulfatophilum
α-Ketobutyrate synthase (EC 1.2.7.2)	propionyl-CoA + CO_2 + ferredoxin (red) → α-ketobutyrate + CoA + ferredoxin (ox)		C. vinosum	

Enzyme	Reaction			
α-Ketoglutarate synthase (EC 1.2.7.3)	succinyl-CoA + CO_2 + ferredoxin (red) → α-ketoglutarate + CoA + ferredoxin (ox)	*R. rubrum*	*C. vinosum; E. shaposhnikovii*	*C. limicola* f. *thiosulfatophilum*
Phenylpyruvate synthase	phenylacetyl-CoA + CO_2 + ferredoxin (red) → phenylpyruvate + CoA + ferredoxin (ox)		*C. vinosum*	*C. limicola* f. *thiosulfatophilum*

Such carboxylases, called "synthases," have been discovered in purple bacteria, green bacteria, and in some chemotrophic bacteria belonging to the strict anaerobes (76, 154, 155). Pyruvate synthase, which catalyzes the formation of pyruvate from acetyl-CoA and CO_2, has the widest distribution. In some purple sulfur bacteria having limited capacity for oxidation of organic compounds such a pathway of acetate assimilation is of particular importance.

The capacity of *C. vinosum* to synthesize phenylpyruvate as a result of phenylacetyl-CoA carboxylation seems to be important for phenylalanine formation. Also, the production by the same bacterium of α-ketobutyric acid by the action of α-ketobutyrate synthase allows the ensuing synthesis of isoleucine by a new pathway without participation of threonine (154, 155). At the same time, it may be of interest to note that in such purple bacteria as *R. rubrum* (152, 153) and *E. shaposhnikovii* (146) propionate metabolism is connected with participation of another carboxylating enzyme, biotin-dependent propionyl-CoA carboxylase.

Thus, it is clear that the presence of different carboxylases in purple bacteria is significant for synthesis of several different metabolites, particularly of some organic acids and amino acids. It is probable, however, that all carboxylases are not always involved in photoassimilation of CO_2; some of them seem to have a decarboxylating function. For instance, the so-called malic enzyme may catalyze production of pyruvate from malate. It is also thought that in *R. rubrum* growing on C_4-dicarboxylic acids, phosphopyruvate carboxykinase effects production of phosphopyruvate from oxaloacetate, which is used for the synthesis of carbohydrates and some other cell components (156).

Special attention should be drawn to several communications concerning properties of enzymes involved in metabolism of CO_2 (128, 151, 157–160). Many of these works are dedicated to ribulose 1,5-diphosphate carboxylase (151, 157, 158) that has been purified from some purple bacteria. According to their molecular weight and subunit composition, ribulose diphosphate carboxylases of *Chr. vinosum* (151, 157, 158) and *Ectothiorhodospira halophila* (161) are similar to the corresponding enzymes of eucaryotes, whereas in some other purple bacteria ribulose diphosphate carboxylases have lower molecular weights. Of all known ribulose diphosphate carboxylases of *C. vinosum* (151, 157, 158) and shown by the enzyme from *R. rubrum*, which consists of only two subunits (151, 157, 158). There is also some evidence that *Rhodopseudomonas sphaeroides* has two forms of ribulose diphosphate carboxylase that differ as to molecular weight (162).

Similar to ribulose diphosphate carboxylases of other organisms, this enzyme in purple bacteria not only catalyzes CO_2 fixation but also has the function of an oxygenase (151, 163, 164):

Ribulose 1,5-diphosphate $+ O_2 \rightarrow$ 3-phosphoglycerate $+$ phosphoglycolate

This may explain the production of glycolate in the presence of O_2 by purple bacteria (165-167) as well as by some other autotrophs, but there are also some data (168) showing that glycolate formation in purple bacteria may be connected with oxidative breakdown of glycolaldehyde-transketolase complex in such a way as:

$$\text{Fructose 6-phosphate} \xrightarrow[\text{transketolase}]{Fe^{3+}} \text{glycolate} + \text{erythrose 4-phosphate}$$

The oxidant involved in this reaction in vivo is not defined.

Metabolism of Reduced One-Carbon Compounds   It has been known for a long time that *R. palustris* can grow on media containing formate (9, 139, 140, 169-172). The same capacity is revealed by some strains of *R. acidophila* and *Chromatium* spp. (11, 128, 141).

Many purple nonsulfur bacteria can use methanol (141, 173, 174). The carbon from formate (139, 140) and from methanol (141) is assimilated by purple bacteria mainly via the ribulose diphosphate cycle. The general result of photometabolism of methanol by *R. acidophila* corresponds to (141, 173):

$$2CH_3OH + CO_2 \xrightarrow{\text{light}} 3(CH_2O) + H_2O$$

Thus it is clear that methanol and formate are used by purple bacteria first of all as electron donors and are oxidized to CO_2:

$$CH_3OH \rightarrow CH_2O \rightarrow HCOOH \rightarrow CO_2$$

Methanol oxidation by *R. palustris* (175) and *R. acidophila* (141) is catalyzed by a methanol dehydrogenase that can be coupled to phenazine methosulfate as electron acceptor. The synthesis of this enzyme is induced in the presence of methanol. The formaldehyde thus produced may be oxidized by the same enzyme as is methanol and also by NAD-linked formaldehyde dehydrogenase, which requires reduced glutathione for activity (141). Similar enzymes catalyze methanol oxidation in many non-photosynthetic methylotrophic bacteria (176-178).

In this respect purple bacteria are similar to *Pseudomonas oxalaticus* and *Micrococcus denitrificans*, which assimilate carbon in the form of CO_2 via the ribulose diphosphate cycle from formate and methanol, respectively (141, 179). In contrast, purple bacteria have not revealed the capacity for formaldehyde assimilation via the ribulose monophosphate or serine cycles (141).

According to some observations, the formate dehydrogenases of purple bacteria are similar to the formaldehyde dehydrogenase in being NAD-linked (128, 141, 180). However, in some strains of *R. palustris* it was active only in the presence of artificial electron acceptors, the best of them being phenazine methosulfate (172).

Activity of formate dehydrogenase has also been discovered in *R. rubrum* and in some other purple bacteria that cannot utilize methanol and formate for growth (128). Some strains of *R. rubrum* (181–183) and *R. palustris* (172) are able to produce CO_2 and H_2 from formate.

Information concerning photoassimilation of other C_1 compounds by purple bacteria ia more limited. There has been a communication about the possibility of growth of *Rhodopseudomonas gelatinosa* in the presence of methane and inclusion of its carbon into cells (184). According to Hirsch's observations (185) several strains of *Rhodopseudomonas* are able to grow in an atmosphere containing carbon monoxide. Recently the ability of some purple bacteria to assimilate CO has been confirmed (186). However, growth of cultures in the presence of CO was observed in the dark and there is as yet no communication about metabolism of CO by purple bacteria in the light.

Metabolism of C_n Organic Compounds   In general, purple bacteria are able to utilize during photosynthesis a considerable number of compounds containing more than one carbon atom, among them alcohols, sugars, organic acids, and amino acids (2, 11, 99, 112, 128, 187). All purple nonsulfur bacteria possess the ability to utilize organic compounds in two ways—as electron donor and as carbon source without obligatory oxidation to CO_2. (11, 123, 125, 127, 128). The same capacity has been demonstrated in some purple sulfur bacteria, in particular in *C. vinosum* and *E. shaposhnikovii* (11, 128, 129), but in many species organic compounds are assimilated only as complementary carbon sources (in relation to CO_2) in the presence of inorganic electron donors (2, 128). The spectrum of organic compounds utilized by such purple sulfur bacteria is usually very limited (2).

Alcohols   Many purple bacteria, including some purple sulfur bacteria, utilize C_2–C_6 aliphatic alcohols in addition to methanol (2, 11, 112, 128, 187). Ethanol, propanol, and glycerol are used most often. Foster (11) showed isopropanol oxidation to acetone in *Rhodopseudomonas* spp. to be coupled with CO_2 assimilation, thus providing incontrovertible evidence for the ability of organic compounds to play the role of electron donors in photosynthesis. But in most cases the alcohols are subjected to further oxidation, with formation of organic acids that may be taken up by the biosynthetic processes of cells and also partly oxidized to CO_2.

Alcohol dehydrogenases of purple bacteria reveal a wide specificity and are able to oxidize some alcohols (C_7–C_8) that do not support growth. In *R. rubrum, R. palustris,* and *Rhodopseudomonas viridis* there are NAD-linked enzymes (175) and in *Rhodomicrobium vannielii,* NADP-dependent enzymes (175–188). In *R. acidophila* alcohol dehydrogenase was active in the presence of phenazine methosulfate and ammonium ions (175).

A mutant of *R. capsulata* has been obtained that, in contrast to parental strain, could grow well on a medium containing glycerol (189, 190). The glycerol oxidation by this mutant and by *R. sphaeroides* (191) proceeds with participation of a soluble glycerokinase and a particulate, pyridine nucleotide–independent glycerophosphate dehydrogenase. In this respect purple bacteria are similar to some aerobic chemotrophs. It has also been shown that glycerophosphate acetyltransferase is present in *R. sphaeroides* and participates in lipid synthesis (192).

Sugars Among different sugars, purple bacteria most often utilize glucose and/or fructose (2, 11, 128, 187); for instance, *R. rubrum* uses only fructose (193). The transport of fructose into the cell occurs through a phosphoenolpyruvate-dependent phosphotransferase system (194).

It is quite possible that the absence of a corresponding transport system prevents some of the purple bacteria, as well as many cyanobacteria (3, 6), from utilizing external sugars. This supposition is supported by the fact that purple bacteria are capable of decomposing endogenous glucan formed by them as reserve product (195).

According to their enzymatic activity many species of purple sulfur and nonsulfur bacteria (*C. okenii, C. vinosum, Chromatium minutissimum, E. shaposhnikovii, T. roseopersicina, R. rubrum, R. palustris, R. viridis,* and *R. vannielii*) catabolize sugar by the Embden-Meyerhof pathway (128–130, 196, 197). The operation of such a pathway was confirmed for *R. rubrum* by the use of ^{14}C-labeled fructose (194). *R. capsulata* metabolizes fructose via the same pathway (198, 199). The growth of this microorganism and of some other purple bacteria on fructose medium is connected with induction of 1-phosphofructokinase (EC 2.7.1.56). This means that metabolism of fructose begins by formation of fructose 1-phosphate instead of fructose 6-phosphate, as produced by many organisms (198). In contrast to fructose, glucose is degraded by *R. capsulata* via the Entner-Doudoroff pathway (198, 199, 200). A similar pathway of degradation of both glucose and fructose exists in *R. sphaeroides* (201).

Organic Acids The utilization of organic acids is a property common to many purple bacteria and for many species they are the best substrates.

Organic acids utilized by purple nonsulfur bacteria include saturated fatty acids (up to C_9), di- and tricarboxylic acids, and hydroxy- and keto acids (from C_3 to C_6). Purple nonsulfur bacteria of different species utilize acetate, pyruvate, malate, succinate, and fumarate; lactate, propionate, and butyrate are also often utilized but glycolate, citrate, gluconate, and aromatic acids are seldom used. (1, 2, 128).

Purple sulfur bacteria are able to utilize acetate and very often pyruvate (1, 2), although some species, e.g., *C. vinosum, C. minutissimum,* and *E. shaposhnikovii,* have greater ranges of substrates (11, 128, 129, 187). Some purple bacteria are able to assimilate amino acids not only as nitrogen source but also as carbon source (1, 112, 115, 127, 128, 193).

Numerous investigations concern acetate metabolism, especially in *R. rubrum* (11, 112, 125, 127, 128, 134, 136, 137, 202, 203). The utilization of acetate begins by formation of acetyl-CoA, catalyzed by acetyl-CoA synthetase (127, 128, 130). Acetyl-CoA may be further involved in different reactions depending on the bacterial species and on the growth medium. The utilization of [^{14}C]acetate with rapid formation of labeled amino acids, especially aspartate and glutamate, is typical for growing cultures of both purple sulfur and nonsulfur bacteria. Besides amino acids, labeled compounds of the tricarboxylic acid (TCA) cycle, succinate, malate, fumarate, and citrate are usually found.

The existence of this cycle is confirmed (11, 127, 128, 137, 204) by the presence of all the necessary enzymes in different purple nonsulfur bacteria (Table 3). In some experiments fluoroacetate inhibited [^{14}C]acetate utilization by *R. rubrum* and *R. palustris* and the formation of labeled glutamate. Simultaneous accumulation of [^{14}C]citrate was observed (11, 128, 205). A glyoxylate cycle (206) may also operate in some purple bacteria (Table 3). The functioning of the TCA and glyoxylate cycles can explain (204, 205) acetate photoassimilation in *R. palustris* (Figure 3).

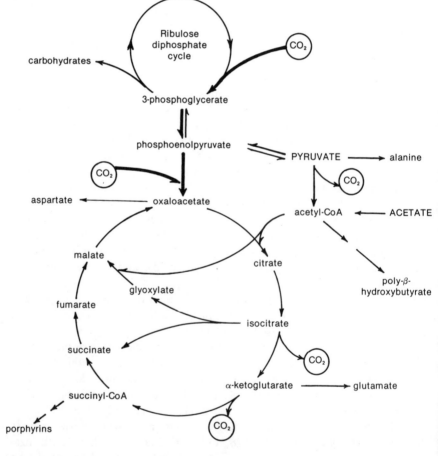

Figure 3. Metabolism of carbon dioxide and some organic compounds in *Rhodopseudomonas palustris* (128).

Table 3. The presence of enzymes of the tricarboxylic acid cycle and glyoxylate cycle in purple and green bacteria (data taken from refs. 127, 128, 204, and 206)

Complete tricarboxylic acid cycle		Incomplete tricarboxylic acid cycle	
Glyoxylate cycle +	Glyoxylate cycle —[a]	Glyoxylate cycle +	Glyoxylate cycle —[a]
R. tenue[b]	R. rubrum	C. vinosum[d]	T. roseopersicina
R. palustris	R. fulvam[b]	C. minutissimum	C. limicola f.
R. gelatinosa[c]	R. molischianum	C. okenii	thiosulfatophilum
R. vannielii[b]	R. sphaeroides	E. shaposhnikovii	P. aestuarii
	R. viridis		
	R. capsulata		

[a]Most bacteria deficient as to glyoxylate cycle do not produce isocitrate lyase.
[b]There are no data for all TCA cycle enzymes.
[c]Malate synthase produced in dark aerobic conditions only.
[d]Bacteria having incomplete TCA cycle do not contain α-ketoglutarate dehydrogenase. In C. vinosum malate dehydrogenase is also lacking.

However, *R. rubrum* and several other purple bacteria have no glyoxylate cycle because they do not contain isocitrate lyase or malate synthase (206). At the same time it was found that alanine produced by *R. rubrum* in the presence of acetate was synthesized with participation of CO_2, since it contained ^{14}C from both these compounds (11, 125, 128, 207). The condensation of $C_2 + CO_2$ fragments in *R. rubrum* and in other purple bacteria is known to be catalyzed by pyruvate synthase (Table 2). The analysis of aspartate and glutamate isolated from proteins of *R. rubrum* after utilization of [14C]acetate in the presence of nonlabeled bicarbonate and vice versa has shown CO_2 to be utilized also in the synthesis of these amino acids (127, 128, 207). Thus it is clear that acetate photometabolism in *R. rubrum* involves not only reactions of the TCA cycle. The same conclusion can be made concerning acetate utilization in the light by *R. sphaeroides* (208).

In all purple sulfur bacteria that have been studied the TCA cycle seems to be incomplete (128–130, 204), owing to the absence of α-ketoglutarate dehydrogenase. Species of purple sulfur bacteria without malate dehydrogenase, or with very low activity (129, 204), also exist, e.g., *C. vinosum* and *C. minutissimum*. However, in *C. vinosum* a modified glyoxylate cycle may operate (129), which explains why this bacterium can grow on acetate medium in the absence of any other compounds capable of playing the role of electron donor. A similar capacity is revealed by *E. shaposhnikovii*, which possesses a glyoxylate cycle (128, 147, 204).

In the absence of inorganic electron donors, acetate is partly oxidized. This makes possible CO_2 fixation via the ribulose diphosphate cycle, but on a restricted scale; considerbly more carbon is assimilated by *E. shaposhnikovii* from acetate as a result of its metabolism via reactions of the TCA and glyoxylate cycles (Figure 4). Assimilation of some of the acetate is possible as a result of acetyl-CoA and phosphoenolpyruvate carboxylation:

$$\text{acetate} \rightarrow \text{acetyl-CoA} \xrightarrow{+CO_2} \text{pyruvate} \rightarrow \text{phosphoenolpyruvate}$$

$$\xrightarrow{+CO_2} \text{oxaloacetate} \rightarrow \text{malate} \rightarrow \text{fumarate} \rightarrow \text{succinate} \rightarrow \text{succinyl-CoA}$$

$$\xrightarrow{+CO_2(?)} \alpha\text{-ketoglutarate}$$

Such a pathway of acetate utilization seems to gain importance if the medium contains sulfide or some other substrate playing the role of electron donor (128, 147).

T. roseopersicina seems to utilize acetate in a somewhat similar way (Figure 5) but this purple sulfur bacterium possesses neither a complete TCA cycle nor a complete glyoxylate cycle (204). That is why its possibilities for metabolism of acetate and some other organic compounds are very

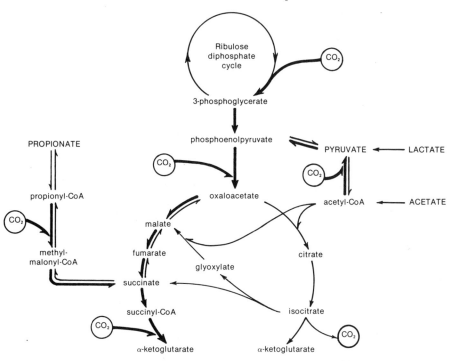

Figure 4. Metabolism of carbon dioxide and some organic compounds in *Ectothiorhodospira shaposhnikovii* (128).

limited and are dependent on the presence of an inorganic electron donor as well as CO_2. The same may be true for several other purple sulfur bacteria, although the necessary investigations have not been made.

Some purple sulfur (*C. vinosum*) and nonsulfur bacteria (*R. rubrum* and *R. sphaeroides*), when utilizing [14C]acetate, produce citramalate as one of the first labeled products (128, 202, 208, 209). This is due to the action of pyruvate transacetylase catalyzing the following reaction:

$$\text{Acetyl-CoA} + \text{pyruvate} \rightarrow \text{citramalate} + \text{CoASH}$$

Citramalate is thought to be utilized by purple bacteria for isoleucine biosynthesis (210).

Finally, there is another pathway of acetate metabolism known to be widely distributed in purple bacteria. This pathway consists of poly-β-hydroxybutyrate production from acetate (127, 128, 203). This process begins by condensation of two molecules of acetyl-CoA, with formation of acetoacetyl-CoA.

Thus acetate photometabolism in purple bacteria seems to be rather varied, differing according to species. Not all pathways of its metabolism are quite clear and they need further study.

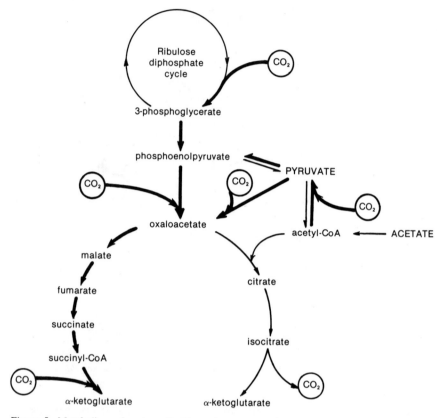

Figure 5. Metabolism of carbon dioxide and some organic compounds in *Thiocapsa rose-opersicina* (128).

There is a clear picture of the initial stages of propionate photometabolism. In such purple bacteria as *R. rubrum* (152, 153) and *E. shaposhnikovii* (146), propionate utilization is connected with propionyl-CoA carboxylation. The general scheme of such a process is as follows:

$$\text{Propionate} \rightarrow \text{propionyl-CoA} \xrightarrow{+CO_2} \text{methylmalonyl-CoA}$$

$$\rightarrow \text{succinyl-CoA} \rightarrow \text{succinate}$$

In contrast to this, in *C. vinosum* (163, 164) propionyl-CoA carboxylation is followed by α-ketobutyrate production (Table 2). Extracts of *R. rubrum* cells also catalyze the following reactions (210, 211):

$$\text{Propionyl-CoA} + \text{glyoxylate} \rightarrow \textit{erythro-}\beta\text{-methylmalyl-CoA}$$

$$\rightarrow \text{mesaconyl-CoA} \rightarrow \text{mesaconate} \rightarrow \text{L-citramalate}$$

Many purple bacteria utilize for their growth lactate that is oxidized to pyruvate by the action of lactate dehydrogenase (11, 128, 212). Pyruvate is one of the compounds utilized by most purple bacteria. Several purple bacteria possess the capacity for its phosphorylation (128, 206, 212). This reaction was earlier thought to be catalyzed by phosphopyruvate synthase, but according to recent evidence (213) pyruvate-orthophosphate dikinase (EC 2.7.9.1) is involved:

$$\text{Pyruvate} + \text{ATP} + P_i \rightarrow \text{phosphoenolpyruvate} + \text{AMP} + PP_i$$

The resulting phosphopyruvate may be utilized for carbohydrate synthesis or be subjected to carboxylation, with formation of oxaloacetate through the action of different enzymes (Table 2). In some species of purple bacteria, carboxylation of pyruvate is possible through the action of pyruvate carboxylase (129, 130, 214) or malic enzyme (129). Finally, production of alanine (128, 215) from pyruvate and citramalate may take place (206, 208, 209).

The breakdown of pyruvate to C_2 and C_1 fragments is of great importance in the metabolism of purple bacteria. Such reactions may proceed with participation of different enzymes.

Purple sulfur and nonsulfur bacteria of many species (*R. palutris, R. sphaeroides, R. viridis, R. gelatinosa, R. capsulata, Rhodospirillum molischianum, C. minutissimum, E. shaposhnikovii,* and *T. roseopersicina*) are able to decarboxylate pyruvate with ensuing production of CO_2 and acetaldehyde, which is converted to acetoin (212, 216). This reaction is catalyzed by pyruvate decarboxylase (EC 4.1.1.1). Some species (*E. shaposhnikovii, R. palustris, R. sphaeroides, R. viridis,* and *R. rubrum*) reveal pyruvate dehydrogenase activity linked with NAD (183, 212). The capacity for pyruvate degradation with participation of pyruvate:ferredoxin: oxidoreductase has also been found in some purple bacteria (*C. vinosum, C. minutissimum,* and *R. rubrum*) (181, 212, 217). *R. rubrum* is able to break down pyruvate in the dark under anaerobic conditions with participation of pyruvate formate lyase (183, 218). It should also be noted that pyruvate kinase (EC 2.7.1.40) has been found in three species of purple bacteria—*R. rubrum, R. capsulata,* and *R. sphaeroides* (219–221).

The assimilation of the dicarboxylic acids, malate, fumarate, and succinate is followed, as a general rule, by CO_2 production. At the same time, experiments with $^{14}CO_2$ show the fixation of carbon dioxide to occur in the presence of these acids, although the greater quantity of carbon assimilated into the cells comes from the dicarboxylic acid substrate (125, 128, 129, 137, 143, 144, 208).

Malate utilization in purple bacteria seems to be connected in many cases with its conversion into oxaloacetic acid, this serving in its turn as a source of phosphopyruvate and other products. However, not all purple bacteria have the capacity for direct production of oxaloacetate from malate. As mentioned above, *C. minutissimum* is without such a capacity, having no malate dehydrogenase activity (129), but the availability of malic enzyme allows the cells of this bacterium to decarboxylate malate to CO_2 and pyruvate. The latter in its turn may be further converted into phosphopyruvate and oxaloacetate, or decarboxylated. *C. minutissimum* seems to have the same possibility (204). Thus malate metabolism in purple bacteria may start in different ways but results in formation of the same products:

$$\text{Malate} \begin{cases} \longrightarrow \text{oxaloacetate} \xrightarrow{-CO_2} \text{phosphoenolpyruvate} \\ \xrightarrow{-CO_2} \text{pyruvate} \rightarrow \text{phosphoenolpyruvate} \end{cases}$$

R. rubrum and *R. sphaeroides* have the capacity not only for synthesis of malate from acetyl-CoA and glyoxylate (206) but also for a reverse process (222–224). According to evidence for *R. sphaeroides*, breakdown of malate with production of glyoxylate and acetyl-CoA occurs with participation of malyl-CoA synthase and malyl-CoA lyase (224, 225):

$$\text{Malate} \xrightarrow{\text{ATP,CoA}} \text{malyl-CoA} \rightarrow \text{acetyl-CoA} + \text{glyoxylate}$$

Glyoxylate and acetyl-CoA formation from malate is very important for the metabolism of methylotrophic bacteria possessing the serine pathway of formaldehyde assimilation (177, 178). Such a process results in regeneration of glycine as the C_1 unit acceptor. In purple bacteria the same process of malate degradation seems to play a somewhat different role. It should be mentioned that neither species of purple bacteria that possess the capacity for malate degradation with acetyl-CoA and glyoxylate production contains isocitrate lyase and hence neither are able to form glyoxylate by cleavage of isocitrate (206).

There is also some evidence for succinate metabolism in purple bacteria. For different species of these microorganisms the presence of succinate dehydrogenase connected with the photosynthetic apparatus is characteristic, although the activity of this enzyme is often low (204). With participation of membrane-linked succinate dehydrogenase, pigment-containing preparations (chromatophores) from *R. rubrum* oxidize succinate in the light with production of reduced NAD (226–228) as a result of reverse electron flow (78, 79). Such a process may ensure CO_2 assimilation by the cells via the ribulose diphosphate cycle (227). Fumarate produced by *R. rubrum* as a result of succinate oxidation may be further converted independently of light into malate and some other compounds (227). Another pathway of succinate utilization by purple bacteria is possible in such microorganisms as *R. rubrum* (154, 155) and *E. shaposhnikovii* (146) due to the availability of α-ketoglutarate synthase, which catalyzes the carboxylation of succinyl-CoA. Such a pathway may also function in *R. palustris* in the light, since utilization of [^{14}C]succinate by this bacterium depends on the presence of CO_2 and results in production of labeled glutamate (170).

Thus some of the purple bacteria dispose of two possibilities of α-ketoglutarate production; either as a result of metabolism via the TCA cycle reactions or by direct carboxylation of succinyl-CoA:

$$\text{Succinate} \begin{cases} \text{succinyl-CoA} \xrightarrow{+CO_2} \text{α-ketoglutarate} \\ \text{fumarate} \to \text{malate} \to \text{oxaloacetate} \xrightarrow{+\text{acetyl-CoA}} \text{citrate} \end{cases}$$

$$\to \text{isocitrate} \xrightarrow{-CO_2} \text{α-ketoglutarate}$$

Cells of *R. rubrum* utilizing 1,4-[^{14}C]succinate in the light produce mainly glutamate labeled in C_1 (228). Such evidence is not consistent with α-ketoglutarate synthesis occurring as a result of succinyl-CoA carboxylation. After 5-10 s of photoassimilation of 2,3-[^{14}C]succinate by cells of *E. shaposhnikovii* (146) in the presence of CO_2, the greatest amount of ^{14}C was found in fumarate and malate; their curves had a negative slope and the content of ^{14}C in glutamate and in phosphorylated products increased with time. Such observations seem to indicate that glutamate generation from succinate proceeds via TCA cycle reactions. The possibility of α-ketoglutarate synthesis by way of succinyl-CoA carboxylation may have importance for some purple bacteria in the presence of sulfide or other substrates capable of functioning as electron donor.

Purple bacteria are also able to condense succinyl-CoA and glycine. This is an important reaction leading to generation of δ-aminolevulinic acid, utilized for synthesis of different porphyrins and corrinoid compounds. There is some evidence that synthesis of δ-aminolevulinate in *R. rubrum* is connected with the function of the so-called succinate-glycine cycle (229).

Investigations of fumarate photometabolism have shown that cells and cell-free preparations of *R. sphaeroides* and *R. rubrum* are capable of converting fumarate into succinate and malate. Extensive production of [^{14}C]succinate from labeled fumarate was observed in the light (208). Fluoroacetate does not influence CO_2 production from fumarate and succinate by *R. sphaeroides*. Such observations allow us to conclude that photometabolism of these compounds is not connected with the TCA cycle (208). According to Gibson's data (230) the transport of succinate, fumarate, and malate into the cells of purple bacteria proceeds with participation of one transport system.

Few purple bacteria possess the capacity to utilize citrate (1,2); one that does is *R. gelatinosa* (231-237). This bacterium grows on citrate medium in the light under anaerobic conditions only. Under the action of citrate lyase (EC 4.1.36) citrate is broken down to acetate and oxaloacetate:

$$\text{Citrate} \xrightarrow{\text{light}} \text{acetate} + \text{oxaloacetate}$$

Oxaloacetate is quickly reduced to malate. Acetate is produced in large amounts but later it may be utilized by the cells.

The molecular weight of citrate lyase of *R. capsulata* purified to homogeneity is 530,000–560,000. It consists of six major and six minor subunits, contains pantothenate, and requires Mg^{2+} for its activity (234). After complete utilization of citrate the enzyme is converted into an inactive HS form owing to its deacetylation (236). The process of deacetylation also occurs under aerobic conditions in spite of citrate availability. Different enzymes are involved in acetylation and deacetylation of citrate lyase (235–237). The first of these enzymes responsible for conversion of citrate lyase to an active acetyl-S form is rapidly inactivated after citrate utilization by the cells. The inactivating enzyme (deacetylase of citrate lyase) transforming citrate lyase to an inactive HS form is strongly substrate-specific and glutamate inhibits its activity. Therefore, it is supposed that as long as there is a high enough concentration of L(+)-glutamate in the cells no activity of deacetylase is evident.

Aromatic Compounds   R. palustris and some other purple bacteria are able to utilize aromatic compounds. The growth of *R. palustris* is possible on media containing benzoate, *p*-hydroxybenzoate, *m*-hydroxybenzoate, mandelate, benzoylformate, and some other cyclic substrates (238–242).

Benzoate can be utilized by this bacterium in the light under anaerobic conditions. Investigations with the use of isotopes, enzymological methods, and different mutant strains have shown that photometabolism of benzoate in *R. palustris* occurs through a unique pathway (240–242). Its principal peculiarity is that benzoate is reduced before being converted to pimelate. Such a phenomenon could be observed only in illuminated cells under anaerobic conditions. There are also data showing that first of all benzoate is converted to benzoyl-CoA, and the whole process of its degradation can be represented as follows (242):

$$\text{Benzoate} \xrightarrow{\text{CoA,ATP}} \text{benzoyl-CoA} \xrightarrow{3(XH_2)} \text{cyclohexanoyl-CoA}$$

$$\xrightarrow{\text{FAD}} \text{cyclohex-1-enoyl-CoA} \xrightarrow{H_2O} \text{2-hydroxycyclohexanoyl-CoA}$$

$$\xrightarrow{\text{NAD}} \text{2-oxocyclohexanoyl-CoA} \xrightarrow{\text{CoA,ATP}} \text{pimelyl-di-CoA} \rightarrow \text{pimelate}$$

This means that photometabolism of benzoate in *R. palustris* is analogous to β-oxidation of fatty acids.

It has also been shown that *R. gelatinosa* is capable of growing in the light on medium containing phloroglucinol, converting it first of all into dihydrophloroglucinol. The same reaction occurs in cell-free extracts in the presence of NADPH (242). In contrast, reduction of benzoate and benzoyl-CoA by cell-free preparations from *R. palustris* in the presence of different electron donors, including reduced ferredoxin, could not be achieved (242). Therefore further studies of this process are needed.

Utilization of Primary Products of Assimilation of CO_2 *and Organic Compounds*   Under conditions of active growth, purple bacteria utilize the products of photoassimilation of CO_2 and organic compounds mainly for synthesis of proteins, nucleic acids, pigments, and other cellular components of vital importance. However, a certain amount of reserve products, carbohydrates included, may be synthesized even in growing cells of purple bacteria (243, 244).

All purple bacteria, excluding some mutants, are capable of synthesizing amino acids using ammonium as nitrogen source (2, 128). Some species have the capacity for assimilative nitrate reduction (128, 245, 246) and several purple bacteria grow on media containing urea and some amino acids (11, 112, 128, 246).

The majority of purple bacteria are known to have the capacity for N_2 fixation (2, 11, 112, 247–249). Amino acid biosynthesis in purple bacteria is possible as a result of transamination (128) or reductive amination of keto acids with participation of glutamate dehydrogenase or alanine dehydrogenase (128, 215). At low concentrations of NH_4^+, glutamine synthetase and glutamate synthase are of special importance (246, 250–253):

$$\text{Glutamate} \xrightarrow{\text{NH}_4^+,\text{ATP}} \text{glutamine} \xrightarrow{\alpha\text{-ketoglutarate,NAD(P)H}} 2 \text{ glutamate}$$

Under nitrogen limitation a considerable part of the photosynthetic products are utilized for generation of nitrogen-free polymer compounds. All purple bacteria are able to accumulate in the cells such reserve polymers as poly-β-hydroxybutyrate and glycogen-like glucan (89, 203). The preferential production of one of these products depends on the carbon source. A considerable amount of poly-β-hydroxybutyrate occurs in purple bacteria utilizing acetate, butyrate, hydroxybutyrate, and some other organic compounds (11, 128, 203). Poly-β-hydroxybutyrate synthesis from acetate consists of the following reactions (3, 89):

$$2 \text{ acetate} \xrightarrow{\text{2CoA, 2ATP}} 2 \text{ acetyl-CoA} \xrightarrow{-\text{CoA}} \text{acetoacetyl-CoA}$$

$$\xrightarrow{+2\text{H}} \beta\text{-hydroxybutyryl-CoA} \xrightarrow[\text{polymerization}]{-\text{CoA}} \text{poly-}\beta\text{-hydroxybutyrate}$$

In the case of butyrate or hydroxybutyrate serving as substrate, the reductive phase is not necessary. Moreover, poly-β-hydroxybutyrate synthesis from butyrate represents an oxidative process proceeding as follows (3):

$$\text{Butyrate} \xrightarrow{\text{CoA,ATP}} \text{butyryl-CoA} \xrightarrow{-2\,H} \text{crotonyl-CoA}$$

$$\xrightarrow{+H_2O} \beta\text{-hydroxybutyryl-CoA} \xrightarrow[\text{polymerization}]{-CoA} \text{poly-}\beta\text{-hydroxybutyrate}$$

Such a process necessitates reoxidation of NADH. Under anaerobic conditions this reductant may be utilized by purple bacteria for CO_2 assimilation via the ribulose diphosphate cycle. The result of this process is accumulation of carbohydrates in resting cells of purple bacteria (3):

$$2n\ C_4H_8O_2\ +\ n\ CO_2 \rightarrow 2(C_4H_6O_2)_n\ +\ (CH_2O)_n\ +\ H_2O$$

| butyrate | poly-β-hydroxy-butyrate | glucan |

The main product of degradation of poly-β-hydroxybutyrate granules by purple bacteria is D($-$)-3-hydroxybutyrate (89, 128).

Reserve carbohydrates may be synthesized by purple bacteria in considerable amounts from CO_2, pyruvate, malate, succinate, propionate, and some other compounds readily converted into phosphopyruvate (128, 203). Different species of purple bacteria (*R. rubrum, R. capsulata, E. shaposhnikovii,* and *C. vinosum*) produce reserve polysaccharide of the same nature, i.e., $\alpha(1 \rightarrow 4)$ polyglucan. Similar to poly-β-hydroxybutyrate, it may form special granules in cells (89, 128, 205, 244).

Like other procaryotes, purple bacteria synthesize reserve glucan from ADP-glucose (89, 254):

$$\text{Glucose 1-phosphate} + \text{ATP} \rightarrow \text{ADP-glucose} + PP_i$$

The reaction is catalyzed by ADP-glucose pyrophosphorylase. The next stage of the process is catalyzed by glycogen synthetase:

$$(\text{Glucan})_n + \text{ADP-glucose} \rightarrow (\text{glucan})_{n+1} + \text{ADP}$$

This enzyme, isolated from *C. vinosum* (255), has been purified to a homogeneous state. The same bacterium forms UDP-glucose (126, 129). It is supposed that UDP-glucose is utilized not for reserve glucan synthesis but for formation of other carbohydrates (89).

Photoproduction of Molecular Hydrogen   In addition to the use of organic compounds for synthesis of different cell components, purple bacteria also possess the capacity for their degradation in the light under anaerobic conditions with generation of CO_2 and of molecular hydrogen (11, 68, 76, 112–116, 123, 182, 256). The photoproduction of H_2 in purple bacteria (115, 256, 257) as well as in cyanobacteria (258) is catalyzed by nitrogenase. This enzyme is contained in cells assimilating N_2 or some amino acids, but in the presence of ammonium its synthesis is repressed

(112, 115, 182, 256). Hydrogen may be evolved by growing cultures and resting cell suspensions of purple bacteria, the latter being more active (112, 115, 116, 123, 256).

Different organic compounds may act as electron donor for H_2 production. The greatest amount of H_2 is usually evolved from pyruvate, lactate, and C_4 dicarboxylic acids (112, 115, 116, 182). *R. rubrum* cells, growing on medium containing glutamate, are able to oxidize completely to CO_2 and H_2 such compounds as acetate, succinate, fumarate, and malate via the TCA cycle (68, 123). Some purple bacteria (*C. vinosum, E. shaposhnikovii,* and *T. roseopersicina*) are capable of producing H_2 as a result of thiosulfate oxidation in the light (68, 76, 182).

Photoproduction of H_2 from thiosulfate and several organic compounds is inhibited by uncouplers and some inhibitors of electron transport (68, 182). Thus it is evident that photoevolution of molecular hydrogen is an energy-dependent process connected with a system of reverse electron flow. Such a system of electron transport may also serve in nitrogen fixation. H_2 production takes place in cases when electrons are not involved in N_2 assimilation. In addition to purple bacteria and cyanobacteria, photoproduction of H_2 has been found in many green and some other algae, but in such eucaryotic microorganisms this process is catalyzed by hydrogenase (118, 259).

Anaerobic Dark Metabolism

It has been known for a long time that purple bacteria have the capacity under anaerobic conditions in the dark for "autofermentation" of endogenous substrates with production of organic acids, CO_2, and H_2 (11, 99). Now it is clear that such substrates are poly-β-hydroxybutyrate and reserve glucan that may be accumulated in the cells in great amounts (3, 195, 260, 261). The utilization of such compounds in purple sulfur bacteria may be followed by reduction of endogenous and exogenous sulfur to hydrogen sulfide (15, 195, 262–264).

In *Chromatium* spp. degradation of endogenous polysaccharide in the presence of sulfur may proceed with production of poly-β-hydroxybutyrate in addition to H_2S and CO_2 (3, 195):

$$(C_6H_{10}O_5)_n + nH_2O + 3nS \rightarrow (C_4H_6O_2)_n + 2nCO_2 + 3nH_2S$$

Thus, metabolism of endogenous reserve compounds in purple bacteria may take the form of fermentation or anaerobic respiration, with utilization of molecular sulfur as electron acceptor.

Such processes can provide cells with energy necessary for their vital activities and even for the synthesis of some compounds, in particular bacteriochlorophyll (195, 260, 261, 265). The cells of *Chromatium* spp. containing no reserve carbohydrates rapidly lose their viability in the dark under anaerobic conditions owing to a sharp decline of ATP level (266).

Purple bacteria are also able to degrade some external organic compounds in the dark under anaerobic conditions, e.g., some sugars, glycerol, pyruvate, and some other organic acids (11, 128, 197, 267–273). There are data showing that some purple nonsulfur bacteria may grow at a low rate in the dark under anaerobic conditions as a result of fermentation of organic compounds. Such a capacity has been found in *R. rubrum* (267–273), *R. palustris, R. viridis,* and *R. sphaeroides* (268). The growth of *R. rubrum* in the dark under anaerobic conditions is possible on media containing fructose or pyruvate.

Fructose is fermented by this bacterium via the fructose diphosphate pathway (Figure 6), with production of pyruvate, acetate, formate, propionate, CO_2, and sometimes H_2 (97, 272).

Pyruvate is metabolized by different strains of *R. rubrum* in the dark under anaerobic conditions mainly to acetate and formate and to a lesser extent to CO_2 and propionate (183, 273). Butyrate, valerate, caproate, and acetoin were also discovered in small amounts. Some strains produce molecular hydrogen (112, 268, 273). In *R. rubrum* the greatest importance for acquisition of energy in the process of fermentation of pyruvate is its cleavage by pyruvate formate lyase, because this process leads to generation of ATP (183, 218):

$$\text{Pyruvate} \xrightarrow{\text{CoA}} \text{acetyl-CoA} + \text{formate}$$

$$\text{Acetyl-CoA} + \text{P}_i \longrightarrow \text{acetylphosphate} + \text{CoA}$$

$$\text{Acetylphosphate} + \text{ADP} \longrightarrow \text{acetate} + \text{ATP}$$

This enzyme plays a similar role in some anaerobic chemotrophic bacteria (274).

The production of H_2 by some strains of *R. rubrum* during fermenta-

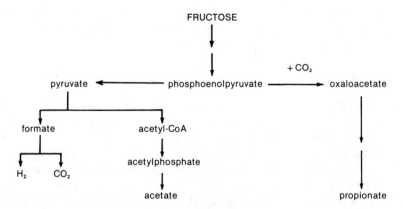

Figure 6. Metabolism of fructose fermentation in *Rhodospirillum rubrum* (272).

tion of fructose and pyruvate seems to be connected with cleavage of formate (181, 183):

$$Formate \rightarrow H_2 + CO_2$$

It is known that in *Escherichia coli* and some other chemotrophic bacteria formate cleavage to CO_2 and H_2 is catalyzed by the formate hydrogen lyase system. The same enzyme system is supposed to be involved in hydrogen production in the dark by *R. rubrum* (183) and by *R. palustris* (172).

In addition to pyruvate formate lyase some *R. rubrum* strains contain pyruvate:ferredoxin oxidoreductase, the participation of which makes production of acetyl-CoA possible (217):

$$Pyruvate \xrightarrow{Fd, CoA} acetyl\text{-}CoA + CO_2 + Fd_{red}$$

The process may be followed by H_2 formation from reduced ferredoxin. The activity of pyruvate:ferredoxin oxidoreductase in *R. rubrum* was much lower than that of pyruvate formate lyase.

As to propionate, all evidence shows it to be produced from methyl-malonyl-CoA under the action of propionyl-CoA carboxylase (152, 153).

It is thought that fructose and pyruvate metabolism in *R. rubrum*, with production of propionate, is also connected with carboxylation of phosphoenolpyruvate (Figure 6). Transcarboxylase activity in *R. rubrum* was not found.

In pyruvate fermentation by other purple nonsulfur bacteria (*R. palustris, R. viridis, R. capsulata, R. acidophila, R. sphaeroides*), acetate, formate, and CO_2 were always found. Propionate, lactate, butyrate, 2,3-butanediol, acetoin, and diacetyl were discovered as well, but the composition and amount of such products were not identical in different species (273).

The purple sulfur bacterium *C. vinosum* strain D also reveals the capacity for pyruvate fermentation. This process was connected with production of acetate, formate, lactate, and CO_2, but no increase of biomass was observed (273). *C. vinosum* may also utilize pyruvate, with simultaneous sulfur reduction to H_2S (262). This fact confirms the ability of some purple bacteria to effect in the dark under anaerobic conditions not only fermentation but also the process of anaerobic respiration. Yet until now no growth of purple bacteria utilizing sulfur as electron acceptor has been observed. However, some strains of purple nonsulfur bacteria similar to *R. sphaeroides* are able to grow in the dark under anaerobic conditions on media with malate or other organic compounds at the expense of reduction of NO_3' to N_2, i.e., as a result of a typical process of denitrification (275).

There is also some evidence that *R. capsulata* is able to grow in the dark under anaerobic conditions on media containing glucose or pyruvate in the presence of dimethyl sulfoxide. The growing cultures reduced this compound with production of dimethyl sulfide. $R.$ *sphaeroides* and $R.$ *palustris* were also able to grow under anaerobic dark conditions with dimethyl sulfoxide (276).

Still more startling is the communication about the capacity of some strains of *Rhodopseudomonas* spp. to grow in dark anaerobic conditions utilizing CO. It was reported as well that suspensions of cells produce equimolar amounts of CO_2 and H_2 from CO according to the equation:

$$CO + H_2O \rightarrow CO_2 + H_2$$

Such a pathway of CO metabolism was confirmed by the use of tritium-labeled water, since $^3H_2$ formation in this case was observed (186).

Thus purple bacteria in the dark under anaerobic conditions reveal the capacity to shift their metabolism to a type proper to different chemotrophs. At the same time they continue to synthesize pigments in considerable amounts and to form a complete photosynthetic apparatus (260, 267–270). This is confirmed by the fact that on illumination of $R.$ *rubrum* cells grown in the dark, photooxidation of c-type cytochromes and NAD reduction becomes evident (268, 269). This means that the reaction centers are in an active state. Purple bacteria are able to accumulate poly-β-hydroxybutyrate in the cells both in the light and in the dark under anaerobic conditions (183, 267, 268).

Aerobic Dark Metabolism

The capacity for growth in the presence of oxygen in the dark is revealed by most purple nonsulfur bacteria (1, 2, 11, 99, 112). $R.$ *sphaeroides* is especially tolerant to high concentrations of O_2; cultures can grow even in an atmosphere of pure oxygen (277, 278, 279). However, optimal oxygen concentrations for this and other purple bacteria are much lower and in different species may vary (11, 278–280). For instance, in the case of $R.$ *capsulata* maximal growth rate was observed at pO_2 in the range of 4–100 mm Hg (280). Some purple nonsulfur bacteria are capable of growing in microaerophilic conditions only (2, 272, 278). Until recently, purple sulfur bacteria were thought to be obligate phototrophs and strict anaerobes (1, 2, 11), but it was known that cells of these microorganisms could survive for a long time in contact with oxygen (11, 279, 281–283). Like purple nonsulfur bacteria, they contain catalase (284) and superoxide dismutase (285, 286).

Some years ago it was shown that some of these microorganisms were capable of growing in the presence of O_2 in the dark, although at a slow rate. Such a capacity was demonstrated in $E.$ *shaposhnikovii* (278, 287, 288), $T.$ *roseopersicina* (132, 289, 290), and *Amoebobacter roseus* (290, 291), but in general aerobic dark metabolism in purple sulfur bacteria is as yet incompletely studied. In contrast, many investigations have been made of dark aerobic metabolism in purple nonsulfur bacteria. Some results of these studies are summarized in special reviews (11, 74, 112).

Respiratory System   Until recently, the growth of purple nonsulfur bacteria under aerobic dark conditions was observed only in the presence of organic compounds (2, 11, 89, 112, 127, 128), although the capacity of some of these microorganisms for oxidation of thiosulfate, sulfide, and molecular hydrogen was conserved (100, 138, 292, 293). Pfennig and Siefert (390) have shown that *R. acidophila* and *R. capsulata* may grow in the dark under microaerophilic conditions on mineral media in the presence of H_2 and CO_2. Thus in the dark under aerobic conditions the behavior of purple nonsulfur bacteria corresponds to that of typical heterotrophs or chemoautotrophs—they get necessary energy as a result of the function of a respiratory electron transport system (74, 127, 128).

The respiratory system of purple bacteria grown in the dark under strict aerobic conditions is located mainly in the cytoplasmic membrane (17, 294). Such a conclusion may be made from evidence showing that oxygen inhibits formation of the intracytoplasmic membrane system and pigments in all known purple bacteria. However, at low oxygen concentration the formation of whole photosynthetic apparatus located in intracytoplasmic membranes is possible both in the light and in the dark (17). Cells grown in the light under anaerobic conditions reveal respiratory activity, although in some cases it is lower than in cells from aerobic dark cultures. Pigment-containing membrane preparations from purple bacteria grown in the light can also take up oxygen in the dark and catalyze phosphorylation of ADP coupled with oxidation of some substrates (17, 295-299).

Since intracytoplasmic and cytoplasmic membranes in purple bacteria are connected to each other, it is thought that there are some common electron carriers participating both in photosynthesis and respiration (17, 74, 298). Their orientations in membranes of both aerobically and photosynthetically grown cells appear to be similar. This is confirmed by respiration of cells of either type grown being coupled to the outward translocation of protons. The same direction is found with proton translocation induced by the light in photosynthetically grown cells (74).

Coupling factors from cells of purple bacteria (*R. capsulata* and *R. sphaeroides*) grown in the light and in the dark contain the same proteins and are interchangeable (74, 300, 301). These facts confirm their identity. Partial reconstruction of systems of electron transport and photophosphorylation has been achieved with membranes from aerobically grown cells of purple bacteria (*R. capsulata* and *R. sphaeroides*) and of their reaction centers (74, 302).

Some of the purple bacteria, in particular *R. rubrum*, when transferred from light anaerobic conditions to dark aerobic conditions, continue to grow without any lag period (3, 303). This indicates the possession by such cells of the complete system of enzymes and electron carriers necessary for the shift from photosynthesis to respiration. However, in the case of *R. palustris*, growth after the transfer of cultures in the dark commenced only after a lag period (297). Such a lag period was not observed in the cultures remaining in the light after the change of condition from anaerobiosis to aerobiosis. This observation indicates that the lag period in the dark is due not to the inhibitory effect of oxygen but to certain changes of cellular metabolism. Such changes may involve different parts of the electron transport chains. According to available information, respiratory systems of purple bacteria contain NAD, flavins, quinones, and cytochromes (17, 74, 294-299, 304-307).

Purple nonsulfur bacteria grown in the dark under aerobic conditions usually possess a stronger NADH-oxidase activity as compared with photosynthetically grown ones (295-297, 304). In some species growing aerobically in the dark, an increase of succinate oxidase activity and activity of dehydrogenases coupled with the electron transport system was noted as well (137, 304), but more often the changes involve the cytochrome part of the respiratory chain, especially of the terminal oxidase (37, 304). All evidence shows c- and b-type cytochromes to be involved in the respiratory system of purple nonsulfur bacteria. They seem to be represented by several components (17, 37, 74, 297, 304, 305). For instance, in the membrane fraction of R. palustris grown in dark aerobic conditions, two c-type cytochromes and three b-type cytochromes were found (304). Three b-type cytochromes have been identified in some membrane preparations of R. capsulata (305).

The composition and the amount of cytochromes in some purple bacteria may vary under different conditions of growth. There is evidence that R. rubrum synthesizes cytochrome c' in the light (27-29). In cells of R. palustris growing in the dark, the amount of c-type cytochromes diminished but the amount of b-type cytochromes increased (299). The function of terminal oxidase in purple nonsulfur bacteria is carried out most often by cytochrome o, one of the b-type cytochromes. Such a cytochrome, capable of interaction with CO, has been found in R. palustris, R. sphaeroides, R. capsulata, R. viridis, and R. rubrum (27, 74, 304, 306, 307). This oxidase of R. palustris has been obtained as a purified preparation (307). There are also data showing that in R. capsulata, in addition to cytochrome o, the function of terminal oxidase may be performed by cytochrome b, not interacting with CO (74, 305). In R. sphaeroides growing under intensive aeration, cytochrome $a + a_3$ is produced (35, 36, 74, 81, 108). An α-type cytochrome has also been found in some strains of R. palustris (37, 81). According to most recent evidence, synthesis of this cytochrome by R. palustris, similarly to R. sphaeroides, depends on the presence of O_2 (37).

Thus it is clear that in purple nonsulfur bacteria different terminal oxidases may function and respiratory electron transport systems (Figure 7) may be branched (74, 304, 305, 308). It was found also that preparations of membranes from R. palustris grown under aerobic conditions may catalyze the synthesis of ATP in the presence of ascorbate and phenazine methosulfate (299). The same preparations from photosynthetically grown cells were deprived of such a capacity. These observations show the existence of an additional point of coupling in the respiratory electron trans-

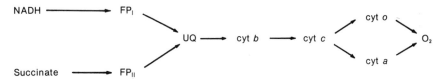

Figure 7. Respiratory electron transport system in *Rhodopseudomonas palustris*: NADH, reduced nicotinamide adenine dinucleotide; FP_I and FP_{II}, different flavoproteins; UQ, ubiquinone; cyt b, cytochrome b; cyt c, cytochrome c; cyt o, cytochrome o; cyt a, cytochrome a.

port system of *R. palustris* grown in aerobic conditions. This is correlated with the appearance in such cells of cytochrome a (37).

Like purple nonsulfur bacteria, different purple sulfur bacteria take up O_2, but often more slowly (11, 132, 278, 283). The rate of respiration of purple sulfur bacteria may increase in the presence of some external substrates (132, 283); for instance, the rate of O_2 consumption by cells of *C. vinosum* strain D was higher in the presence of malate, H_2, sulfide, or thiosulfate (283). The oxidation of thiosulfate by aerated suspensions of *C. vinosum* in the dark resulted in an increased amount of ATP in the cells (283), but not as much as in the light (309). It was also shown that oxidation of thiosulfate by *C. vinosum* in the dark under aerobic conditions was followed by sulfur and sulfate production (101, 282, 283). There is also evidence for *Chromatium* sp. strain L being able to oxidize sulfur in the dark with production of sulfate and with very slow CO_2 fixation (281), but no growth of this purple sulfur bacterium in the dark under aerobic conditions was observed.

By contrast, different strains of *T. roseopersicina* are capable of growing in the dark in the presence of oxygen (132, 289, 290). The growth is possible in autotrophic conditions and depends on sulfide or thiosulfate oxidation with consumption of O_2 (132, 288, 290). Organic compounds (acetate, glucose, and some others) may be utilized by this bacterium, and increase the cell yield but do not stimulate O_2 uptake in the dark (132, 288, 290).

Dark aerobic metabolism of thiosulfate in *T. roseopersicina* is identical with that in the light (106, 107, 288). The process begins with cleavage of thiosulfate to S^{2-} and SO_3^{2-}, with participation of thiosulfate reductase (106). The oxidation of sulfite to sulfate is catalyzed by APS-reductase and APS-sulfurylase (107). This allows the cells to synthesize ATP as a result of phosphorylation at substrate level (310):

$$SO_3^{2-} + AMP \xrightarrow{\text{APS-reductase}} APS + e^-$$

$$APS + P_i \xrightarrow{\text{APS-sulfurylase}} ADP + SO_4^{2-}$$

$$2ADP \xrightarrow{\text{adenylate kinase}} AMP + ATP$$

Part of the sulfite may also be oxidized with participation of sulfite: cytochrome c oxidoreductase (107). Both these processes of oxidation of thiosulfate are membrane-linked. Cell-free preparations of *T. roseopersicina* also exhibit NADH-oxidase activity, although less than that observed in purple nonsulfur bacteria. Malate and succinate dehydrogenase activities also are very low (204). However, ATP formation linked with the electron transport system has been demonstrated.

Common to all purple bacteria containing bacteriochlorophyll is inhibition of respiration by light (11, 283, 311), but in the case of repression of bacteriochlorophyll synthesis by O_2 (17) some of these microorganisms may shift their metabolism to respiration even in the presence of light. It is interesting to note that oxygen depresses first the formation of lightharvesting (antenna) bacteriochlorophyll. Pigment included in reaction centers is conserved even at high pO_2 (312). This seems to have importance for the ability of purple bacteria to switch from respiration to photosynthesis under changed conditions.

Carbon Metabolism   The shift of purple bacteria from photosynthesis to aerobic respiration also results in changed carbon metabolism. Even in the light the presence of O_2 may change the capacity of these microorganisms for synthesis of some enzymes. Moreover, oxygen may inhibit the activity of some enzymes. The effect of oxygen on CO_2 assimilation is quite evident. In the presence of considerable amounts of O_2, ribulose diphosphate carboxylase synthesis in different species of these microorganisms is usually repressed (128, 148, 313–315). As a result, there can be no CO_2 assimilation via the ribulose diphosphate cycle in aerobic conditions. Oxygen is also capable of inhibiting CO_2 fixation catalyzed by ribulose diphosphate carboxylase and of shifting this enzyme to oxygenase function (151, 165–168).

The only known exception concerns *T. roseopersicina* synthesizing ribulose diphosphate carboxylase even in conditions of strong aeration (132, 133). This seems to allow *T. roseopersicina* to change from photosynthesis to dark chemolithoautotrophic metabolism, like some thiobacilli. There is also evidence that *R. capsulata* (138) assimilates CO_2 in the dark in $H_2 + O_2$ atmosphere, like some microalgae (316).

As was indicated earlier, this bacterium and *R. acidophila* are also capable of growing using H_2 and O_2 in the dark under microaerophilic conditions (390). Under such conditions *R. acidophila* may also grow in the dark on media with methanol or formate and N_2 as source of nitrogen (390), but at high pO_2 the growth of cultures of *R. acidophila* in the presence of H_2 or methanol is not possible. Along with repression of ribulose diphosphate carboxylase synthesis, in the presence of a high amount of O_2, hydrogenase and methanol dehydrogenase function is not observed (390). The repression of ribulose diphosphate carboxylase and hydrogenase synthesis may also prevent the growth of *E. shaposhnikovii* in the dark aerobic conditions on mineral media (287, 288).

Oxygen also seems to inhibit carboxylases whose action is connected with ferredoxin. But at the same time the capacity of purple bacteria for some carboxylation reactions in the presence of oxygen may be conserved (128, 144). In *R. sphaeroides,* activity of pyruvate carboxylase was even higher in darker aerobic conditions than in the light under anerobiosis (214). According to Clayton's data (11), in *R. rubrum* the utilization of propionate both in the light under anaerobic conditions and in the dark in the presence of O_2 proceeds through carboxylation of propionyl-CoA. As was mentioned earlier, however, such compounds as citrate (231, 232) and benzoate (239–242) support growth of purple bacteria in the light under anaerobic conditions. It is known that in the dark under aerobic conditions *R. palustris* oxidizes *p*-hydroxybenzoate by the same pathway as in some aerobic species of *Pseudomonas,* according to the following sequence (238):

$$p\text{-Hydroxybenzoate} \xrightarrow{\text{NADPH, FAD, O}_2} \text{protocatechuate}$$

$$\xrightarrow{\text{O}_2} \alpha\text{-hydroxy-}\ \gamma\text{-carboxymuconic semialdehyde} \xrightarrow{\text{NADP}} \alpha\text{-hydroxy-}$$

$$\gamma\text{-carboxymuconic acid} \rightarrow \rightarrow \rightarrow \text{pyruvate}$$

The first and the second reactions are catalyzed by *p*-hydroxybenzoate hydroxylase and protocatechuate-4,5-oxygenase, respectively. In *R. palustris* cells grown in the light under anaerobic conditions, the activity of these enzymes is very low even in the presence of *p*-hydroxybenzoate. The synthesis of them is induced only in the presence of O_2. This fact explains a certain lag period in the growth of *R. palustris* culture on media containing *p*-hydroxybenzoate after transfer from anaerobic light to aerobic dark conditions (238).

As a rule, metabolism of different organic compounds in the dark under aerobic conditions is connected to terminal stages by way of the TCA cycle (11, 127, 128). The importance of the TCA cycle for utilization of acetate and other organic acids by purple bacteria under dark aerobic conditions is demonstrated by inhibition of their assimilation by fluoroacetate (11, 127, 128, 208). In certain purple bacteria the activity of some enzymes of the TCA cycle is higher under dark aerobic conditions (137, 204). In such bacteria as *R. capsulata* the level of citrate synthase is especially increased (200). Along with the TCA cycle the glyoxylate shunt may be involved in dark metabolism of organic compounds in some purple bacteria (206). In *R. gelatinosa* this shunt is only active in the cells growing under aerobic dark conditions, because only in such conditions is malate synthase formation observed (206).

The possibility of growth of some purple nonsulfur bacteria (in particular *R. viridis*) by glyoxylate bypass on acetate medium only in the light confirms the difference of its metabolism in the light and in the dark (206). But there are some purple nonsulfur bacteria species (for instance, *R.*

sphaeroides) lacking the glyoxylate shunt (Table 3) but capable of growing in dark aerobic conditions on acetate medium (206). This means the availability of some other mechanism of synthesis of C_4 acids from acetate in such microorganisms.

In contrast to purple nonsulfur bacteria, in purple sulfur bacteria grown in the dark (*E. shaposhnikovii* and *T. roseopersicina*) the TCA cycle remains incomplete as these bacteria do not contain α-ketoglutarate dehydrogenase (204, 317). However, the availability of the glyoxylate bypass (147, 204) seems to allow *E. shaposhnikovii* to utilize acetate and other organic compounds in the absence of light not only as source of carbon but as source of energy (electron donors) as well.

The ability of *T. roseopersicina* to utilize acetate and other organic compounds in the dark remains as limited as it is in the light. This seems to be connected with lack both of a complete TCA cycle and of glyoxylate bypass (204, 323). Moreover, respiratory systems of purple sulfur bacteria seem not to be perfect enough to provide cells with energy for adequate growth (283). In this respect they are similar to cyanobacteria that have no complete TCA cycle and often reveal only slight respiratory activity (3, 6).

Therefore, although some purple sulfur bacteria are able to grow in the dark under aerobic conditions, their possibilities in this respect are more limited than in nonsulfur bacteria. The absence of growth of many purple sulfur bacteria in aerobic dark conditions may be also caused by inhibitory effects of oxygen, although these microorganisms contain catalase (289) and superoxide dismutase (285, 286). The inhibitory effect of oxygen on purple sulfur bacteria is confirmed by the absence of growth of most of these microorganisms in aerobic conditions in the light.

GREEN BACTERIA

General Characteristics

Green bacteria, as purple bacteria, are included in the order Rhodospirillales (2). They are united into one suborder, chlorobiinae, which is subdivided into two families, chlorobiaceae and chloroflexaceae, comprising seven genera (Table 4).

Table 4.   Some genera of green bacteria (refs. 1, 2)

Genus	Cell shape, arrangement, and motility
Chlorobium	ovoid, rod or vibrioid, single and in short chains
Prosthecochloris	irregular sphere with prosthecae
Pelodictyon	ovoid or rod, may form nets
Clathrochloris	sphere and ovoid, united in characteristic aggregates
Chloroflexus	filaments composed of rods, gliding motility

They grow under anaerobic photoautotrophic conditions and some of them need vitamin B_{12}. All species oxidize hydrogen sulfide and sulfur. Some of them can use thiosulfate and molecular hydrogen as electron donors. They are known by the name green sulfur bacteria.

Some of these green bacteria live in consortia with other microorganisms (2, 11). One such complex (318, 319) was described as a particular species—*Chloropseudomonas ethylica* (11).

Recently, filamentous green bacteria have been discovered (320–323). Like some cyanobacteria, these microorganisms have a capacity for gliding motility. They are included in the new family Chloroflexaceae (324), with type species *Chloroflexus aurantiacus* isolated in pure culture. On the basis of investigations of natural material it has been proposed that two more genera and three more species of green bacteria be included into this family (325, 326). One of these microorganisms, *Oscillochloris chises*, was described earlier as a cyanobacterium (i.e., blue-green alga). Therefore one may suppose the existence of other green bacteria wrongly considered as cyanobacteria.

In contrast to representatives of Chlorobiaceae, different strains of *Chloroflexus* are facultative anaerobes and are able to grow in the light and in the dark on media containing different organic substrates and B group vitamins (320–322, 32;7). However, in one strain the capacity for growth in photoautotrophic conditions was found (328).

The members of the Chloroflexaceae (320–322, 324) are similar to other green bacteria in the structure of their photosynthetic apparatus, their chlorophylls, and their DNA base ratios (% guanine + cytosine). Moreover, green bacteria belonging to the genera *Chlorobium* and *Chloroflexus* have similar lipid compositions (329). In this respect they are more closely related to cyanobacteria than to purple bacteria.

As do most purple bacteria, all green bacteria studied contain bacteriochlorophyll *a*, but in a small amount. The main bulk of their pigments is represented by bacteriochlorophylls *c*, *d*, and *e* (22). Depending on the content of these bacteriochlorophylls, green bacteria have absorption maxima in the long wave region from 715 to 755 nm (1, 2, 22).

With respect to carotenoid composition, green bacteria differ from purple bacteria. The main carotenoids of green bacteria belonging to Chlorobiaceae are mono or diaryl pigments, such as chlorobacten, β-isorenieratene,, and isorenieraten (1, 2, 12). *Chloroflexus* have another composition of carotenoids. Here the major carotenoids are β- and γ-carotenes. Among minor carotenoids of this microorganism, echinenone (similar to β-carotene) was found; this pigment is also found in cyanobacteria (330).

Data concerning compounds capable of playing a role as electron carriers in photosynthesis of green bacteria exist for *Chlorobium limicola* forma *thiosulphatophilum* and *Prosthecochloris aestuarii*. Both these microorganisms contain *c*-type cytochromes. In *Chlorobium* cells, three such cytochromes were found (c_{551}, c_{553}, c_{555}) and in *Prosthecochloris* two were found ($c_{551.5}$ and c_{555}) (26–31, 331, 332). c_{555} Cytochromes of these bacteria are the most studied (333). There are also data about *C. limicola* f. *thiosulfatophilum* containing cytochrome *b* (34,334).

Contrary to purple bacteria, investigated strains of green bacteria do not contain ubiquinones, but menaquinones are found, among them a particular compound named chlorobiumquinone (335).

The greatest part of nicotinamide nucleotide coenzymes in green bacteria are in the form of NAD. Preparations of NAD(P)-reductase bound with flavins have been obtained from *C. limicola* f. *thiosulfatophilum* (336) and *P. aestuarii* (337).

The green bacteria also contain several high potential ($E_m = +430$ mV) and low potential ($E_m = -600$ mV) iron-sulfur proteins, including rubredoxin and ferredoxins. The latter, isolated from *C. limicola* f. *thiosulfatophilum*, have 2–4Fe:4S, 1–4Fe:4S, and 2Fe:2S centers (45, 46, 338–342).

A characteristic feature of green bacteria is the presence in all species of so-called chlorobium vesicles (1–3, 17, 18, 343). They have an oval form, are surrounded by a thin (30–40 Å) membrane, probably of a protein nature, and are located close to the cytoplasmic membrane. According to the latest evidence, chlorobium vesicles contain only such light-harvesting pigments as bacteriochlorophyll *c, d,* or *e*. Bacteriochlorophyll *a* and the reaction centers are located in the cytoplasmic membrane, with which chlorobium vesicles are in some way connected (344–347).

Thus the photosynthetic apparatus in green bacteria has a more complex structure than that of purple bacteria. It more resembles the photosynthetic apparatus of cyanobacteria, in which phycobilisomes may have the same function as chlorobium vesicles (6).

The Initial Stages of Photosynthesis

Photochemical Process and Electron Transport　On the basis of investigations of intact cells and different pigment-containing complexes isolated from green bacteria (30, 31, 49, 346–350), the reaction centers of these microorganisms are supposed to contain the pigment P_{840}, which seems to represent a special form of bacteriochlorophyll *a*. Light energy absorbed by other pigments migrates to P_{840} through the main part of bacteriochlorophyll *a*.

The photochemical reaction consists of oxidation of reaction center pigment with its ensuing reduction as a result of interaction with *c*-type cytochromes (+165 mV). The role of electron acceptor in photooxidation of P_{840} (+250 mV) may be played by bacteriopheophytin and later by iron-sulfur protein, with a midpoint redox potential of about −550 mV (30, 31, 49, 331, 332, 346–350).

As with purple bacteria, the photochemical process taking place in reaction centers of green bacteria causes a further cyclic electron transport.

In such a system of electron flow quinones and c-type cytochromes take part (30, 332), but in addition a noncyclic ("open") photosynthetic electron transport system seems to operate (Figure 8). The possibility of such a system is indicated by several observations. Pigment-containing preparations from *Chlorobium* have the capacity for ferredoxin photoreduction (31, 76, 351). In its turn ferredoxin stimulates the photoreduction of NAD by such preparations (76, 352). The possibility of the occurrence of these reactions corresponds to the measured value of the redox potential of the primary electron acceptor (-550 mV). It has also been shown that uncouplers do not prevent photoreduction of NAD(P) in intact cells of green bacteria, in contrast to purple ones (82). The noncyclic electron transport system in green bacteria seems to contain, besides iron-sulfur proteins and NAD, flavins and c-type cytochromes, differing from cytochromes participating in cyclic electron flow (30, 49, 332, 337).

The site of cytochrome b action in electron transport systems of green bacteria is still uncertain (34, 334). Thus, although neither green nor purple bacteria utilize water as electron donor in photosynthesis, they seem to have differences in organization of photosynthetic apparatus and its functions. The photosystem of green bacteria is closer to the first photosystem of cyanobacteria and of green plants than is the photosystem of purple bacteria (334).

Energy Conservation   Illumination of green bacteria results in inorganic phosphate consumption by the cells (11) and in increase of their ATP level (353), but most attempts to attain photophosphorylation in cell-free preparations of these microorganisms have failed. This failure seems to be caused by peculiarities of the structure of photosynthetic apparatus and damage to it during the cells' disruption. There are only two reports about the utilization of inorganic phosphate by illuminated pigment-containing preparations from green bacteria (354) and about synthesis of ATP by them (355).

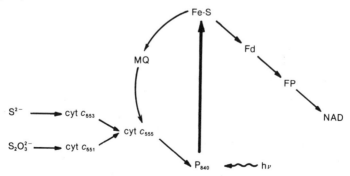

Figure 8. Photosynthetic electron transport system in green sulfur bacteria (332, 360, 362): P_{840}, pigment of reaction center; Fe-S, iron-sulfur protein; Fd, ferredoxin; FP, flavoprotein; NAD, nicotinamide adenine dinucleotide; cyt c, cytochrome c (different); MQ, menaquinone.

The addition of uncoupler (carbonyl-m-chlorphenylhydrazone) inhibits the formation of ATP by such preparations (355). A similar effect of uncoupler was observed on generation of ATP in illuminated intact cells (353). This fact confirms that ATP synthesis is connected with the action of the photosynthetic electron transport system. It was found that C. limicola f. thiosulfatophilum reveals an ATPase activity associated with a membrane fraction (353).

Like purple bacteria and other microorganisms, green bacteria produce in a considerable amount long chain polyphosphates and seem to synthesize pyrophosphate also (11, 86, 89, 356, 357). The existence was also shown in green bacteria of polyphosphate phosphotransferase catalyzing the reaction (89, 357):

$$(\text{Polyphosphate})_n + \text{ATP} \rightleftharpoons (\text{polyphosphate})_{n+1} + \text{ADP}$$

Electron Donors

Green bacteria belonging to the Chlorobiaceae utilize as electron donor only inorganic compounds (1, 2, 10, 12, 13, 15). Sulfide oxidation by different species of green sulfur bacteria is connected with production of sulfur, which accumulates in the media:

$$2\text{H}_2\text{S} + \text{CO}_2 \xrightarrow{\text{light}} 2\text{S} + \text{H}_2\text{O} + (\text{CH}_2\text{O})$$

Then the sulfur is usually oxidized to sulfates, but in some cases this oxidation is incomplete.

Growing cultures of some species of green sulfur bacteria, as mentioned above, may oxidize thiosulfate with production of sulfate (2, 11, 15, 358):

$$\text{Na}_2\text{S}_2\text{O}_3 + 2\text{CO}_2 + 3\text{H}_2\text{O} \xrightarrow{\text{light}} 2\text{NaHSO}_4 + 2(\text{CH}_2\text{O})$$

The oxidation of sulfide in C. limicola f. thiosulfatophilum seems to proceed with participation of cytochrome c_{553} bound with flavin (FMN). Such a cytochrome is believed to have a function of sulfide–cytochrome c reductase (15, 359, 360).

Oxidation of thiosulfate in the same green bacterium is catalyzed by thiosulfate–cytochrome c reductase, which has been highly purified (15, 360–362). The oxidation of thiosulfate catalyzed by this enzyme results in reduction of cytochrome c_{551} and then of cytochrome c_{555}. Thus, depending on electron donor, cytochrome c_{555} of C. limicola f. thiosulfatophilum may interact with either cytochrome c_{553} or cytochrome c_{551} (Figure 8). There is evidence that thiosulfate oxidation results first of all in the production of tetrathionate, which may be utilized by Chlorobium cells as electron donor in CO_2 photoassimilation (358).

The capacity of green bacteria for thiosulfate cleavage with participation of thiosulfate reductase has not been demonstrated (15), but it is certain that these microorganisms are similar to purple bacteria in containing APS-reductase (15, 363). The properties of this enzyme of *C. limicola* f. *thiosulfatophilum* differ somewhat from those of the purple bacterium *T. roseopersicina*. It was also shown that suspensions of cells of two species of *Chlorobium* in anaerobic conditions in the light in the absence of CO_2 metabolized elemental sulfur with production of sulfide and thiosulfate (364, 365):

$$4S^0 + 3H_2O \longrightarrow 2H_2S + H_2S_2O_3$$

This means that sulfur may serve both as electron donor and as electron acceptor.

Like many purple bacteria, green sulfur bacteria also need reduced sulfur compounds for biosynthetic processes, since they possess no capacity for assimilatory sulfate reduction (2, 15).

In contrast to the Chlorobiaceae, green bacteria belonging to the genus *Chloroflexus* seem to be able to utilize organic compounds in photosynthesis as electron donor (321, 322, 327). However, some of them also possess the capacity to oxidize sulfide, with production mainly of sulfur (328).

Many green bacteria can use molecular hydrogen as electron donor. The growth of these microorganisms in the presence of H_2 and CO_2 is possible if cysteine or some other reduced compound as source of sulfur are available (366).

Photometabolism of Carbon

Assimilation of Carbon Dioxide   Pathways of carbon dioxide assimilation in green sulfur bacteria, especially in *C. limicola* f. *thiosulfatophilum*, have been investigated for more than 10 years (11, 76, 127, 128, 154, 155, 367–381). There is evidence indicating that green sulfur bacteria possess all the enzymes of the ribulose diphosphate cycle, although their activity, in particular that of ribulose diphosphate carboxylase and ribose 5-phosphate isomerase, is less than in purple bacteria (127, 367, 377). However, some studies, including the most recent ones, have failed to identify the availability of ribulose diphosphate carboxylase and phosphoribulokinase in *C. limicola* f. *thiosulfatophilum* and some other green sulfur bacteria (127, 315, 370, 371, 375).

After short-term exposure of cells of *C. limicola* f. *thiosulfatophilum* to $^{14}CO_2$ the labeled products were pyruvate, malate, oxaloacetate, α-ketoglutarate, aspartate, and glutamate (369, 370, 375). It was also found that photoassimilation of CO_2 by cells of *C. limicola* f. *thiosulfatophilum* results in production of branched-chain keto acids, mainly β-methyl-α-ketovalerate (372).

Similar results were obtained for other strains of green sulfur bacteria in the genera *Chlorobium* and *Prosthecochloris* (378, 379). Only in one experiment with *Chlorobium* when the CO_2 content was small (0.8 mM) was the incorporation of radioactivity found in 3-phosphoglycerate after 5 sec (370). At CO_2 contents of 8 and 40 mM this compound did not appear among early labeled products (375).

Thus the composition and kinetics of labeled products of $^{14}CO_2$ fixation of green bacteria are not typical for ribulose diphosphate cycle function. On the contrary, the results of most experiments indicate that carboxylation reactions other than ribulose diphosphate carboxylase are taking place. Such a possibility was confirmed by the capacity of *C. limicola* f. *thiosulfatophilum* for acetyl-CoA and succinyl-CoA carboxylation (153, 154), with participation of the corresponding ferredoxin-linked enzymes (Table 2). The carboxylation of phosphoenolpyruvate by this bacterium is possible too.

Different strains of *C. limicola* f. *thiosulfatophilum* and of *P. aestuarii* have been shown to contain several enzymes of the TCA cycle (204, 369, 380).

All this evidence led to the conclusion that autotrophic CO_2 photoassimilation in *C. limicola* f. *thiosulfatophilum* proceeds via a particular system named the reductive carboxylic acid cycle (76, 153, 154, 369). It is also thought that such a cycle may operate in two variants. The first of them represents a complete reductive carboxylic acid cycle, including carboxylation of acetyl-CoA, phosphoenolpyruvate, succinyl-CoA, and α-ketoglutarate. The second short variant of this cycle includes only acetyl-CoA and succinyl-CoA carboxylation. This variant is to be considered as a reverse TCA cycle (369). According to the opinion of some authors, the identical cyclic system of CO_2 assimilation may function along with the ribulose diphosphate pathway in some purple bacteria, in particular in *R. rubrum* (76, 153, 369) and *R. palustris* (170).

However, all available evidence has not yet given definite proof of the existence of such a cycle, although the function of some part of it is quite certain. One of the controversial points concerns the existence in green bacteria and in *R. rubrum* of citrate lyase, which is a key enzyme of such a cycle. The early communications indicated a slight activity of this enzyme in named microorganisms (369), but repeated investigations have failed to confirm its presence (381). The ability of green and purple bacteria for α-ketoglutarate carboxylation is not determined either (157).

Thus there is no doubt that CO_2 metabolism in those green sulfur bacteria studied differs significantly from that in purple bacteria. However, it is as yet difficult to determine the functioning mechanism as a whole.

If it is confirmed that at least some of the green bacteria do not synthesize ribulose diphosphate carboxylase, then it will be impossible to consider this enzyme as characteristic for all organisms assimilating CO_2 in autotrophic conditions (157, 382, 383). In this connection the investigation of different species of green bacteria, especially of filamentous forms belonging to the Chloroflexaceae, would be very important.

Metabolism of Organic Compounds With respect to metabolism of organic compounds, green sulfur bacteria are similar to some purple sulfur bacteria (2). They are unable to use organic compounds as electron donors, but may assimilate them as additional carbon sources. The utilization of organic compounds by growing cultures and often by cell suspensions of green bacteria usually depends on the availability of inorganic electron donors and occurs simultaneously with active CO_2 assimilation (1, 2, 203, 368, 384, 385).

The limited capacity of green sulfur bacteria for utilization of organic compounds in their metabolism is, according to the available evidence (204, 380) for *Chlorobium* and *Prosthecochloris*, due to an incomplete TCA cycle (the lack of α-ketoglutarate dehydrogenase) and to the absence of the glyoxylate bypass (Table 3). The permeability factor seems not to be of great importance because the possibility of penetration of different organic compounds into the cells of *C. limicola* f. *thiosulfatophilum* has been proved (385).

Acetate is considered to be utilized by different species of green sulfur bacteria (1, 2). The possibility of utilization of pyruvate, propionate, glucose, formate, and some amino acids in small amounts has also been shown (11, 128, 373, 384, 385). The addition of acetate to the medium increased production by *C. limicola* f. *thiosulfatophilum* of α-ketoglutarate and other compounds synthesized from CO_2 (370, 372, 376). Distribution of ^{14}C in alanine isolated from the protein fraction of *C. limicola* f. *thiosulfatophilum* after growth in the presence of labeled acetate or bicarbonate allowed the conclusion as to its synthesis from acetyl-CoA and CO_2. Such a conclusion corresponds with the presence of pyruvate synthase (Table 2). The same analysis has shown the synthesis of aspartate to be connected with carboxylation of C_3 compounds (368). Thus the pathway of generation of C_3 and C_4 organic acids from acetate in *C. limicola* f. *thiosulfatophilum* may be represented as follows:

$$\text{Acetate} \rightarrow \text{acetyl-CoA} \xrightarrow{+CO_2} \text{pyruvate} \xrightarrow{P_i} \text{phosphoenolpyruvate} \xrightarrow{+CO_2} \text{oxaloacetate}$$

There also exists the possibility of α-ketoglutarate formation from acetate via carboxylation of succinyl-CoA (154, 155, 369, 372, 376, 386), but distribution of ^{14}C from acetate in glutamate does not give a clear idea about the pathway of biosynthesis of the latter (372).

Photoassimilation of acetate by *C. limicola* f. *thiosulfatophilum* significantly increased synthesis of polyglucose in the presence of CO_2. The use of [^{14}C]acetate has shown that its carbon, like that from CO_2, is incorporated into this polysaccharide (373, 376). Therefore it is clear that pathways of acetate and CO_2 metabolism in green sulfur bacteria are closely related.

In the presence of pyruvate as well as acetate, the character of the products of CO_2 assimilation of *C. limicola* f. *thiosulfatophilum* does not change, but formation of branched-chain keto acids increases (379). The utilization of pyruvate by cell suspensions in the absence of CO_2 also results in considerable formation of storage glucan (372). Evidently this process must be connected with production of phosphoenolpyruvate from pyruvate, with participation of pyruvate-orthophosphate dikinase (213). In the presence of CO_2, carboxylation of phosphoenolpyruvate with production of oxaloacetate is also possible (Table 2). The cells of *C. limicola* f. *thiosulfatophilum* can also decompose pyruvate with production of CO_2 and other acid products (358). Probably such a process is catalyzed by pyruvate:ferredoxin oxidoreductase, which was found in this microorganism (387).

The utilization of propionate by cell suspensions of *C. limicola* f. *thiosulfatophilum* in the presence of $^{14}CO_2$ is connected with production of succinate containing ^{14}C in the carboxyl carbon (358). This observation indicates the presence of propionyl-CoA carboxylase, which has also been found in some purple bacteria (153).

Like acetate, succinate significantly increases α-ketoglutarate formation in *Chlorobium*. Hence one may suppose it to be synthesized from succinyl-CoA with participation of α-ketoglutarate synthase (Table 2), but after short-term (5 sec) photoassimilation of [^{14}C]succinate by cells of *C. limicola* f. *thiosulfatophilum* in the presence of unlabeled CO_2, the greatest percentage of radiocarbon was contained in malate and fumarate. Only later was increase of its content in glutamate and glutamine observed (370). Such kinetics of appearance of labeled products corresponds not to production of α-ketaglutarate from succinyl-CoA but to reactions of the TCA cycle.

It has been shown for *C. limicola* f. *thiosulfatophilum* that ^{14}C from formate is incorporated mainly into purines in spite of inhibition of cell growth (385).

In contrast to members of the Chlorobiaceae, green bacteria of the *Chloroflexus* genus are able to utilize different organic compounds for their growth both under anaerobic and aerobic conditions (320–322, 327). These organic compounds include acetate, pyruvate, lactate, butyrate, C_4 dicarboxylic acids, ethanol, glycerol, mannitol, some sugars, and amino acids. Metabolism of all these compounds has not yet been studied.

Like purple bacteria, different species of green bacteria are capable of using ammonium as source of nitrogen (11, 128). The capacity of some green bacteria for assimilation of N_2 has also been determined (2, 112, 128, 247, 249). Such enzymes as glutamate dehydrogenase, glutamine synthetase, and glutamate synthase may be involved in the formation of amino acids in *Chlorobium* (128, 250). Transamination with participation of other NH_2 donors may also take place (128, 369).

Dark Metabolism

Of all green bacteria isolated in pure cultures, the capacity for growth in the dark has only been shown in *Chloroflexus* (320–322, 327). The growth of different strains of this bacterium in the dark was observed under aerobic conditions on media containing the same organic compounds that are utilized in the light. Both in the light and in the dark under aerobic conditions, the synthesis of bacteriochlorophyll in this bacterium is repressed (322).

In contrast to *Chloroflexus,* green sulfur bacteria related to Chlorobiaceae not only fail to grow in the presence of oxygen but rapidly lose their vital capacity (11, 278) in spite of containing catalase (284) and superoxide dismutase (285, 286). The uptake of O_2 by cells of different strains of *Chlorobium* is very slight. In some cases it cannot be detected at all (358).

In contrast to purple bacteria, the oxygen uptake by *Chlorobium* spp. cells is not inhibited but stimulated by light (388). The mechanism of such an action of light remains to be determined.

All available evidence shows green sulfur bacteria to be very sensitive to oxygen and unable to survive in the presence of O_2.

In the dark under anaerobic conditions the cells of *C. limicola* f. *thiosulfatophilum* are capable of fermenting (373, 376) endogenous compounds and added pyruvate (358), with production of CO_2, organic acids, and hydrogen sulfide. Endogenous storage polyglucose is utilized by green bacteria in dark anaerobic conditions (373, 376).

As mentioned above, such polysaccharide may be synthesized by green bacteria in great amounts during assimilation of CO_2, acetate, and pyruvate in the absence of a source of nitrogen (128, 337, 376). Fermentation of this product proceeds via the fructose diphosphate pathway (204) and is followed by production of acetate in considerable amount. Pyruvate, propionate, succinate, and caproate were also found (373, 376).

Washed suspensions of *C. limicola* f. *thiosulfatophilum* also utilize endogenous glucan in the light, but in such cases a lesser production of volatile acids occurs (376). Thus there exists a definite similarity of dark endogenous metabolism in green and purple bacteria, but purple bacteria are capable of synthesizing as storage products and utilizing glucan and poly-β-hydroxybutyrate. In green bactaria the capacity for poly-β-hydroxybutyrate synthesis has not been found.

CONCLUSION

Purple and green bacteria include microorganisms possessing different pathways of carbon metabolism and of energy production. In the light all purple bacteria are capable of assimilating carbon dioxide via the ribulose diphosphate cycle and through other carboxylation reactions. Moreover, they photoassimilate other carbon compounds, and several species also utilize organic substrates as electron donors.

Greater metabolic versatility is revealed by purple nonsulfur bacteria.

Such a capacity allows many of them to switch over from photosynthesis to energy generation through aerobic and anaerobic respiration or fermentation. Some purple sulfur bacteria are also capable of growing in the dark under heterotrophic and even chemoautotrophic conditions, but most of these microorganisms are obligate phototrophs possessing only limited possibilities for utilization of organic compounds.

The majority of known green bacteria are obligate photosynthetic forms, but there are among them strains growing only in the presence of organic compounds and capable of dark aerobic heterotrophic metabolism. A significant peculiarity of green sulfur bacteria is their autotrophic pathway of CO_2 utilization. This aspect of their metabolism needs further investigation.

ACKNOWLEDGMENTS

I wish to thank Professor J. R. Quayle for his great help in the preparation of this chapter.

REFERENCES

1. Pfennig, N., and Trüper, H. G. (1973). In A. I. Laskin and H. A. Lechevalier (eds.), Handbook of Microbiology, Vol. 1, Organismic Microbiology, p. 17. C.R.C. Press, Cleveland, Ohio.
2. Pfennig, N., and Trüper, H. G. (1974). In Bergey's Manual of Determinative Bacteriology, 8th ed., p. 24. The Williams & Wilkins Company, Baltimore.
3. Stanier, R. Y., Adelberg, E. A., and Ingraham, J. L. (1976). General Microbiology, 4th ed. Prentice-Hall, Inc., Englewood Cliffs, New Jersey.
4. Murray, R. G. E. (1974). In Bergey's Manual of Determinative Bacteriology, 8th ed., p. 21. The Williams & Wilkins Company, Baltimore.
5. Fogg, G. E., Stewart, W. D. P., Fay, P., and Walsby, A. E. (1973). The Blue-Green Algae. Academic Press, New York and London.
6. Carr, N. G., and Whitton, B. A. (1973). The Biology of Blue-Green Algae. Blackwell Scientific Publications, Oxford, England.
7. Lewin, R. A. (1976). Nature 261:697.
8. Schreckenback, Th. (1976). In H. G. Schlegel and J. Barnea (eds.), Microbial Energy Conversion, p. 245. Erich Goltze KG, Göttingen.
9. Oesterhelt, D., Gottschlich, R., Hartmann, R., Michel, H., and Wagner, G. (1977). Symp. Soc. Gen. Microbiol. 27:333.
10. Van Niel, C. B. (1931). Arch. Mikrobiol. 3:1.
11. Kondratieva, E. N. (1965). Photosynthetic Bacteria. Oldbourne Press, London.
12. Pfennig, N. (1967). Annu. Rev. Microbiol. 21:285.
13. Pfennig, N. (1975). Plant Soil 43:1.
14. Kondratieva, E. N., and Gorlenko, V. M. (1977). Usp. Mikrobiol. 13:8.
15. Trüper, H. G. (1975). Plant Soil 43:29.
16. Hansen, T. A., Sepers, A. B. J., and van Gemerden, H. (1975). Plant Soil 43:17.
17. Oelze, J., and Drews, G. (1972). Biochim. Biophys. Acta 265:209.
18. Echlin, P. (1970). Symp. Soc. Gen. Microbiol. 20:221.

19. Hurlbert, R. E., Golecki, J. R., and Drews, G. (1974). Arch. Microbiol. 101:169.
20. Golecki, J. R., and Oelze, J. (1975). J. Gen. Microbiol. 88:253.
21. Lascelles, J. (1968). Adv. Microb. Physiol. 2:1.
22. Gloe, A., Pfennig, N., Brockmann H., Jr., and Trowitzsch, W. (1975). Arch. Microbiol. 102:103.
23. Jensen, L. S., and Andrewes, A. G. (1972). Annu. Rev. Microb. 26:225.
24. Schmidt, K. (1976). In G. A. Codd and W. D. P. Stewart (eds.), Proceedings of the Second International Symposium on Photosynthetic Prokaryotes, p. 58. Federation of European Microbiological Societies, Dundee.
25. Hind, G., and Olson, J. M. (1968). Annu. Rev. Plant Physiol. 19:249.
26. Vernon, L. P. (1968). Bacteriol. Rev. 32:243.
27. Bartsch, R. G. (1968). Annu. Rev. Microbiol. 22:181.
28. Horio, T., and Kamen, M. D. (1970). Annu. Rev. Microbiol. 24:399.
29. Kamen, M. D., and Horio, T. (1970). Annu. Rev. Biochem. 39:673.
30. Frenkel, A. W. (1970). Biol. Rev. 45:569.
31. Parson, W. W. (1974). Annu. Rev. Microbiol. 28:41.
32. Ambler, R. P., Meyer, T. E., Bartsch, R. G., and Kamen, M. D. (1976). In G. A. Codd and W. D. P. Stewart (eds.), Proceedings of the Second International Symposium on Photosynthetic Prokaryotes, p. 248. Federation of European Microbiological Societies, Dundee.
33. Saunders, V. A., and Jones, O. T. G. (1975). Biochim. Biophys. Acta 396:220.
34. Knaff, D. B., and Buchanan, B. B. (1975). Biochim. Biophys. Acta 376:549.
35. Kikuchi, G., Saito, Y., and Motokawa, Y. (1965). Biochim. Biophys. Acta 94:1.
36. Saunders, V. A., and Jones, O. T. G. (1973). Biochim. Biophys. Acta 333:439.
37. Ivanovskii, R. N., and Rodova, N. A. (1975). Mikrobiologiia, 44:16.
38. Oelze, J., Pahlke, W., and Bohm, S. (1975). Arch. Microbiol. 102:65.
39. Higuti, T., Erabi, T., Kakuno, T., and Horio, T. (1975). J. Biochem. (Japan), 78:51.
40. Okamura, M. Y., Isaacson, R. A., and Feher, G. (1975). Proc. Natl. Acad. Sci. USA 72:3491.
41. Yoch, D. C., and Valentine, R. C. (1972). Annu. Rev. Microbiol. 26:139.
42. London, J., and Knight, M. (1966). J. Gen. Microbiol. 44:241.
43. Buchanan, B. B., and Arnon, D. I. (1970). Adv. Enzymol. 33:119.
44. Orme-Johnson, W. H. (1973). Annu. Rev. Biochem. 42:159.
45. Hall, D. O., Rao, K. K., and Cammack, R. (1975). Sci. Prog. (Oxford) 62:285.
46. Rogers, L. J., Hutson, K. G., Hastel. B. G., and Boulter, D. (1976). In G. A. Codd and W. D. P. Stewart (eds.), Proceedings of the Second Symposium on Photosynthetic Prokaryotes, p. 245. Federation of European Microbiological Societies, Dundee.
47. Evans, M. C. W., Lord, A. V., and Reeves, S. G. (1974). Biochem. J. 138:177.
48. Mizrahi, I. A., Wood, F. E., and Cusanovich, M. A. (1976). Biochemistry 15:343.
49. Sybesma, Chr. (1970). In P. Halldal (ed.), Photobiology of Microorganisms, p. 57. Wiley-Interscience, London, New York.
50. Clayton, R. K. (1973) Annu. Rev. Biophys. Bioeng. 2:131.
51. Parson, W. W., and Cogdell, R. J. (1975). Biochim. Biophys. Acta 416:105.
52. Cogdell, R. J. (1976). In G. A. Codd and W. D. P. Stewart (eds), Proceed-

ings of the Second International Symposium on Photosynthetic Prokaryotes, p. 66. Federation of European Microbiological Societies, Dundee.

53. Pucheu, N. L., Kerber, N. L., and Garcia, A. F. (1976). Arch. Microbiol. 109:301.
54. Feher, G., and Okamura M. Y. (1976). Brookhaven Symp. Biol. 28:183.
55. Barsky, E. L., Borisow, A. Y., and Samuilov, V. D. (1976). Usp. Sovrem. Biol. 82:222.
56. Parson, W. W., and Monger, T. G. (1976). Brookhaven Symp. Biol. 28:195.
57. Dutton, P. L., Prince, R. C., Tiede, D. M., and Petty, K. M. (1976). Brookhaven Symp. Biol. 28:213.
58. Fajer, J., Davis, M. S., Brune, D. C., Spaulding, L. D., Borg, D. C., and Forman, A. (1976). Brookhaven Symp. Biol. 28:74.
59. Prince, R. C., Leigh, J. S., and Dutton, P. L. (1976). Biochim. Biophys. Acta 440:622.
60. Vermeglio, A., and Clayton, R. K. (1976). Biochim. Biophys. Acta 449:500.
61. Kononenko, A. A., Knox, P. P., Adamova, N. P., Paschenko, V. Z., Timofeev, K. N., Rubin, A. B., and Morita, S. (1976). Stud. Biophys. 55:183.
62. Shuvalov, V. A., Krakhmaleva, I. N., and Klimov, V. V. (1976). Biochim. Biophys. Acta 449:597.
63. Wraight, C. A. (1977). Biochim. Biophys. Acta 459:525.
64. Grondelle, R., Romijn, F. C., and Holmes, N. G. FEBS Lett. 72:187.
65. Thiede, D. M., Prince, R. C., and Dutton P. L. (1976). Biochim. Biophys. Acta 449:447.
66. Samuilov, V. D. (1969). Usp. Sovrem. Biol. 68:232.
67. Evans, M. C. W., and Whatley, F. R. (1970). Symp. Soc. Gen. Microb. 20: 203.
68. Gest, H. (1972). Adv. Microb. Physiol. 7:243.
69. Evans, E. H., Crofts, A. R. (1974). Biochim. Biophys. Acta 357:89.
70. Evans, E. H., and Gooding, D. A. (1976). Arch. Microbiol. 111:171.
71. Prince, R. C., Baccarini-Melandri, A., Hauska, G. A., Melandri, B. A., and Crofts A. R. (1975). Biochem. Biophys. Acta 387:212.
72. Prince, R. C., and Dutton, P. L. (1976). International Conference of Primary Electron Transport and Energy Transduction in Photosynthetic Bacteria, TB4. Vrije Universitet, Brussels.
73. Cromet-Elhanan, Z., and Gest, H. (1976). In G. A. Codd and W. D. P. Stewart (eds.), Proceedings of the Second International Symposium on Photosynthetic Prokaryotes, p. 77. Federation of European Microbiological Societies, Dundee.
74. Jones, O. T. G. (1977). Symp. Gen. Microbiol. 27:151.
75. Avron, M. (1975). Bioenergetics of Photosynthesis, p. 373. Academic Press, New York, London.
76. Arnon, D. I., and Yoch, D. C. (1974). In A. Quespel (ed.), Biology of Nitrogen Fixation, p. 168. North-Holland Publishing Co., Amsterdam.
77. Van Grondelle, R., Duysens, N. M., and van der Wal, H. N. (1976). International Conference of Primary Electron Transport and Energy Transduction in Photosynthetic Bacteria, MB8 Brussels.
78. Keister, D. L., and Yike N. J. (1967). Arch. Biochem. Biophys. 121:415.
79. Keister, D. L., and Minton, N. J. (1969). Biochemistry 8:167.
80. Jones, O. T. G., and Whale, F. R. (1970). Arch. Microbiol. 72:48.
81. Knobloch, K., Eley, J. H., and Aleem, M. I. H. (1971). Arch. Microbiol. 80:97.
82. Ivanovsky, R. N. (1975). Mikrobiologiia 44:965.
83. Baltscheffsky, M. (1967). Nature 216:241.
84. Baltscheffsky, H., Baltscheffsky, M., and von Stedingk, L.-V. (1969). In H.

Mezner (ed.), Progress Photosynthesis Research, Vol. 3, p. 1313, H. Laupp Jr., Tübingen.
85. Guillory, R. J., and Fisher, R. R. (1972). Biochem. J. 129:471.
86. Kulaev, I. S. (1975). Biochemistry of High Molecular Polyphosphates. Moscow State University.
87. Keister, D. L., and Minton, N. J. (1971). Arch. Biochem. Biophys. 147:330.
88. Weber, H. (1965). Z. Allg. Mikrobiol. 5:315.
89. Dawes, E. A., and Senior, P. J. (1973). Adv. Microb. Physiol. 10:139.
90. Crofts, A. R. (1974). In O. S. Estrada and C. Gitler (eds.), Perspectives in Membrane Biology, p. 373. Academic Press, New York, London.
91. Skulachev, V. P. (1977). FEBS Lett. 74:1.
92. Rinehart, C. A., and Hubbard, J. S. (1976). J. Bacteriol. 127:1255.
93. Baltscheffsky, M. (1969). Arch. Biochem. Biophys. 130:646.
94. Baccarini-Melandri, A., Gest, H., and San Pietro, A. (1970). J. Biol. Chem. 245:1224.
95. Gromet-Elhanan, Z. (1974). J. Biol. Chem. 249:2522.
96. Johansson, B. C., and Baltscheffsky, M. (1975). FEBS Lett. 53:221.
97. Kerber, N. L., Pucheu, N. L., and García, A. F. (1976). FEBS Lett. 72:63.
98. Trüper, H. G., Lorenz, C., and Fischer, U. (1976). In G. A. Codd and W. D. P. Stewart (eds.), Proceedings of the Second International Symposium on Photosynthetic Prokaryotes, p. 41. Federation of European Microbiological Societies, Dundee.
99. van Niel, C. B. (1944). Bacteriol. Rev. 8:1.
100. Rolls, J. P., and Lindstrom, E. S. (1967). J. Bacteriol. 94:860.
101. Smith, A. J., and Lascelles, J. (1966. J. Gen. Microbiol. 42:357.
102. Trüper, H. G., and Pfennig, N. (1966). Antonie van Leeuwenhoek 32:261.
103. Trudinger, P. A. (1969). Adv. Microb. Physiol. 3:111.
104. Siegel, L. M. (1975). In D. M. Greenberg (ed.), Metabolic Pathways, Vol. VII, Metabolism of Sulfur Compounds, p. 217. Academic Press, New York.
105. Hashwa, F. (1975). Plant Soil 43:41.
106. Petushkova, Yu. P., and Ivanovsky, R. N. (1976). Mikrobiologiia 45:96.
107. Petushkova, Yu. P., and Ivanovsky, R. N. (1976). Mikrobiologiia 45:592.
108. Eley, J. H., Knobloch, K., and Aleem, M. I. H. (1971). Arch. Biochem. Biophys. 147:419.
109. Hensel, G., and Trüper, H. G. (1976). Arch. Microbiol. 109:101.
110. Schmidt, A. (1977). Arch. Microbiol. 112:263.
111. Pfennig, N. (1974). Arch. Microbiol. 100:197.
112. Gest, H., and Kamen, M. D. (1960). Handbuch der Planzenphysiologie, Vol. 5/2, p. 568. Springer Verlag, Berlin.
113. Schick, H. J. (1971). Arch. Microbiol. 75:102.
114. Schick, H. J. (1971). Arch. Microbiol. 75:110.
115. Hillmer, P., and Gest, H. (1977). J. Bacteriol. 129:724.
116. Hillmer, P., and Gest, H. (1977). J. Bacteriol. 129:732.
117. Klemme, J. H. (1969). Z. Naturforsch. [B] 24:67.
118. Mortenson, L. E., and Chen. J.-S. (1974). Microbiology of Iron Metabolism, p. 231. Academic Press, New York.
119. Gitlitz, P. H., and Krasna, A. I. (1975). Biochemistry 14:2561.
120. Kakuno, T., Kaplan, N. O., and Kamen, M. D. (1977). Proc. Natl. Acad. Sci. USA 74:861.
121. Gogotov, I. N., and Kondratieva, E. N. (1976). In H. G. Schlegel, G. Gottschalk, and N. Pfenning (eds.), Microbial Production and Utilization of Gases, p. 255. Erich Goltze KG, Göttingen.
122. Gogotov, I. N., Zorin, N. A., and Kondratieva, E. N. (1976). Biochemija 41:836.

123. Ormerod, J. G., and Gest, H. (1962). Bacteriol. Rev. 26:51.
124. Elsden, S. R. (1962). *In* I. C. Gunsalus and R. Y. Stanier (eds.), The Bacteria, Vol. 3, p. 1. Academic Press, New York.
125. Elsden, S. R. (1962). Fed. Proc. 21:1047.
126. Fuller, R. C. (1969). *In* H. Metzner (ed.), Progress in Photosynthesis Research, Vol. 3, p. 1579. H. Laupp Jr., Tübingen.
127. Wiessner, W. (1970). *In* P. Halldal (ed.), Photobiology of Microorganisms, p. 95. Wiley-Interscience, New York, London.
128. Kondratieva, E. N. (1974). Usp. Mikrobiol. 9:44.
129. Fuller, R. C., Smillie, R. M., Sisler, E. C., and Kornberg, H. L. (1961). J. Biol. Chem. 236:2140.
130. Trüper, H. G. (1964). Arch. Mikrobiol. 49:23.
131. Firsov, N. N., Cherniadjev, I. I., Ivanovskii, R. N., Kondratieva, E. N., Vdovina, N. V., and Donan, N. G. (1974). Mikrobiologiia 43:214.
132. Kondratieva, E. N., Zhukov, V. G., Ivanovsky, R. N., Petushkova, Yu. P., and Monosov, E. Z. (1976). Arch. Microbiol. 108:287.
133. Zhukov, V. G. (1976). Mikrobiologiia 45:915.
134. Glover, J., Kamen, M. D., and van Genderen, H. (1952). Arch. Biochem. Biophys. 35:384.
135. Anderson, L., and Fuller, R. C. (1967). Plant Physiol. 42:487.
136. Anderson, L., and Fuller, R. C. (1967). Plant Physiol. 42:491.
137. Anderson, L., and Fuller, R. C. (1967). Plant Physiol. 42:497.
138. Stoppani, A. O. M., Fuller, R. C., and Calvin, M. (1955). J. Bacteriol. 69: 491.
139. Stokes, J. E., and Hoare, D. S. (1969). J. Bacteriol. 100:890.
140. Cherniadiev, I. I., Kondratieva, E. N., and Doman, N. G. (1970). Izv. Akad. Nauk SSSR 6:895.
141. Sahm, H., Cox, R. B., and Quayle, J. R. (1976). J. Gen. Microbiol. 94:313.
142. Porter, J., and Merrett, M. J. (1972). Plant Physiol. 50:252.
143. Slater, J. H., and Morris, I. (1973). Arch. Microbiol. 88:213.
144. Slater, J. H., and Morris, I. (1973). Arch. Microbiol. 92:235.
145. Cherniadiev, I. I., Kondratieva, E. N., and Doman, N. G. (1968). Izv. Akad. Nauk SSSR [Biol] 6:670.
146. Firsov, N. N., and Ivanovsky, R. N. (1974). Mikrobiologiia 43:400.
147. Firsov, N. N., and Ivanovsky, R. N. (1975). Mikrobiologiia 44:197.
148. Lascelles, J. (1960). J. Gen. Microbiol. 23:499.
149. Hurlbert, R. E., and Lascelles, J. (1963). J. Gen. Microbiol. 33:445.
150. Cherniadiev, I. I., Kondratieva, E. N., and Doman, N. G. (1974). Mikrobiologiia 43:949.
151. McFadden, B. C., and Tabita, F. R. (1974). Biosystems 6:93.
152. Knight, M. (1962). Biochem. J. 84:170.
153. Olsen, I., and Merrick, J. M. (1968). J. Bacteriol. 95:1774.
154. Buchanan, B. B. (1972). *In* P. D. Boyer (ed.), The Enzymes, 3rd ed., Vol. 6, p. 193. Academic Press, New York, London.
155. Buchanan, B. B. (1973). Iron-Sulfur Proteins, Vol. 1, p. 129. Academic Press, New York, London.
156. Klemme, J. -H. (1976). Arch. Microbiol. 107:189.
157. McFadden, B. A. (1973). Bacteriol. Rev. 37:289.
158. McFadden, B. A. (1976). *In* H. G. Schlegel, G. Gottschalk and N. Pfennig (eds.), Microbial Production and Utilization of Gases, p. 267. Erich Goltze KG, Göttingen.
159. Priess, J., and Kosuge, T. (1970). Annu. Rev. Plant Physiol. 21:433.
160. Romanova, A. K. (1975). Usp. Mikrobiol. 10:27.
161. Tabita, F. R., and McFadden, B. A. (1976). J. Bacteriol. 126:1271.

162. Gibson, J., and Tabita, F. R. (1976). Abst. Annu. Meet. Amer. Soc. Microbiol., p. 159.
163. Takabe, T., and Akazawa, T. (1973). Biochem. Biophys. Res. Comm. 53: 1173.
164. Asami, S., and Akazawa, T. (1975). Plant Cell Physiol. 16:631.
165. Asami, S., and Akazawa, T. (1974). Plant Cell Physiol. 15:571.
166. Codd, G. A., and Turnbull, G. 1975. Arch. Microbiol. 104:155.
167. Codd, G. A., Sabbal. A. K. J., and Stewart, R. In G. A. Codd and W. D. P. Stewart (eds.), Proceedings of the Second International Symposium on Photosynthetic Prokaryotes, p. 193. Federation of European Microbiological Societies, Dundee.
168. Asami, S., and Akazawa, T. (1975). Plant Cell Physiol. 16:805.
169. Rolls, J. P., and Lindstrom, E. S. (1966). Fed. Proc. 25:739.
170. Yoch, D. C., and Lindstrom, E. S. (1967). Biochem. Biophys. Res. Comm. 28:65.
171. Yoch, D. C., and Lindstrom, E. S. (1967). Bacteriol. Proc., p. 121.
172. Qadri, S. M. H., and Hoare, D. S. (1968). J. Bacteriol. 95:2344.
173. Quayle, J. R., and Pfennig, N. (1975). Arch. Microbiol. 102:193.
174. Douthit, H. A., and Pfennig, N. (1976). Arch. Microbiol. 107:233.
175. Krassilnikova, E. N. (1975). Mikrobiologiia 44:795.
176. Quayle, J. R. (1972). Adv. Microb. Physiol. 7:119.
177. Quayle, J. R. (1976). In H. G. Schlegel, G. Gottschalk, and N. Pfennig (eds.),Microbial Production and Utilization of Gases, p. 353. Erich Goltze KG, Göttingen.
178. Anthony, C. (1975). Sci. Prog. Oxf. 62:167.
179. Cox, R. B., and Quayle, J. R. (1975). Biochem. J. 150:569.
180. Yoch, D. C., and Lindstrom, E. S. (1969). Arch. Microbiol. 67:182.
181. Bennet, R., Rigopoulos, N., and Fuller, R. C. (1974). Proc. Natl. Acad. Sci. USA 52:762.
182. Gogotov, I. N. (1973). In G. Drews, N. Pfennig, and R. Y. Stanier (eds.), Abstracts of the Symposium on Prokaryotic Photosynthetic Organisms, p. 118. Deutsche Forschunggemeinschaft and International Union of Biological Sciences, Freiburg.
183. Schön, G., and Voelskow, H. (1976). Arch. Microbiol. 107:87.
184. Wertlieb, D., and Vishniac, W. (1967). J. Bacteriol. 93:1722.
185. Hirsch, P. (1968). Nature 217:555.
186. Uffen, R. L. (1976). Proc. Natl. Acad. Sci. USA 73:3298.
187. Thiele, H. H. (1968). Arch. Microbiol. 60:124.
188. Sandhu, G. R. and Carr, N. G. (1970). Arch. Microbiol. 70:340.
189. Leuking, D., Tokuhisa, D., and Sojka, G. (1973). J. Bacteriol. 115:897.
190. Leuking, D. R., Pike, L., and Sojka, G. (1976). J. Bacteriol 125:750.
191. Pike, L., and Sojka, G. (1975). J. Bacteriol. 124:1101.
192. Leuking, D. R., and Goldfine, H. (1975). J. Biol. Chem. 250:8530.
193. Gibson, M. S., and Wang, C. H. (1968). Can. J. Microbiol. 14:493.
194. Saier, M. H., Feucht, B. U., and Roseman, S. (1971). J. Biol. Chem. 246: 7819.
195. van Gemerden, H. (1968). Arch. Microbiol. 64:118.
196. Krassilnikova, E. N. (1975). Mikrobiologiia 44:5.
197. Schön, G., and Biedermann, M. (1972). Arch. Microbiol. 85:77.
198. Conrad, R., and Schlegel, H. G. (1974). Biochim. Biophys. Acta 358:221.
199. Conrad, R., and Schlegel, H. G. (1976). In G. A. Codd and W. D. P. Stewart (eds.), International Symposium on Photosynthetic Prokaryotes, p. 206. Federation of European Microbiological Societies, Dundee.
200. Eidels, L., and Preiss, J. (1970). J. Biol. Chem. 245:2937.

201. Szymona, M., and Doudoroff, M. (1960). J. Gen. Microbiol. 22:167.
202. Losada, M., Trebst, A. V., Ogata, S., and Arnon, D. I. (1960). Nature 186:753.
203. Stanier, R. Y. (1961). Bacteriol. Rev. 25:1.
204. Krassilnikova, E. N., Pedan, L. V., Firsov, N. N., and Kondratieva, E. N. (1973). Mikrobiologiia 42:995.
205. Cherniadiev, I. I., Kondratieva, E. N., and Doman, N. G. (1970). Mikrobiologiia 39:24.
206. Albers, H., and Gottschalk, G. (1976). Arch. Microbiol. 111:45.
207. Hoare, D. S. (1963). Biochem. J. 87:284.
208. Kikuchi, G., Tsuiki, S., Muto, A., and Yamada, H. (1963). Studies in Microalgae and Photosynthetic Bacteria, p. 547. University of Tokyo Press, Tokyo.
209. Bennedict, C. R., and Rinne, R. W. (1964). Biochem. Biophys. Res. Comm. 14:474.
210. Osumi, T., Ebisuno, T., Nakano, H., and Katsuki, H. (1975). J. Biochem. 78:763.
211. Osumi, T., and Katsuki, H. (1975). J. Biochem. 81:771.
212. Krassilnikova, E. N., and Kondratieva, E. N. (1974). Mikrobiologiia 43:776.
213. Buchanan, B. B. (1974). J. Bacteriol. 119:1066.
214. Payne, J., and Morris, J. G. (1969). J. Gen. Microbiol. 59:97.
215. Johansson, B. C., and Gest, H. (1976). J. Bacteriol. 128:683.
216. Qadri, S. M. H., and Hoare, D. S. (1967). J. Bacteriol. 95:2344.
217. Uffen, R. L. (1973). In G. Drews, N. Pfennig, and R. Y. Stanier (eds.), Abstracts of the Symposium on Prokaryotic Phototrophic Organisms, p. 155. Deutsche Forschungsgemeinschaft and the International Union of Biological Sciences, Freiburg.
218. Jungermann, K., and Schön, G. (1974). Arch. Microbiol. 99:109.
219. Klemme, J. -H. (1973). Arch. Microbiol. 90:305.
220. Klemme, J. -H. (1974). Arch. Microbiol. 100:57.
221. Schedel, M., Klemme, J. H., and Schlegel, H. G. (1975). Arch. Microbiol. 103:237.
222. Stern, J. R. (1963). Biochim. Biophys. Acta 69:435.
223. Tuboi, S., and Kikuchi, G. (1966). J. Biochem. (Tokyo) 59:456.
224. Tuboi, S., and Kikuchi, G. (1965). Biochim. Biophys. Acta 96:148.
225. Mue, S., Tuboi, S., and Kikuchi, G. (1964). J. Biochem. (Tokyo) 56:545.
226. Evans, M. C. W. (1965). Biochem. J. 95:661.
227. Evans, M. C. W. (1965). Biochem. J. 95:669.
228. Ehrensvärd, G., and Gatenbeck, S. (1965). Acta Chem. Scand. 19:2006.
229. Schigesada, K. (1972). J. Biochem. 71:961.
230. Gibson, J. (1975). J. Bacteriol. 123:471.
231. Schaab, C., Giffhorn, F., Schoberth, S., Pfennig, N., and Gottschalk, G. (1972). Z. Naturforsch. 27b:962.
232. Giffhorn, F., Beuscher, N., and Gottschalk, G. (1972). Biochem. Biophys. Res. Comm. 49:467.
233. Giffhorn, F., Beuscher, N., and Gottschalk, G. (1973). In G. Drews, N. Pfennig, and R. Y. Stanier (eds.), Abstracts of the Symposium on Prokaryotic Photosynthetic Organisms, p. 147. Deutsche Forschungsgemeinschaft and International Union of Biological Sciences, Freiburg.
234. Beuscher, N., Mayer, F., and Gottschalk, G. (1974). Arch. Microbiol. 100:307.
235. Giffhorn, F., and Gottschalk, G. (1975). J. Bacteriol. 124:1052.
236. Giffhorn, F., and Gottschalk, G. (1975). J. Bacteriol. 124:1046.
237. Giffhorn, F., and Gottschalk, G. (1976). In G. A. Codd and W. D. P. Stewart

(eds.), Proceedings of the Second International Symposium on Photosynthetic Prokaryotes, p. 204. Federation of European Microbiological Societies, Dundee.

238. Hegeman, G. D. (1967). Arch. Mikrobiol. 59:143.
239. Dutton, P. L., and Evans, W. C. (1969). Biochem. J. 113:525.
240. Dutton, P. L., and Evans, W. C. (1970). Arch. Biochem. Biophys. 136:228.
241. Dagley, S. (1971). Adv. Microb. Physiol. 6:1.
242. Whittle, P. J., Lunt, D. O., and Evans, W. C. (1976). Biochem. Soc. Trans. 4:490.
243. van Gemerden, H. (1968). Arch. Mikrobiol. 64:111.
244. Hara, F., Akazawa, T., and Kojima, K. (1973). Plant Cell Physiol. 14:737.
245. Malofeeva, I. V., Kondratieva, E. N., and Rubin, A. B. (1975). FEBS Lett. 53:188.
246. Bast, E. (1977). Arch. Microbiol. 113:91.
247. Stewart, W. D. P. (1973). Annu. Rev. Microb. 27:283.
248. Stewart, W. D. P. (1975). Nitrogen Fixation by Free-Living Microorganisms. Cambridge University Press, Cambridge, England.
249. Yoch, D. C., and Arnon, D. I. (1974). In A. Quispel (ed.), Biology of Nitrogen Fixation, p. 687. North-Holland Publishing Co., Amsterdam.
250. Nagatani, H., Shimizu, M., and Valentine, R. C. (1971). Arch. Microbiol. 79:164.
251. Slater, J. H., and Morris, I. (1974). Arch. Microbiol. 95:337.
252. Brown, C. M., and Herbert, R. A. (1976). In G. A. Codd and W. D. P. Stewart (eds.), Proceedings of the Second International Symposium on Photosynthetic Prokaryotes, p. 152. Federation of European Microbiological Societies, Dundee.
253. Weare, N. M., and Shanmugan, K. T. (1976). Arch. Microbiol. 110:207.
254. Furlong, C. C., and Preiss, J. (1969). J. Biol. Chem. 244:2539.
255. Hara, F., and Akazawa, T. (1974). Plant Cell Physiol. 15:545.
256. Kondratieva, E. N. (1976). In H. G. Schlegel and J. Boznea (eds.), Microbial Energy Conversion, p. 205. Erich Goltze KG, Göttingen.
257. Wall, J. D., Weaver, P. F., and Gest, H. (1975). Nature 258:630.
258. Benemann, J. R., and Weare, M. M. (1974). Science 184:174.
259. Kessler, E. (1973). In W. D. P. Stewart (ed.), Algae Physiology and Biochemistry, Vol. 10, p. 456. Blackwell Scientific Publications, Oxford.
260. Schön, G., and Drews, G. (1966). Arch. Mikrobiol. 54:199.
261. Schön, G. (1968). Arch. Mikrobiol. 61:187.
262. Hendley, D. D. (1955). J. Bacteriol. 70:625.
263. Trüper, H. G., and Schlegel, H. G. (1964). Antonie van Leeuwenhoek 30:225.
264. Krassilnikova, E. N. (1976). Mikrobiologiia 45:372.
265. Schön, G., and Ladwig, R. (1970). Arch. Microbiol. 74:356.
266. van Gemerden, H. (1976). In G. A. Codd and W. D. P. Stewart (eds.), Proceedings of the Second International Symposium on Photosynthetic Prokaryotes, p. 50. Federation of European Microbiological Societies, Dundee.
267. Schön, G. (1968). Arch. Mikrobiol. 63:362.
268. Uffen, R. L., and Wolfe, R. S. (1970). J. Bacteriol. 104:462.
269. Uffen, R. L., Sybesma, C., and Wolfe, R. S. (1971). J. Bacteriol. 108:1348.
270. Uffen, R. L. (1973). J. Bacteriol. 116:874.
271. Uffen, R. L. (1973). J. Bacteriol. 116:1086.
272. Schön, G., and Biedermann, M. (1973). Biochem. Biophys. Acta 304:65.
273. Gürgün, V., Kirchner, G., and Pfennig, N. (1976). Z. Allg. Mikrobiol. 16:573.
274. Thauer, R. N., Jungermann, K., and Becker, K. (1977). Bacteriol. Rev. 41:100.
275. Satoh, T., Hoshino, Y., and Kitamura, H. (1976). Arch. Microbiol. 108:265.
276. Yen, H. C., and Marrs, B. (1977). Arch. Biochem. Biophys. 181:411.

277. Gorchein, A., Neuberger, A., and Tait, G. H. (1968). Proc. R. Soc. 170:229.
278. Gusev, M. V., Shendereva, L. V., and Kondratieva, E. N. (1969). Mikrobiologiia 38:787.
279. Gusev, M. V., Shenderova, L. V., and Kondratieva, E. N. (1970). Mikrobiologiia 39:562.
280. Drews, G., Lampe, H. H., and Ladwig, R. (1969). Arch. Mikrobiol. 65:12.
281. Breäcker, E. (1964). Zentralbl. Bakteriol. 118:561.
282. Hurlbert, R. E. (1967). J. Bacteriol. 93:1346.
283. Gibson, J. (1967). Arch. Mikrobiol. 59:104.
284. Uspenskaia, V. E., Rodova, N. A., and Kondratieva, E. N. (1971). Mikrobiologiia 40:455.
285. Lumsden, J., and Hall, D. O. (1975). Nature 257:670.
286. Henry, E. A., Lumsden, J., and Hall, D. O. (1976). In G. A. Codd and W. D. P. Stewart (eds.), Proceedings of the Second International Symposium on Photosynthetic Prokaryotes, p. 242. Federation of European Microbiological Societies, Dundee.
287. Kondratieva, E. N., Krassilnikova, E. N., and Pedan, L. V. (1975). Mikrobiologiia 45:172.
288. Kondratieva, E. N. (1976). In G. A. Codd and W. D. P. Stewart (eds.), Proceedings of the Second International Symposium on Photosynthetic Prokaryotes, p. 47. Federation of European Microbiological Societies, Dundee.
289. Pfennig, N. (1970). J. Gen. Microbiol. 61:ii.
290. Bogorov, L. V. (1974). Mikrobiologiia 43:326.
291. Gorlenko, V. M. (1974). Mikrobiologiia 43:729.
292. Hansen, T. A. (1976). In G. A. Codd and W. D. P. Stewart (eds.), Proceedings of the Second International Symposium on Photosynthetic Prokaryotes, p. 43. Federation of European Microbiological Societies, Dundee.
293. Keppen, O. I., Nozhevnikova, A. N., and Gorlenko, V. M. (1976). Mikrobiologiia 45:15.
294. Throm, E., Oelze, J., and Drews, G. (1970). Arch. Microbiol. 72:361.
295. Thore, A., Keister, D. L., and San Pietro, A. (1969). Arch. Mikrobiol. 67:378.
296. Klemme, J.-H., and Schlegel, H. G. (1969). Arch. Mikrobiol. 68:326.
297. Rodova, N. A., and Krassilnikova, E. N. (1974). Mikrobiologiia 43:208.
298. Irschik, H., and Oelze, J. (1976). Arch. Microbiol. 109:307.
299. Ivanovsky, R. N., and Rodova, N. A. (1976). Mikrobiologiia 45:197.
300. Baccarini-Melandri, A., and Melandri, B. A. (1972). FEBS Lett. 21:131.
301. Lien, S., and Gest, H. (1973). Arch. Biochem. Biophys. 159:730.
302. Garcia, A. F., Drews, G., and Kamen, M. D. (1975). Biochim. Biophys. Acta 387:129.
303. Cohen-Bazire, G., Sistrom, W. R., and Stanier, R. Y. (1957). J. Cell. Comp. Physiol. 49:25.
304. King, M. T., and Drews, G. (1975). Arch. Microbiol. 102:219.
305. Zannoni, D., Baccarini-Melandri, A., Melandri, B. A., Evans, E. H., Prince, R. C., and Crofts, A. R. (1974). FEBS Lett. 48:153.
306. Jurtshuk, P., Jr., Mueller, T. J., and Acord, W. C. (1975). CRC Crit. Rev. Microbiol. 3:399.
307. King, M. T., and Drews, G. (1976). Eur. J. Biochem. 68:5.
308. Zannoni, D., Melandri, B. A., and Baccarini-Melandri, A. (1976). Biochim. Biophys. Acta 423:413.
309. Gibson, J., and Morita, S. (1977). J. Bacteriol. 93:1544.
310. Ivanovskii, R. N., and Petushkova, I. P. (1976). Mikrobiologiia 45:1102.
311. Oelze, J., and Weaver, P. (1971). Arch. Microbiol. 79:108.
312. Dierstein, R., and Drews, G. (1975). Arch. Microbiol. 106:227.
313. Lascelles, J., and Wertlabl, W. D. (1971). Biochim. Biophys. Acta 226:328.

314. Hicughi, M., and Kikuchi, G. (1969). Plant Cell Physiol. 10:149.
315. Cherniadiev, I. I., Kondratieva, E. N., and Doman, N. G. (1974). Mikrobiologiia 43:949.
316. Bishop, N. I. (1966). Annu. Rev. Plant Physiol. 17:185.
317. Krassilnikova, E. N. (1977). Mikrobiologiia 46:217.
318. Gray, B. H., Fowler, C. F., Nugent, N. A., Rigopoulos, N., and Fuller, R. C. (1973). Int. J. System. Bacteriol. 23:256.
319. Pfennig, N., and Biebl, H. (1976). Arch. Microbiol. 110:3.
320. Pierson, B. K., and Castenholz, R. W. (1971). Nature New Biol. 233:25.
321. Pierson, B. K., and Castenholz, R. W. (1974). Arch. Microbiol. 100:5.
322. Pierson, B. K., and Castenholz, R. W. (1974). Arch. Microbiol. 100:283.
323. Bauld, J., and Brock, T. D. (1973). Arch. Microbiol. 92:267.
324. Trüper, H. G. (1976). Int. J. System. Bacteriol. 26:74.
325. Dubinina, G. A., and Gorlenko, V. M. (1975). Mikrobiologiia 44:511.
326. Gorlenko, V. M., and Pivovarova, T. A. (1977). Izv. Akad. Nauk SSSR N3:396.
327. Madigan, M. T., Peterson, S. R., and Brock, T. D. (1974). Arch. Microbiol. 100:97.
328. Madigan, M. T., and Brock, T. D. (1975). J. Bacteriol. 122:782.
329. Kenyon, C. N., and Gray, A. M. (1974). J. Bacteriol. 120:131.
330. Halfen, L. N., Pierson, B. K., and Francis, G. W. (1972). Arch. Microbiol. 82:240.
331. Knaff, D. B., Buchanan, B. B., and Malkin, R. (1973). Biochem. Biophys. Acta 325:94.
332. Shioi, Y., Takamiya, K., and Nishimura, M. (1976). J. Biochem. (Tokyo) 80:811.
333. van Beeumen, J., Ambler, R. P., Meyer, T. E., Kamen, M. D., Olson, J. M., and Shaw, E. K. (1976). Biochem. J. 159:757.
334. Fowler, C. F. (1974). Biochim. Biophys. Acta 357:327.
335. Powls, R., and Redfearn, E. R. (1969). Biochim. Biophys. Acta 172:429.
336. Kusai, A., and Yamanaka, T. (1973). Biochim. Biophys. Acta 292:621.
337. Shioi, Y., Takamiya, K., and Nishimura, M. (1976). J. Biochem. (Tokyo) 79:361.
338. Buchanan, B. B., Matsubara, H., and Evans, M. C. W. (1969). Biochim. Biophys. Acta 189:46.
339. Meyer, T. E., Sharp, J. J., and Bartsch, R. G. (1971). Biochim. Biophys. Acta 234:266.
340. Tanaka, I., Haniu, M., Yasunobu, K. T., Evans, M. C. W., and Rao, K. K. (1974). Biochemistry 13:2953.
341. Knaff, D. B., and Malkin, R. (1976). Biochim. Biophys. Acta 430:244.
342. Jennings, J. V., and Evans, M. C. W. (1976). International Conference on Primary Electron Transport and Energy Transduction in Photosynthetic Bacteria, MB3. Vrije Universitet, Brussels.
343. Cruden, D. L., and Stanier, T. Y. (1970). Arch. Microbiol. 72:115.
344. Boyce, C. O. L., Oyewole, S. H., and Fuller, R. C. (1976). Brookhaven Symp. Biol. 28:365.
345. Fuller, R. C., Boyce, C. O., and Oyewole, S. H. (1976). International Conference on Primary Electron Transport and Energy Transduction in Photosynthetic Bacteria, MB7. Vrije Universitet, Brussels.
346. Olson, J. M., Prince, R. C., and Brune, D. C. (1976). International Conference on Primary Electron Transport and Energy Transduction in Photosynthetic Bacteria, MB6. Brussels.
347. Olson, J. M., Prince, R. C., and Brune, D. C. (1976). Brookhaven Symp. Biol. 28:238.

348. Olson, J. M., Ke, B., and Thompson, K. H. (1976). Biochim. Biophys. Acta 430:524.
349. Prince, R. C., and Olson, J. M. (1976). Biochim. Biophys. Acta 423:357.
350. Fowler, C. F., Nugent, N. A., and Fuller, R. C. (1971). Proc. Natl. Acad. Sci. USA 68:2278.
351. Buchanan, B. B., and Evans, M. C. W. (1969). Biochim. Biophys. Acta 180:123.
352. Evans, M. C. W. (1969). In H. Metzner (ed.), Progress in Photosynthesis Research, Vol. III, p. 1474. H. Laupp Jr., Tübingen.
353. Burns, D. D., and Midgley, M. (1976). Eur. J. Biochem. 67:323.
354. Williams, A. M. (1956). Biochim. Biophys. Acta 19:571.
355. Sykes, J., and Gibbon, J. A. (1967). Biochim. Biophys. Acta 143:173.
356. Hughes, D. E., Conti, S. F., and Fuller, R. C. (1963). J. Bacteriol. 85:577.
357. Cole, J. A., and Hughes, D. E. (1965). J. Gen. Microbiol. 38:65.
358. Larsen, H. (1953). K. Nor. Vidensk. Selak Srk 1:1.
359. Kusai, A., and Yamanaka, T. (1973). FEBS Lett. 34:235.
360. Kusai, A., and Yamanaka, T. (1973). Biochim. Biophys. Acta 325:304.
361. Mathewson, J. H., Burger, L. J., and Millstone, H. G. (1968). Fed. Proc. 27:774.
362. Kusai, A., and Yamanaka, T. (1973). Biochem. Biophys. Res. Comm. 51:107.
363. Kirchhoff, J., and Trüper, H. G. (1974). Arch. Microbiol. 100:115.
364. Paschinger, H., Paschinger, J., and Gaffron, H. (1974). Arch. Microbiol. 96:341.
365. Trüper, H. G., Lorenz, C., and Fischer, U. (1976). In G. A. Codd and W. D. P. Stewart (eds.), Proceedings of the Second International Symposium on Photosynthetic Prokaryotes, p. 41. Federation of European Microbiological Societies, Dundee.
366. Lippert, K. -D., and Pfennig, N. (1969). Arch. Mikrobiol., 65:29.
367. Smillie, R. M., Rigopoulos, N., and Kelly, H. (1962). Biochim. Biophys. Acta 56:612.
368. Hoare, D. S. (1963). Biochem. J. 87:284.
369. Evans, M. C. W., and Buchanan, B. B. (1965). Proc. Natl. Acad. Sci. USA 53:1420.
370. Buchanan, B. B., Schürmann, P., and Shanmugan, K. T. (1972). Biochim. Biophys. Acta 283:136.
371. Buchanan, B. B., and Sirevåg, R. (1976). Arch. Microbiol. 109:15.
372. Sirevåg, R., and Ormerod, J. G. (1970). Biochem. J. 120:399.
373. Sirevåg, R., and Ormerod, J. G. (1977). Arch. Microbiol. 111:239.
374. Sirevåg, R., Buchanan, B. B., Berry, J. A., and Troughton, J. H. (1977). Arch. Microbiol. 112:35.
375. Sirevåg, R. (1974). Arch. Microbiol. 98:3.
376. Sirevåg, R. (1975). Arch. Microbiol. 104:105.
377. Tabita, F. R., McFadden, B. A., and Pfennig, N. (1974). Biochim. Biophys. Acta 341:187.
378. Doman, N. G., Kondratieva, E. N., and Simisker, Ya. A. (1967). Dokl. Akad. Nauk SSSR 172:396.
379. Trozenko, Y. A., Simisker, J. A., Kondratieva, E. N., and Doman, N. G. (1970). Izv. Akad. Nauk SSSR 3:415.
380. Callely, A. G., Rigopoulos, N., and Fuller, R. C. (1968). Biochem. J. 106:615.
381. Beuscher, N., and Gottschalk, G. (1972). Z. Naturforsch. 27b:967.
382. Whittenbury, R., and Kelly, D. P. (1977). Symp. Soc. Gen. Microb. 27:121.
383. Kelly, D. P. (1971). Annu. Rev. Microbiol. 25:177.

384. Sadler, W. R., and Stanier, R. Y. (1960). Proc. Natl. Acad. Sci. USA 46: 1328.
385. Kelly, D. P. (1974). Arch. Microbiol. 100:163.
386. Ormerod, J. G., and Sirevåg, R. (1976). *In* G. A. Codd and W. D. P. Stewart (eds.), Proceedings of the Second International Symposium on Photosynthetic Prokaryotes, p. 187. Federation of European Microbiological Societies, Dundee.
387. Bothe, H., and Nolteernsting, U. (1975). Arch. Microbiol. 102:53.
388. Shuvalov, V. A., Kondratieva, E. N., and Litvin, F. F. (1968). Dokl. Akad. Nauk SSSR, 178:711.
389. Adams, M. W. W., and Hall, D. O. (1977). Biochem. Biophys. Res. Comm. 77:730.
390. Pfennig, N., and Siefert, E. (1977). *In* Abstracts of the Second International Symposium on Microbial Growth on C_1 Compounds, p. 146. Puschino.

International Review of Biochemistry
Microbial Biochemistry, Volume 21
Edited by J. R. Quayle
Copyright 1979   University Park Press   Baltimore

5
Continuous Culture Applications to Microbial Biochemistry

A. T. BULL AND C. M. BROWN

Department of Applied Biology,
UWIST, Cardiff, Wales
and Department of Biological Sciences
University of Dundee, Scotland

The foundations of continuous culture theory laid by Monod and by Novick and Szilard, the subsequent extensions of the theory, and developments in

Research reported from our own laboratories was supported by the Science Research Council and National Environment Research Council.

instrumentation during the 30 intervening years are becoming increasingly known to microbiologists. These early studies, many of which were concerned with experimental testing of chemostat theory, kinetics, and basic chemical analyses, have been reviewed on several occasions and are omitted from the present discussion. The reader can readily gauge the directions and pace of continuous culture research by reference to the series of International Symposia on Continuous Culture that reports on the meetings held since 1958 in Czechoslovakia and England; the most recent of these Symposia was published in 1976 (1). Other major sources of information on continuous culture theory and practice include the books by Malek and Fencl (2), Kubitschek (3), and Pirt (4). In this chapter we assume that the reader is familiar with basic chemostat theory.

Continuous culture techniques are now regarded as established laboratory procedures and for the past several years only relatively minor developments have occurred in relation to their theory and technology. By contrast, use of continuous cultures is being adopted to provide answers to an extremely wide spectrum of microbiological problems, a situation that has been encouraged by the publication of good practical information and advice (5), and the increasing availability of reliable and often inexpensive equipment. A number of reviews published in the 1970s by Tempest (6), Bull (7), and Veldkamp (8) have emphasized the research potential of continuous cultures and canvassed their use as the method of choice in numerous areas of microbiology and biochemistry. Similarly, the case for an integral position for continuous cultures in the teaching of microbial physiology has been argued by Bull and Slater (9).

Because of the diversity of application of continuous culture techniques, we concentrate in this chapter on recent research; most of the work that is considered here has been published since 1970. Moreover, we concentrate on areas of microbial biochemistry whose study is especially suited to continuous culture experimentation and where a steady flow of publications is beginning to appear. In particular the reader may discern a tenuous theme of ecophysiology running through this article and touching upon such topics as the modulation of metabolism in response to changing environments, mixed substrate utilization, mixed culture activities, and experimental microbial evolution. Developments in this area within the past few years have been very exciting. As Tempest and Neijssel (10) have remarked, "In natural environments, conditions for the growth of microorganisms are often unstable and rarely ideal," and clearly successful species, or communities, have evolved adaptive responses that have allowed them to exploit such environments in the most effective manner. One of our objectives in this chapter is to explore some of these adaptive responses in metabolic terms.

PHENOTYPIC VARIATION OF MICROBIAL CHEMISTRY

According to Herbert (11), "it is virtually meaningless to speak of the chemical composition of a microorganism without at the same time specifying the environmental conditions producing it." The use of continuous culture to control the growth rate under conditions of known substrate limitation, temperature, pH, and oxygen tension has been exploited by many researchers in relation to the cellular contents and composition of most macromolecules. As might be anticipated, initial experiments illustrated changes in gross composition [Herbert (12) has summarized much of these data in a recent article on the stoichiometric aspects of microbial growth] while more recent work has centered on fine changes in macromolecular structure or replacement.

The alternative to the use of the chemostat for studies of chemical composition was set out in the report of Schaechter et al. (13) in which 22 media of varying complexity were used to study the influence of growth rate on the macromolecular composition of *Salmonella typhimurium*. Their results indicated that the faster the specific growth rate (μ) the higher the content of RNA, DNA, cell mass, and cell size, while protein content was fairly constant. Because the increase in RNA was accounted for as ribosomes it was suggested that the rate of protein synthesis/ribosome was constant and that an increased rate of protein synthesis (at higher μ) could be accomplished only by increasing the amount of ribosomal RNA.

Later chemostat experiments described by Herbert (11) extended these observations to *Klebsiella aerogenes* and *Bacillus subtilis* and further demonstrated that these effects were due to μ rather than to the varying complexity of the media used, a point not excluded in the batch culture experiments. However, Schaechter et al. (13) also varied the growth temperature in their batch culture experiments and found that the RNA and DNA contents and the cell size did not vary between 25 °C and 37 °C: they concluded that "the size and chemical composition of the cells are related to the growth rate only in so far as it depends on the medium." Tempest and Hunter (14) subsequently reported that when grown in a chemostat (at a fixed dilution rate D) the RNA (and hence ribosome) content of *K. aerogenes* increased markedly with decreasing growth temperature, results clearly at variance with those of Schaechter et al. (13).

In the batch experiments, a decrease in temperature results in a decrease in both growth rate and temperature, while temperature is the only variable in a chemostat. Therefore in the batch experiments it seems likely that a potential rise in RNA, etc. due to decreased temperature was compensated for by the decreased μ. This may be viewed more finitely if it is assumed that cell size, RNA content, etc. are a function of the relative growth rate of the organism. Relative growth rate (15, 16) is the ratio μ/μ_{max} and obviously batch cultures in the same medium grow at a relative growth rate of 1 during so-called balanced growth, irrespective of temperature. Since μ_{max} decreases with temperature, then, in the chemostat (at fixed dilution rate D), the lower the temperature the higher the relative growth rate. Variations in the RNA content and cell size of the yeast *Candida utilis* with μ and temperature (17,

18) may also be interpreted on this basis. This should be borne in mind in all studies in which μ_{max} varies with, for example, varying pH and pO_2 and different carbon and nitrogen sources, when it may be misleading to perform experiments at constant D and assume that cultures are in the same physiological state. Perhaps the use of the turbidostat should be more widespread in this type of study.

The use of the chemostat to grow organisms under specific substrate limitations has been of particular use in the study of phenotypic variations in the composition and structure of the bacterial cell envelope. Much of this work stems from the observation of Tempest (19) that Mg^{2+}-limited and K^+-limited cultures of *B. subtilis* contained non-nucleic acid phosphorus, which was absent from phosphate-limited cultures. Because the cell walls of this organism were known to contain teichoic acids, it was reasoned that these polymers might be present in phosphate-sufficient and absent in phosphate-limited cultures (19). This was shown to be the case with phosphate-limited cultures containing teichuronic acid polymers (glucuronic and *N*-acylated galactosaminuronic acids) replacing the glycerol teichoic acids.

These results demonstrated that cell wall composition was a function of substrate limitation. Further experiments (20) demonstrated, by studying wall composition during the transition between steady states, that these changes were due to a phenotypic response rather than to selection of different genotypes. For example, it was shown that the synthesis of teichoic acids during the transition from phosphate to K^+ limitation occurred faster than could be accounted for by replacement of an organism synthesizing teichuronic acids with one synthesizing teichoic acids. Ellwood and Tempest (20) stated that, because teichoic acids could account for over 50% of the *Bacillus* cell wall and yet could be replaced rapidly and totally by teichuronic acids during phosphorus deprivation, the cell wall must be in a dynamic state during growth and moreover must be a phenotypically variable structure.

Archibald has exploited these observations in his work on phage binding and mode of wall growth in *B. subtilis* (21, 22, 23, 24). In *B. subtilis* W23 the receptor for phage SP50 consists of wall material containing a ribitol teichoic acid (TA) with the integrity of both TA and peptidoglycan components being essential for phage adsorption. Phosphate-limited bacteria contained teichuronic acid and did not bind, nor were they infected by, the phage. In contrast, K^+-limited organisms contained TA, bound the phage, and were lysed by it. During transition from phosphate to K^+ limitation the bacteria developed a capacity to bind phage that increased exponentially in relation to the TA content of the cell walls. During transition from K^+ to phosphate limitation the bacteria retained a near-maximum phage binding capacity until the wall TA content fell to a very low level. Archibald suggested that these results could be explained if newly synthesized wall material was not immediately exposed at the cell surface in a structure to which phage could adsorb.

Using heat-killed bacteria (not lysed by phage) it was possible to locate the sites of phage adsorption by electron microscopy and it was shown that newly incorporated receptor material appeared at the cell surface first along the whole length of the cell but not at the end caps. This suggested that this receptor material was intercalated into a large number of sites distributed along the length of the cell. Further experiments included 'pulsing' a phosphate-limited culture with phosphate—maximum phage adsorption occurred half a generation time after the pulse, after the effects of the pulse and the incorporation of TA had ceased.

In some elegant experiments showing phage attachment to isolated cell walls all these effects were accounted for. Walls from organisms grown in phosphate sufficiency had phage adsorption sites on both inner and outer wall sufaces. Walls from phosphate-limited organisms pulsed with phosphate had phage attachment sites only on the inside of the wall in the early stages of the pulse, while TA was synthesized to the outside of the wall only later (after incorporation of TA had ceased). Thus this experimental system yielded valuable information that TA in *B. subtilis* is intercalated into the wall along the length of the cell, initially at sites on the inside of the wall.

Another area of interest is the influence of growth conditions on the resistance of *Pseudomonas aeruginosa* to antibacterial agents. Melling and Brown (25) have summarized the results of batch culture experiments in which resistance to EDTA and polymyxin was shown to be dependent upon the culture Mg^{2+} concentration. Melling et al. (26) and Finch and Brown (27) extended these observations to chemostat cultures and reported that, at all values of μ used, Mg^{2+}-limited cultures were more resistant to polymyxin and EDTA than phosphate- or carbon-limited cultures. Attempts to correlate resistance with envelope composition revealed no clear pattern, although the phosphorus and hexose contents of envelopes were shown to vary. It was suggested that Mg^{2+}-limited organisms were modified to deny access of EDTA and polymyxin to their sites of action (27). Dean et al. (28) reported changes in sensitivity to carbenicillin, actinomycin, streptomycin, and gentomycin with μ and substrate limitation.

Finch and Brown (27) tested the susceptibility of organisms grown under carbon- or Mg^{2+}-limitation at D 0.05 h^{-1} and D 0.5 h^{-1} to killing by rabbit phagocytes in the presence of rabbit serum. While carbon-limited organisms and Mg^{2+}-limited organisms grown at D 0.5 h^{-1} were killed rapidly ($<5\%$ surviving after 120 min treatment), Mg^{2+}-limited organisms grown at D 0.05 h^{-1} were markedly resistant. Similarly, resistance to cationic proteins derived from rabbit phagocytes was shown by Mg^{2+}-limited cultures at slow growth rates, but not in organisms grown under the other conditions listed.

Gilbert and Brown (29) tested the effect of halogenated phenols on *P. aeruginosa*. At low concentrations 3-chlorphenol and 4-chlorphenol uncoupled oxidative phosphorylation and increased membrane permeability to protons. The drug concentration required to produce similar levels of proton translocation varied with μ and substrate limitation. In general, fast-growing cultures were most sensitive, with carbon-limited organisms more sensitive than Mg^{2+}-limited organisms. The authors suggested that these differences in sensitivity were due to variations in drug penetrability and reported a correlation between lipopolysaccharide (LPS) content and drug resistance, i.e., higher LPS in Mg^{2+}-limited organisms at low μ. This also offers an explanation for the results with phagocytes outlined above.

While the precise nature of envelope modification causing resistance to antibacterials remains unclear, it is apparent that resistance is a function of growth environment. Brown (30, 31) drew attention to the pharmaceutical importance of these results and questioned the relevance of the normal practice of determining resistance patterns on stationary phase cultures grown in nutrient broth. He further speculated on the nature of nutrient limitations in natural environments, such as the human body, pointing to the need for drug testing to be correlated with in vivo bacterial growth. The usefulness of continuous culture in this area is self-evident.

The lipid components of bacterial membranes also vary with growth environment (32). Minnikin's group has proposed that bacterial lipids fall into two categories: 1) acidic polar lipids, including phosphatidyl (P), glycerol (PG), diphosphatidyl glycerol (DPG), P-serine, P-inositol, acidic peptidolipids, and acidic glycolipids; and 2) neutral or amphoteric polar lipids, including P-ethanolamine (PE), P-choline, ornithine-containing lipids (OL), and neutral glycolipids. Within the limits of retaining some acidic and some neutral or amphoteric component(s), gross variations in composition may occur. Batch-grown *B. subtilis* (Marburg) contained DPG, PE, PG, the glycolipid diglucosyl diacyl glycerol (DG), and traces of the lysyl ester of PG. In Mg^{2+}-limited chemostat cultures little PE was detected and the only components present in significant amounts were PG and DG. In complete contrast, phosphate-limited cultures contained lower amounts of the acidic PG and DPG but major amounts of DG and an acidic peptidolipid containing leucine, valine, aspartate, and glutamate (33). Variation in pH also resulted in changes in lipid composition in both Mg^{2+} and phosphate-limited cultures (34).

Even more marked effects of nutrient limitation were evident in chemostat cultures of *Pseudomonas diminuta* (35) and a marine strain of *Pseudomonas fluorescens* (36). *P. diminuta* envelopes usually contain only small quantities of phosphorus and do not contain PE. Lipids extracted from batch cultures contained both PG and phosphatidyl glucosyl diacyl glycerol (PGD), together with some glycolipids. In phosphate-limited cultures there was no detectable PGD and PG constituted only a trace (0.3%) of total polar lipid (37). Thus phosphate-limited cultures contained only trace amounts of phospholipids and it was evident that PE was replaced by glucosyl diacyl glycerol and that acidic glycolipids (with glucuronosyl residues) replaced PG. Lipids from batch cultures of the marine *P. fluorescens* contained PE, PG, DPG, and OL. Mg^{2+}-limited chemostat cultures contained the three phospholipids but no OL, while under phosphate-limitation at D 0.1 h^{-1} OL was the major component present, along with small quantities of phospholipid. At D 0.2 h^{-1}, however, no traces of phospholipids could be detected; they had been completely replaced by OL and small quantities of an acid glycolipid (probably a glucuronosyl diacyl glycerol). In this organism, therefore, an acidic glycolipid replaced the acidic PG and DPG, while PE is interchangeable with the zwitterionic OL. This strain of *P. fluorescens* represents one organism that can exist without detectable quantities of phospholipids in its cytoplasmic membrane. The yeast *Ca. utilis* shows an equivalent loss of phospholipid during phosphate-limited growth, but while some glycolipid is produced under these conditions the other lipids present have not been characterized.

Changes in cell size with μ and temperature were mentioned earlier in this section. Cell shape, which is the gross manifestation of chemical composition, is also environmentally determined. Brown and Hough (38) reported changes in the morphology of *Saccharomyces cerevisiae* with substrate limitation and Johnson (39) briefly reported the effects of growth on the morphology of the very interesting "triangular" yeast *Trigonopsis variabilis*. One of the best characterized changes in morphology, however, is observed in *Arthrobacter* spp. In batch cultures on media allowing rapid growth rates a 'life cycle' is observed in which coccoid cells change to rods that change back to cocci. It was found that, while cocci alone were produced in a glucose–mineral salts medium, if this medium was supplemented with an additional carbon source, such as succinate, then rods were formed. These effects of medium supplementation could be interpreted as induction of the rod form by specific chemicals or by related changes in growth rate.

Luscombe and Gray (40), using glucose-limited chemostats (with no medium supplementation), reported that cell shape was μ-dependent. For example, at 25 °C rods were produced at D >0.25 h^{-1}, and cocci were produced at lower growth rates. It was also reported that the lower the growth temperature the lower the D at which the coccus/rod transition occurred, which suggests that this transition may be a function of the relative growth rate of the culture. In a later study of a number of *Arthrobacter* spp., the coccus/rod transition was shown to be a μ-determined feature in all organisms, but the actual "μ transition" was species dependent (41). It is noteworthy that the higher the μ_{max} the higher the μ transition, with transition occurring at relative growth rates between 0.5 and 0.7 h^{-1}.

Cell walls of both rods and cocci are lysozyme-resistant, while containing high amounts of polysaccharide relative to peptidoglycan. The walls of cocci were up to twice as thick as those from rods and contained more galactose and rhamnose, while the muramic acid content of rods was higher than that of cocci. In addition, walls from cocci contained about three times more phosphorus, although the significance of this is unclear. It will be interesting to determine how wall composition and morphology vary under different substrate limitations, with phosphate limitation being of particular note (42).

The phenotypic variability of fungal chemistry has recently been reviewed by Bull and Trinci (43).

MODIFICATION OF MICROBIAL METABOLISM IN RESPONSE TO FLUCTUATING ENVIRONMENTS

A number of examples of the study of metabolic regulation using continuous cultures are covered elsewhere in this book, in the articles on microbial growth yields, energy metabolism, and the uptake of nitrogen, and in other recent publications (43); consequently only certain aspects of the subject are mentioned here. As in the applications to microbial chemistry, the chemostat provides defined environments with known substrate limitations and enables changes in enzyme and metabolite levels to be followed kinetically during transition in a way not possible in batch culture. A few examples are considered below.

Two pathways of glycerol assimilation are known to occur in members of the Enterobacteriaceae. In *Escherichia coli* the following pathway is found:

$$
\begin{array}{ccc}
\text{ATP} & & \text{flavoprotein} \\
\searrow & & \searrow \\
\text{Glycerol} \longrightarrow \text{glycerol 3-phosphate} \longrightarrow \text{dihydroxyacetone phosphate} \\
\text{glycerol} & \text{glycerol-3-phosphate} \\
\text{kinase} & \text{dehydrogenase}
\end{array}
$$

with mutants lacking glycerolkinase being unable to grow on glycerol and anaerobic growth of the wild type being possible only in the presence of an inorganic electron acceptor.

In *Klebsiella* spp. an alternative pathway is known:

$$
\begin{array}{ccc}
\text{NAD}^+ & & \text{ATP} \\
\searrow & & \searrow \\
\text{Glycerol} \longrightarrow \text{dihydroxyacetone} \longrightarrow \text{dihydroxyacetone phosphate} \\
\text{glycerol} & \text{dihydroxyacetone} \\
\text{dehydrogenase} & \text{kinase}
\end{array}
$$

The overall reaction is the same but the phosphorylation and oxidation steps are reversed. The *Klebsiella* system employs NAD^+ as immediate electron acceptor, rather than a flavoprotein, and can be used anaerobically in the absence of an alternative terminal electron acceptor. The results of Lin et al. (44) indicated that some strains of *K. aerogenes* possessed both pathways, with the glycerol kinase system operating aerobically and the glycerol dehydrogenase system operating anaerobically. This proved to be an over-simplification because the route employed appears to be dependent on the glycerol concentration. Chemostat grown glycerol- (or glucose-) limited cultures of *K. aerogenes* contained high levels of glycerol kinase but little glycerol dehydrogenase when grown aerobically, while when grown anaerobically the cultures contained little kinase but high levels of the dehydrogenase (45), results supporting the findings of Lin et al. (44). However, when glycerol excess (SO_4^{2-}-or ammonia-limited) cultures were studied there was little kinase activity when grown aerobically but de-hydrogenase activity was present. While the dehydrogenase activity was lower than with anaerobic glycerol-limited cultures, presumably there was sufficient enzyme present for growth since it would be saturated with substrate. Therefore, with cultures grown aerobically the route of glycerol assimilation depends upon the concentration of glycerol available and this may be explained in terms of the affinities of glycerol kinase (K_m 1-2 $\times$ 10^{-6} M) and glycerol dehydrogenase (K_m 2-4 $\times$ 10^{-2} M) for glycerol. Washed suspensions of glycerol-limited organisms had a K_m for glycerol of $< 4 \times$ 10^{-4} M, while with SO_4^{2-}- or ammonia-limitation the K_m was $> 1 \times 10^{-2}$ M. The reasons for this switch in assimilatory pathways are not clear, but obviously the change to a low affinity system at high substrate concentration effectively regulates the rate of catabolism of glycerol. Neijssel et al. (45) speculate that this prevents the accumulation of "traumatic substrates" (46) that might otherwise repress adenyl cyclase, with a resultant disruption of metabolic control, with the production of a catabolite-repressed culture.

Another dual pathway system, this time for glucose catabolism, has been elucidated in chemostat cultures of *P. aeruginosa* (47, 48). One pathway begins with a high affinity system of glucose uptake with a K_m of 8 $\times$ 10^{-6} M. Intracellularly glucose is converted via glucose 6-phosphate to gluconate 6-phosphate, which in turn is metabolized by either the Entner-Doudoroff pathway or the hexose monophosphate pathway:

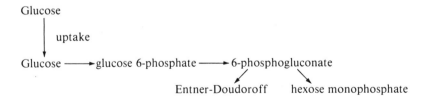

The alternative system involves the oxidation of glucose to gluconic and 2-oxogluconic acids by periplasmic enzymes, i.e., extracellular oxidation mediated by periplasmic glucose and gluconate dehydrogenases. Gluconic and 2-oxogluconic acids are transported into the cell on specific uptake systems and are converted intracellularly to gluconate 6-phosphate:

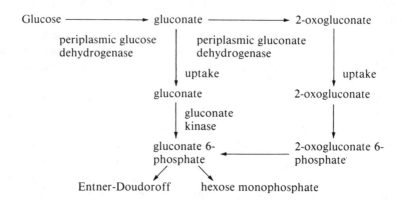

The K_m of the periplasmic glucose dehydrogenase is about 1 mM and in the presence of excess glucose a significant portion is metabolized by the extracellular pathway, with gluconate and 2-oxogluconate accumulating in the culture medium since their rate of production exceeds the rate of uptake. There are, therefore, high and low affinity pathways of glucose catabolism apparently controlled by the environmental concentration of glucose. During nitrogen-limited growth (excess glucose) the activity of the high affinity glucose uptake system was low while the activities of the uptake systems for gluconate and 2-oxogluconate, the periplasmic dehydrogenases, gluconate kinase, and the 2-oxogluconate-metabolizing enzymes were high, i.e., the low affinity system was fully expressed and the high affinity system partly repressed. During glucose-limited growth, however, very little gluconate or 2-oxogluconate was detected, and the glucose uptake system was present with a high activity while the enzymes and uptake systems of the low affinity system were repressed. Perhaps this dual pathway is needed to prevent traumatic substrates from accumulating intracellularly during growth with an excess of glucose; an alternative explanation proposed by Dawes et al. (47) is that the periplasmic production of gluconate and 2-oxogluconate effectively sequesters the carbon source, making it less available to other organisms.

Dual pathways are not confined to the dissimilation of sources of carbon and energy, and Tempest and Neijssel (10) have listed a number of nutrients whose metabolism is mediated by alternative systems, depending upon the prevailing environmental conditions. One particular system, the discovery of which was completely dependent on the ability to exploit substrate-limited growth in chemostats, was the high affinity ammonia assimilation mechanism first found in $K.$ $aerogenes$ (49). In studies of the biosynthesis of glutamate, the first cultures studied were K$^+$-limited, with ammonia as nitrogen source. These organisms contained a NADP-dependent glutamate dehydrogenase in adequate amounts to account for glutamate synthesis, but it was obvious from the K_m for ammonia of this enzyme that it would function poorly under conditions of ammonia limitation. It was anticipated that ammonia-limited growth would lead either to the increased synthesis of glutamate dehydrogenase (as occurs in many yeasts) or to the synthesis of a system with a higher affinity for ammonia. Ammonia-limited cultures contained no detectable glutamate dehydrogenase. The high affinity system was demonstrated by pulsing an ammonia-limited culture with ammonia and following changes in the amino acid pool over time (50). Such an experiment showed an initial rise in the glutamine pool content followed by an increase in the glutamate concentration. From this information the glutamine synthetase levels of ammonia- and K^+-limited cultures were compared and the biosynthetic activity of this enzyme was found to be high in ammonia limitation and not detectable with ammonia in excess. Since glutamine synthetase alone could not account for net ammonia assimilation, extracts of ammonia-limited organisms were incubated with glutamine and possible nitrogen acceptors and the products analyzed. Using this method it was found that glutamine and 2-oxoglutarate, when incubated with NADPH, yielded 2 molecules of glutamate, a reaction catalyzed by the previously unknown enzyme glutamate synthase. Glutamate synthase synthesis in $K.$ *aerogenes* occurred in ammonia-limited but not in ammonia-excess cultures.

The overall scheme is as follows:

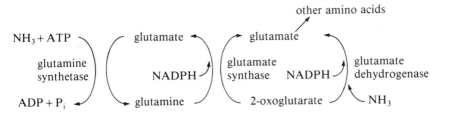

with the high affinity glutamine synthetase/glutamate synthase route owing its role to the low K_m of glutamine synthetase for ammonia. This route has been shown to operate in all bacteria studied that are able to assimilate ammonia, in some yeasts, in algae, and in higher plants (51, 52, 53).

Sims and Folkes used turbidostats in a detailed study of nitrogen assimilation in the yeast $Ca.$ *utilis*, making use of $^{15}NH_3$ as a tracer. Continuous culture was particularly useful here because it allowed a full kinetic analysis to be made (54, 55, 56, 57). For example, steady-state conditions were disturbed by withholding the nitrogen supply for 10 to 20 min and the incorporation of $^{15}NH_3$ into amino acids followed with time. It was found that 75% of the ammonia assimilated appeared first as glutamate [which is in keeping with the role of NADP-linked glutamate dehydrogenase in this organism (57a)], much of the remainder being assimilated via glutamine. The rate of glutamate synthesis was found to be inversely related to the size of the intracellular amino acid pool and in vitro experiments confirmed that amino acids acted as allosteric inhibitors of glutamate dehydrogenase. The rate of synthesis of glutamine correlated with the intracellular glutamine concentration and it was found that the activity of glutamine synthetase was modulated according to the biosynthetic requirements of the culture. A sudden increase in the ammonia supply or a decrease in the glucose supply resulted in a rapid decrease in enzyme activity that was faster than could be accounted for by repression of enzyme synthesis followed by dilution-out during growth. In fact, the enzyme was irreversibly inactivated and the control of activity is by enzyme inactivation and resynthesis. A similar system appears to be present in the fission yeast *Schizosaccharomyces pombe* (58).

Another area in which continuous culture methods have been exploited is the study of sugar uptake in bacteria, and both chemostats and turbidostats have been employed. Glucose and fructose are transported into *E. coli* via phosphoenolpyruvate-dependent phosphotransferase (PT) systems and since these sugars are used sequentially in batch cultures without an intermediate lag phase, it seemed likely that the synthesis and/or activity of the fructose PT system was modulated by the presence or absence of glucose and vice versa. Clark and Holms (59) reported that turbidostat cultures grown on fructose had a μ of 0.73 h^{-1} and showed high fructose PT and low glucose PT activities. The addition of 0.1 mM glucose to the medium increased the μ to 0.83 h^{-1}, and all the glucose was fully utilized and there was little change in either PT activity. The addition of 0.4 mM glucose increased the μ to 0.91 h^{-1}. Again, all the glucose was fully utilized and while the fructose PT activity remained high the glucose PT activity was increased to an equivalent level—thus the two sugars were used simultaneously and full PT activities expressed. The addition of higher concentrations of glucose did not further increase the μ but gradually decreased the level of fructose PT and the amount of fructose utilized, until with 3.3 mM glucose addition no fructose was utilized and fructose PT level activity was some fivefold less than during growth with fructose alone. Time course studies showed that fructose PT was apparently diluted-out by growth after addition of glucose to a fructose culture with a concomitant rise in glucose PT. When glucose addition ceased the fructose PT activity increased very rapidly with some overshoot before settling to a steady-state value.

Herbert and Kornberg (60) reported that the rate of glucose transport in *E. coli* was the growth-limiting step in metabolism. For instance, the rate of glucose utilization (in these experiments, equivalent to the glucose PT activity) at D 0.4 h^{-1} was more than twice the activity at D 0.1 h^{-1}, i.e., the glucose PT activity was related to the μ of a glucose-limited culture. Herbert and Kornberg calculated the required rates of glucose utilization as a function of μ and compared these with observed rates of utilization and found a marked correlation at D > 0.2 h^{-1}. Thus in the range of D 0.2-0.5 h^{-1} the rate of growth was closely related to, if not limited by, the glucose PT activity. At D < 0.2 h^{-1} some other factor (presumably glucose concentration) was involved, because glucose PT activity was greater than the specific rate of glucose utilization. These results give a useful insight into the way in which metabolism is modulated to suit the prevailing environmental conditions.

The uptake of glucose via a PT system raises a question about the function of hexokinase in organisms growing on glucose. Carter and Dean (61) carried out a comparative study on the control of glucose PT and hexokinase in a number of strains of *K. aerogenes* and found that hexokinase synthesis, in some strains, was repressed during growth on glucose but not on lactose (in chemostats at D 0.5 h^{-1}). Conversely, the glucose PT activity was repressed during growth on lactose but not on glucose. Thus hexokinase functions in glucose catabolism when the substrate is produced intracellularly but not when it is provided exogenously. In glucose-grown organisms the glucose PT activity increased linearly with μ in a number of organisms and was close to the calculated rates of glucose uptake. This confirms and extends the results of Herbert and Kornberg (60) for cultures of *E. coli* discussed above.

A correlation between μ and uptake rate of the amino acids (amino isobutyrate, alanine, glutamate, and arginine) was reported to occur in chemostat-grown *Streptomyces hydrogenans* (62). The actual number of sites for transport appears to increase, since the rate of both uptake and efflux of [^{14}C] cycloleucine increased with increasing μ.

There are a large number of reports of enzyme synthesis varying with substrate limitation, μ, etc.; some have already been discussed and others are included in the paper of Dean (63). Two further examples will be outlined here because they illustrate different aspects of the use of continuous culture. Herbert and Phipps (64) reported some particularly elegant experiments on the control of catalase synthesis in *Bacillus megaterium*. Catalase is usually considered to be a constitutive enzyme and studies on induction with hydrogen peroxide were hampered by the fact that in batch culture large amounts of the substrate sterilize the inoculum, while small amounts are rapidly decomposed by background levels of enzyme (produced in the absence of peroxide). Herbert and Phipps (64) used a mannitol-limited culture with hydrogen peroxide added to the medium. The residual culture levels were very low ($< 10^{-4}$M) even though up to 0.3 M concentration was used in the medium feed. Catalase was reported to be induced up to 35- to 60-fold and as such constituted about 1% cell protein. This is an example of the use of a metabolizable yet toxic substrate that can be studied effectively in continuous culture because its residual level in the culture was low.

The second example of modulation of enzyme synthesis concerns the control of methanol oxidase during the oxidation of methanol by the yeast *Hansenula polymorpha*. Methanol catabolism is initiated by methanol oxidase, which catalyzes the oxygen-dependent formation of formaldehyde and hydrogen peroxide. The latter product is contained within peroxisomes that are produced during growth on methanol. In *H. polymorpha* the peroxisomes are cubic in form, with a completely crystalline or striated substructure (65). Organisms harvested from methanol-limited cultures at D 0.03 h^{-1} had a maximum rate of oxygen uptake (Qo_2^{max}) some 20 times the culture Qo_2, while at high D (0.16 h^{-1}) the Qo_2^{max} was 1.6 Qo_2 (66). The basis for this vast oxidative 'overcapacity' at low growth rates was the intracellular content of methanol oxidase, which accounted for about 20% of cell protein at D 0.03 h^{-1}. These results were based on the poor affinity of methanol for its substrate—the K_m for methanol in air-saturated reaction mixtures is 1.3 mM and the K_m for oxygen (even with 100 mM methanol) is 0.4 mM. The high content of enzyme at low μ was required presumably because of the low concentration of methanol present. Methanol oxidase appeared to be the limiting enzyme for methanol oxidation because the levels of the associated catalase, formaldehyde dehydrogenase, and formic dehydrogenase were much lower and not influenced by μ. This high methanol oxidase content is reminiscent of the "hyper" levels of "substrate-capturing enzymes" discussed under "Continuous Culture and Microbial Evolution," below.

Another example of metabolic control concerns the role of oxygen as a modulator of metabolism. Harrison (67, 68) has recently reviewed this subject in detail. Oxygen tension is particularly difficult to control in batch cultures because oxygen demand increases exponentially with growth and, while MacLennan and Pirt (69) have described equipment for automatic control, most workers have applied this or similar systems to chemostat cultures. Over a wide range of oxygen tensions the oxygen uptake rate remains constant over time in chemostats, although oscillations have been reported for *K. aerogenes*. In general, dissolved oxygen tensions above 5–15 mm Hg and below hyperbaric levels have little effect on metabolism. Between 5–15 and 0 mm Hg, metabolism changes toward fermentation or the use of alternative inorganic electron acceptors. For example, in *K. aerogenes* grown in the presence of excess glucose, the main excreted metabolic product at > 15 mm Hg was pyruvate, with smaller amounts of βdiol and ethanol. At 1 to 15 mm Hg (when oscillations occur) more diol was produced at the expense of pyruvate, while between <1 and 0 all pyruvate was converted to βdiol, ethanol, and volatile acids, products reflecting the use of organic electron acceptors and substrate-level phosphorylation.

Dawes and co-workers (70, 71, 72) have made a detailed study of the biosynthesis of poly-β-hydroxybutyrate (PHB) as a function of oxygen tension. During batch growth of *Azotobacter beijerinckii* on 0.5% (wt/vol) glucose, PHB formation commenced at the end of the exponential phase and reached 35% dry weight, but when the glucose was exhausted synthesis ceased and the PHB level, now being used for cellular synthesis and as a source of energy for nitrogen fixation, fell. With 2% (wt/vol) glucose, PHB synthesis again occurred at the end of the exponential phase and accounted for over 70% dry weight. Oxygen monitoring of these cultures showed that PHB synthesis began at about the time when the dissolved oxygen tension reached zero. Thus PHB synthesis occurred at the end of exponential growth at near zero oxygen tension, and was dependent to some extent on carbon availability and on whether or not the organisms were fixing nitrogen. It was then uncertain whether the cultures were oxygen- or nitrogen-limited. This is of course the normal dilemma of interpreting results from batch cultures.

In carbon-limited nitrogen-fixing chemostat cultures, PHB synthesis occurred at low growth rates, but the cellular content never exceeded 3% dry weight. Under nitrogen-limitation the PHB content never exceeded 1.5% dry weight. Under oxygen limitation, however, the PHB content was highest at $D < 0.12 \text{ h}^{-1}$ but still accounted for 20% dry weight at D 0.25 h^{-1}. Thus oxygen availability was the most important factor influencing PHB synthesis. PHB has two functions—it serves as a store of reusable carbon and as an electron sink in this obligate aerobe (70). The biosynthesis of PHB represents a flow of acetyl-CoA away from oxidation in the tricarboxylic acid (TCA) cycle to polymer synthesis:

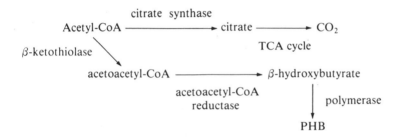

In nitrogen-fixing cultures the NADH/NAD ratio increased following oxygen limitation, but rapidly readjusted with the onset of synthesis of PHB as electron sink (71). On relaxation of oxygen limitation the activity of some oxidative enzymes (NADH oxidase, isocitrate dehydrogenase, and 2-oxoglutarate dehydrogenase) increased while those of β-ketothiolase and acetoacetyl-CoA reductase decreased.

Similar overall results were obtained with ammonia-grown cultures (72), with the PHB content increasing to about 70% dry weight under oxygen limitation. On transition from oxygen to ammonia limitation, however, (with an O_2/argon gas phase) the PHB content fell to 5–10% dry weight with a gas phase containing 1.25% O_2 (vol/vol with argon); a further increase of oxygen in the gas phase caused the PHB content to increase again to 20 to 40% dry weight. The difference from nitrogen-fixing cultures may be due to the requirements for an electron sink in the absence of reductant utilization during nitrogen fixation. At still higher oxygen levels the polymer level fell to about 3% dry weight, due presumably to uncoupled electron transport.

The final example in this section is a further instance where specific substrate limitations were exploited in order to determine the function of that substrate in metabolism. Light and Garland (73) and Light (74) have summarized experiments on the role of iron and sulfur in the mitochondrial function of *Ca. utilis* in which they showed that site I energy conservation was absent from iron- and sulfate-limited organisms. This correlated with a loss of rotenone and piericidin sensitivity and with the assumption that an iron-sulfur protein was lacking from these cultures.

Aiking and Tempest (75) have described the role of K^+ in mitochondrial function in *Ca. utilis*. This study began with the observation that the potassium content of K^+-limited organisms was some 10 times less than those grown with a K^+ excess. In K^+-limited cultures the cellular potassium level increased with increasing D and correlated both with the growth yield and the oxygen uptake rate of the culture. These results suggested a role for potassium in oxidative metabolism and further experiments were carried out with isolated mitochondria. Mitochondria were prepared from cultures harvested directly from a 6-liter capacity chemostat, thus assuring a minimum delay between harvesting and assay. Using this method intact mitochondria with good respiratory control ratios were produced after treatment of cultures with Helicase and without prior thiol treatment. A comparison of mitochondria from glucose- and K^+-limited organisms grown at D 0.1 h^{-1} showed the K^+-limited system to have a lower P:O ratio, consistent with the loss of one site (either site II or III) of energy conservation. At D of 0.2 h^{-1} and higher, however, K^+-limited organisms had 3 sites of energy conservation. Glucose-, ammonia-, phosphate,- and sulfate-limited organisms contained 3 energy conservation sites at all D values studied between 0.1 and 0.45 h^{-1}. While some variations in cytochrome contents were observed, the K^+ effect remains unexplained (76).

RESPONSES OF MICROORGANISMS TO MIXED SUBSTRATES

In the operation of chemostats it is usual for nutrient-limited growth to be established by a single constituent of the culture medium. Energy-limited growth of heterotrophs, for example, is frequently arranged by feeding a single carbohydrate or carboxylic acid to the culture in the complete absence of alternative energy-yielding substrates. However, there are compelling reasons for the study of mixed substrate fermentations and for extending such analyses to continuous culture systems. Our use of the term "mixed substrate" implies a multiplicity of sources of a given nutrient and chemically these sources may compromise homologous or nonhomologous compounds. Culture media for commercial fermentations often contain complex mixtures of energy, carbon, and nitrogen sources; indeed, empirical adoption of mixed carbon sources for antibiotic and amino acid fermentations is commonly undertaken to improve yields. Similarly, in natural environments the growth of microorganisms takes place in the presence of a diversity of mixed substrates and it is unfortunate, therefore, that most laboratory-fixed microbial ecologists fail to take cognizance of this fact. Moreover, while continuous culture methods have been encouraged as model systems for investigating ecological principles, remarkably few studies have focused on mixed substrate situations.

Much of the literature on mixed substrate utilization has been assessed recently in an excellent review by Harder and Dijkhuizen (77). These authors comment that the majority of reports in this field discuss carbon and energy substrate interactions. Clearly there is considerable need for comparable studies of other mixed substrates, particularly nitrogen and phosphorus. It is not our intention to retrace the ground covered by Harder and Dijkhuizen; rather, we will concentrate on a selection of data that are concerned with regulatory mechanisms and have relevance to ecophysiology and fermentation.

In batch cultures the pattern of mixed carbon source utilization varies with the combination of substrates and with the organism; a unitary response to given substrate combinations does not occur. Substrate utilization may be sequential, during the course of which metabolism of a secondary source is preceded by a lag in growth [diauxie (78)]. The diauxic growth of *E. coli* on glucose plus lactose has been analyzed exhaustively. During growth on glucose the utilization of the secondary sugar is believed to be prevented by repression of *lac* operon protein synthesis and catabolite inhibition of the extant lactose-utilizing system. Alternatively, substrate utilization may occur sequentially in the absence of a growth lag (79, 80) or simultaneously. In either case the specific growth rate may remain at its initial value or be reduced as secondary or tertiary carbon sources are utilized.

Mixed carbon utilization by chemostat populations is influenced primarily by the nature of the growth-limiting substrate and by the dilution rate and, as in batch cultures, preferential or simultaneous utilization patterns have been observed (77, 80). Some of the most revealing studies of this type have been reported by Mateles and his colleagues (81, 82, 83). The utilization of sugar mixtures by bacteria under carbon limitation was subject to distinct on-off control at high dilution rates, whereas simultaneous utilization resulted at low dilution rates. The classic glucose-lactose-*E. coli* system was among those examined by Mateles in carbon-limited continuous cultures (83). Typical on-off control of lactose utilization occurred in wild-type organisms at a dilution rate rquivalent to 72% of μ_{max}, but eventually β-galactosidase mutants that were constitutive for the enzyme arose in the population. Simultaneous utilization of both sugars by these mutants was evident over the whole growth rate range, a result clearly indicating that catabolic repression of β-galactosidase alone is insufficient to prevent lactose utilization. Silver and Mateles (83) concluded that the inhibition of lactose utilization by wild-type *E. coli* at high dilution rates was caused by the competitive inhibition of lactose transport into the bacterium by glucose and the consequent inability to induce β-galactosidase. In contrast, constitutive mutants, even under the most severely repressive conditions, could synthesize adequate levels of β-galactosidase to enable total assimilation of lactose. Experiments of this type, in which unequivocal effects of growth rate can be determined, amply demonstrate the utility of chemostasis in the elucidation of complementary transport-level and catabolite repression controls.

Edwards (84) examined the response of *K. aerogenes* to mixed feeds of glucose and maltose in carbon-limited chemostats. When glucose cultures were switched to glucose plus maltose, the response was related to the dilution rate. At low dilution rates maltose was utilized without a lag, at intermediate dilution rates a lag in maltose utilization was observed (its duration being related linearly to the actual value of D), and maltose remained unmetabolized at dilution rates approaching the critical value.

On the basis of the results of Silver and Mateles (83) it is tempting to conclude that diauxic growth in batch culture is analogous to preferential utilization of sugars in chemostat cultures. However, even the relatively few studies that have been reported invalidate this generalization and many variations are found in utilization patterns. *E. coli*, for example, grows diauxically on glucose-xylose mixtures, whereas in a mixed substrate carbon-limited chemostat the sugars are used simultaneously (79). Tsuchiya and his co-workers (79) attempted to gain further understanding of this system by analyzing the transient behavior of chemostat populations challenged with dual substrates. For example, switching feeds from glucose to xylose at a constant dilution rate caused a transient reduction in population density and a concomitant accumulation of xylose in the chemostat; subsequently the culture progressed to a complete utilization of the pentose. Complementary switches from xylose to glucose feeds did not perturb the population in any detectable way. However, stepwise increases in dilution rate of glucose-xylose continuous cultures created similar, transient changes in population density and xylose concentration, while the glucose concentration remained at a very low value. The magnitude of the dilution rate change had an expected graded effect on the size and duration of the transient state. Unfortunately enzyme data are not available for this system. However, it is reasonable to postulate that the inducible xylose enzymes are synthesized only in minimum amounts commensurate with total utilization of xylose; the glucose enzymes, in contrast, are constitutive and are likely to be synthesized somewhat in excess of essential requirements. Such a situation would enable organisms to respond immediately to an increase in glucose concentration.

Different responses to mixed substrates can be seen in members of the same genus. Thus, *Saccharomyces fragilis* growing on glucose-fructose mixtures shows a significant on-off effect on fructose uptake at a dilution rate equivalent to approximately 75% μ_{max} (80), while fructose uptake was impaired at all dilution rates in similar chemostat cultures of *Sa. cerevisiae* (81). It should be noted that the utilization of nonhomologous mixed carbon substrates, such as sugar–organic acid comingations, also has been analyzed in chemostats. E. A. Dawes and his collaborators have made a comprehensive study of glucose and citrate utilization by *P. aeruginosa* (85, 47) following the interesting initial observation that the organic acid was the preferred substrate in batch cultures (86). *P. aeruginosa* was grown under ammonia-limited conditions in a chemostat with citrate (75 mM) as the source of carbon and, with glucose, fed to the culture at varying concentrations. Under these conditions a threshold steady-state concentration of 4.2 mM glucose was required to promote glucose catabolism, a result that argued that induction of a glucose uptake system was a necessary prelude to glucose assimilation.

Further work substantiated the view that the preferential utilization of citrate by this organism was defined by the dual regulation of glucose uptake and glucose metabolism. Glucose catabolism in *P. aeruginosa* proceeds via the Entner-Doudoroff and pentose phosphate pathways and recently Dawes (47) has found that addition of gluconate to chemostat populations growing on citrate plus glucose represses glucose uptake and induces gluconate and 2-oxogluconate transport systems. Thus, in the presence of excess glucose, *P. aeruginosa* metabolizes a proportion of it extracellularly to gluconate and 2-oxogluconate, which in turn leads to repression of sugar transport and growth occurs at the expense of the organic acids. Interestingly, under conditions of glucose limitation the extracellular metabolism is switched off via enzyme and transport repression while a high affinity glucose transport system is induced (see section above).

Some findings from Higgins' laboratory also are noteworthy in this context of carboxylic acid utilization. Sariaslani et al. (87) have been concerned with the control of isocitrate lyase synthesis and activity in *Nocardia salmonicolor* and have described a number of batch and chemostat experiments involving mixed carbon substrates. The effects of adding equimolar concentrations of acetate to batch populations growing on glucose were very unusual and quite contrary to those previously noted with *E. coli*: both carbon sources were utilized simultaneously, while the specific growth rate, glucose utilization rate, isocitrate lyase activity, and dehydrogenase activity all increased dramatically. Point addition of very low concentrations of acetate to glucose-limited chemostat cultures of *N. salmonicolor* caused a rapid and large synthesis of isocitrate lyase, a result that the authors reasonably ascribe to enzyme induction. Additions even of large amounts of acetate to nitrogen-limited populations did not elicit synthesis of the lyase. It appears that fumarate is the likely catabolite repressor of isocitrate lyase in this organism, not phosphoenolpyruvate, as in *E. coli*. The *Nocardia* grows more rapidly on acetate (μ 0.14 h^{-1}) than on glucose (μ 0.06 h^{-1}), results that are fully consistent with the observed pattern of lyase regulation. As Higgins and his colleagues conclude, the different manner in which key anaplerotic enzymes seem to be regulated in soil-inhabiting *Nocardia* and enterobacteria may be expected, and is certainly consistent with a consideration of potential carbon substrates available in their respective habitats.

An understanding of the utilization of mixed substrates has special importance in the design and operation of industrial fermentations and waste treatment processes, in which complex mixtures of substrates are frequently found. The first extensive commercial adoption of continuous fermentation was made by the brewing industry (88) and beer is the product of a classic mixed-sugar fermentation. Wort, the product of the malting process, contains sucrose, maltose, and maltodextrins. It has been known for many years that in batch fermentations these sugars are utilized sequentially by yeast: first sucrose is hydrolyzed by invertase and the glucose metabolized, while subsequently fructose, maltose, and maltotriose are consumed. Initially the synthesis of α-glucosidase (maltase) is repressed by glucose (89), and maltose utilization is further depressed by the inhibition of α-glucosidase and maltose transport by glucose (90, 91). In homogeneous continuous cultures on wort, the capacity of yeast to utilize maltose is much less than it is in batch cultures, and α-glucosidase probably is rate-limiting (92).

It can be predicted that undamped oscillations may occur in the concentration of inducible enzymes regulated by feedback controls in continuous cultures. Such oscillations—manifested as variations in maltose concentration—have been observed in wort fermentations. Portno (93) experimented with a single-stage chemostat operated with partial feedback and reported increasing maltose concentrations with successive oscillations. Yeast flocculation is affected both by maltose (negatively) and ethanol (positively); thus, as the maltose concentration increased the majority of organisms eventually deflocculated and washed out of the fermenter. Subsequent behavior was dependent on the dilution rate: 1) when sufficient yeast was retained to reduce maltose to a concentration permitting reflocculation the population again increased; and 2) when only a few organisms were retained in the fermenter population recovery did not occur. The definition of substrate utilization patterns obviously has considerable bearing on fermenter design and as far as most fermentation is concerned the use of a heterogeneous continuous culture system[1] appears to be advantageous. An experimental fermenter of this type was successfully operated by Portno (94).

The production of single cell protein from materials such as sulfite liquor and whey similarly involves carbon utilization from mixed substrates. Thus, *Ca. utilis* assimilates only glucose from sulfite liquor when grown at high dilution rates in a chemostat, but xylose and acetate are also utilized as the dilution rate is lowered (95). Recently Smith and Bull (80) have reported a feasibility study of single cell protein production from waste coconut water, a material that contains glucose, fructose, sucrose, and sorbitol in total concentration up to 8% (wt/vol). Chemostat culture of

---

[1] A system characterized by *gradients* of organisms, substrates, and products in which temporal changes that occur in closed cultures may be mimicked.

Sa. fragilis on a mixture of these four carbohydrates revealed that sorbitol was utilized only at very low dilution rates, a result ascribed, at least in part, to feedback inhibition of the sorbitol-utilizing system. Sucrose was completely utilized only at dilution rates of less than 0.2 h^{-1} and invertase activity was zero until the total, steady-state, reducing sugar concentration had been diminished to 0.10 g/liter. Invertase exhibited a sharp peak in activity at a dilution rate of 0.08 h^{-1}, reflecting a balance point between induction (by sucrose) and repression (by hexoses). Consequently, it becomes desirable to use a multistage chemostat or a heterogeneous continuous fermenter in order to maximize biomass productivity from fermentation feedstocks of this nature.

It has been suggested (96) that to convert diauxic fermentations to continuous processes it is necessary only to "adapt" the population once at a low dilution rate. However, as we have seen, this approach to simultaneous mixed substrate utilization does not have general applicability. Instead, two alternative fermentation strategies may be considered; the first concerns process designs of the sort briefly referred to above and the second involves the use of mixed cultures. Study of the latter has obvious relevance to waste treatment processes and a number of model mixed-substrate/mixed-population continuous cultures have been analyzed. Some early work on activated sludge populations growing on glucose-sorbitol mixtures revealed that sorbitol utilization was inhibited by glucose over a wide range of dilution rates (97), a pattern that was very similar to that found in monocultures (98). Likewise, similar patterns of lactose and glucose utilization have been reported for mono- and mixed-chemostat cultures of bacteria (81). Generally, therefore, mixed substrate utilization by mixed populations resembles that of monocultures, with only small quantitative differences being evident. However, significant differences occur when, for example, one or another of the substrates are growth inhibitory, as in the case of hydrocarbon oxidation and pesticide degradation products (see next section). Mixed substrate utilization by chemostat populations of mixotrophs and photoheterotrophs has been discussed recently by Harder and Dijkhuizen (77).

ANALYSIS OF INTERACTING MICROBIAL COMMUNITIES BY CONTINUOUS CULTURE TECHNIQUES

The argument that continuous cultures provide the most appropriate laboratory models of microbial ecosystems has been tacitly accepted by many microbiologists. The common "open" character of both systems is the main ground for promulgating this view: the continuous culture, while it cannot reproduce microbial ecosystems, enables many of their features to be mimicked and uniquely allows nutrient-limited and very slowly growing populations (99) to be analyzed. One almost universal attribute of natural ecosystems is the occurrence of mixed populations, or communities, of microorganisms. However, it is only very recently that microbiologists have sought to break with the Koch tradition of pure (i.e., mono) culture studies and, in keeping with this change of approach, we contend that the analysis of microbial community structure and behavior is most effectively made in continuous cultures.

Given the widespread existence of microbial communities, the advantages that they have over monospecies populations need to be defined. Little experimental information is available with which to support our thinking but the following points are pertinent. Stability is likely to be enhanced in a multispecies community and, although the results of mathematical analyses run contrary (100, 101, 102), experimental support for this hypothesis is accumulating (some examples are considered below). A probable solution to this apparent paradox has been suggested by May (101): ecosystems have evolved to their present states over enormous time spans and presumably stable rather than unstable communities have been selected. We hope to reveal certain community characteristics that result in their stability, but one proposition might be considered here. Clearly microorganisms have evolved such that genetic information resides in a heterogeneous multispecies gene pool rather than being concentrated within single species, a condition that may benefit individual species by sparing energy and substrates for biomass synthesis and higher growth rates. (Preliminary support for such a view can be found in the significantly increased specific growth rate of an F^- $E.$ $coli$ population following loss of a plasmid; see "Continuous Culture and Microbial Evolution"). Of course, the corollary to this situation is the cooperative action of a community of microorganisms in the degradation of chemically complex substrates, such as pesticides, petroleum, and lignin. However, following the rapid developments in microbial genetics, researchers are currently attempting to circumvent natural evolution and to combine desirable genetic information from several microorganisms into the genome of a single species, e.g., the construction of a hydrocarbonoclastic "superbug" (103).

The description of two-species microbial interactions has been extensive and quite successful in mechanistic terms and many such systems have been amenable to sophisticated mathematical analysis. Moreover, the chemostat has been used to considerable advantage in these studies. However, it is not our intention to dwell on these simple, and frequently synthetic (i.e., gnotobiotic), systems, since the interested reader has recourse to several comprehensive reviews for information (104, 105–111). Rather, we discuss the behavior of more complex naturally existing communities under the following headings: 1) ecosystem modeling; 2) pollution effects on microorganisms; 3) biodegradation; and 4) microbial technology.

Ecosystem Modeling

Closed or batch culture studies are unsatisfactory for revealing ecological principles—they invariably impose nutrient concentrations that are much

in excess of those found in the environment, while batch enrichment iso-
lations result in selection of organisms very largely on the basis of maximum specific growth rates. Unfortunately, species having high nutrient
affinities and/or low growth rates may be ones with decisive roles in the
ecosystem under investigation—these populations are unlikely to be isolated
via traditional batch enrichments. Adoption of the latter also precludes to
a great extent the isolation of communities because of the disrupted spatial
arrangement of microorganisms following plating operations. Consequently,
the use of continuous-flow methods for community isolation becomes obligatory. Such methods enable an infinite variety of selection conditions to
be imposed, although, of course, the microbiologist concerned with ecosystem modeling will relate these conditions to those that appear to have a
determinant role in the environment.

The case for continuous-flow enrichment culture has been compell-
ingly stated by Veldkamp (112). Enrichment isolation is a crucial phase
of any investigation and lack of attention probably explains the failure of
many programs intended to explore, for example, the biodegradation of
xenobiotic compounds. Thus, success in isolating Amitrole-degrading (3-
amino-1,2,4-triazole) (113) and Lontrel-degrading (3,6-dichloropicolinic
acid) (D. Lovatt, A. T. Bull, and J. H. Slater, unpublished results) communities was consequent upon using these herbicides as growth-limiting
sources of nitrogen rather than of carbon.

The single most important interaction between microbial populations
is competition, and the chemostat has been deployed widely for its study
(104, 111). In this context continuous culture experiments have provided
substantial understanding of adaptation and evolution within microbial
communities, and some genetic and biochemical aspects of evolutionary
processes are discussed later (see "Continuous Culture and Microbial
Evolution"). The classic analysis of competition was made 50 years ago by
Volterra, who concluded that competition in a common limited environment leads to extinction of all species but one—the so-called competitive
exclusion principle. Experience shows, however, that free competition between different microorganisms can be removed and stable mixed populations established in continuous culture. Competition may be eliminated
as a consequence of film development (see "Heterogenous Continuous Culture Systems," below) and via population interactions other than those of
a negative type. These noncompetitive interactions frequently involve a
hierarchy of nutrient-consumer relationships referred to as food chains or
food webs, depending on their complexity (111).

When grown in continuous culture, interacting populations of this
type rarely attain steady states *sensu stricto*; instead, stable population
oscillations become established, the amplitude and periodicity of which are
system dependent (see refs. 114 and 115 for examples). On perturbing the
stable chemostat community, the values of 1 to n measured parameters: 1)
will return to their original equilibrium states; 2) will establish new equili-

brium states, not necessarily identical to the original states; or 3) will remain unstable. In the latter case, community disintegration may result in the loss of one or all of the constituent populations. The stability of microbial communities in the face of environmental perturbations will, in part, be related to the degree of integration between populations. Relatively lax associations may exist in simple food chains involving commensalism (115), for example, whereas tight or even obligatory associations appear possible when the interactions involve the mutual sustaining of energy generating processes or the alleviation of toxic conditions. Thus, *"Methanobacillus omelianskii"* (116) and *"Chloropseudomonas ethylica"* (117) each represent pairs of bacteria whose metabolisms are so tightly coupled that, until quite recently, they were accepted as being bona fide species.

Mixed microbial populations have a ubiquitous distribution in nature, being found in such widely different contexts as biogeochemical transformations, pathological conditions, digestive metabolism, and native fermented food processes. The exploitation of chemostat culture is beginning to provide useful predictive models for eutrophication, pollution effects, and biodegradation, processes that are discussed in subsequent sections. At this stage, however, the value of model chemostat experiments in revealing relationships between a disease situation (dental caries) and the physiology of the microorganisms causing the infection is assessed.

The mouth is an open system through which food and saliva flow; moreover, the submaximal growth rates of the bacteria present strongly suggest that growth is nutrient-limited in this ecosystem. Consequently, chemostat cultures should be very appropriate for modeling pathological conditions like dental caries. Several groups of researchers (118, 119) have designed "artificial mouths" containing extracted teeth and other surfaces for experimental caries research and the special composition and structural organization of these artificial plaques closely resemble naturally occurring colonies on teeth in vivo (119). A somewhat more artificial system has been advocated recently by Sudo (120). She has used hydroxyapatite-coated glass beads to model the tooth surface and found that they are selectively colonized by salivary bacteria from which human dental plaque develops. Systems of this type have the additional advantages of a large surface area for growth and ease of regular sampling during the course of long-term experiments.

Coulter and Russell (118) have used an "artificial mouth" chemostat to explore the relationship between pH and E_h during plaque development; similar results were obtained when either saliva or defined mixed populations were used as inocula. Plaque that developed in an artificial saliva medium reduced the E_h at the tooth surface to negative values within 24 h and had a high pH (7–7.5). The plaque flora contained large numbers of anaerobes. In contrast, plaque that developed in the presence of sucrose had a high E_h (approximately +300 mV) and a pH of about 5. Anaerobes were absent from such plaque and streptococci were dominant. Furthermore, the plaque produced in sucrose medium was tenacious, the tenacity being attributable to the presence of streptococci.

The question of adherence of plaque to surfaces has been studied by Ellwood with chemostat populations of the critical organism in this context, *Streptococcus mutans* (121, 122). "Stickiness" has been correlated with the production of various extracellular α-glucans, the synthesis of which is controlled by a constitutive or semiconstitutive glycosyl transferase. Maximum transferase production occurred at low dilution rates in the chemostat when the growth rate of *S. mutans* was similar to that pertaining in the mouth. However, Ellwood and his colleagues (121) caution against a too ready acceptance of this polysaccharide matrix hypothesis. They observed that even rapidly growing carbon-limited populations of *S. mutans* adhered strongly to surfaces and that sucrose was not an essential factor in adherence.

Apart from "stickiness," the ability to produce acid is the other major activity associated with the development of tooth decay. Work in several laboratories has shown that *S. mutans* ferments sugars to acetate, ethanol, and lactate at low dilution rates, whereas at high dilution rates lactate becomes the sole product (122). The overall pattern of fermentation by the mixed plaque flora is more complex; *Veillonella alcalescens* for example, which colonizes the interior of plaque deposits, derives energy from the metabolism of lactate to propionate and is unable to utilize glucose (123). The Porton group (122) analyzed fatty acid production by complete plaque communities growing under glucose limitation. Total acid production increased with dilution and the proportions of acetate, propionate, and lactate at low, intermediate, and high dilution rates were 51:25:0, 35:33:12, and 19:19:43, respectively. Finally, the advantages of working with model systems of this type are well illustrated by a study of fluoride effects on *S. mutans* (124). Acid production at low dilution rates was completely repressed by 15 ppm F^-, whereas acid production at high rates was unaffected even by concentrations of F^- as high as 100 ppm. Hunter et al. (124) were able to relate this differential susceptibility to F^- ions to the glucose uptake mechanisms of *S. mutans*. Glucose uptake at low dilution rates is via a PEP transferase system that was sensitive to 10 ppm F^-, presumably via enolase inhibition and the consequent suppression of PEP synthesis. The glucose uptake system operative at high dilution rates was fluoride resistant.

Pollution Effects on Microorganisms

Microcosms have been used increasingly to reveal how pollutants can affect microbial communities and their activities. Essentially, a microcosm should be *functionally* similar to the ecosystem being mimicked and it may be related to a subset of a natural ecosystem (e.g., lysimeter or stream section) or to an artificial laboratory system (e.g., soil column, water channel, or chemostat) (125). Obviously the fate of chemical pollutants may be determined in field tests, but there are several drawbacks to the use of the latter for extensive screening purposes (see ref. 125). Laboratory microcosms, including continuous cultures, have many advantages in this context: control of environmental conditions over very wide ranges, ease of replication, avoidance of large scale environmental contamination, rapidity of screening. However, the choice of microcosm will be determined by the type of information being sought for predictive modeling, and invariably the degree of biological complexity is the critical factor.

A discussion of multitrophic microcosms of the kind originally designed by Metcalf and his colleagues (126) is outside the scope of this chapter. However, two points of specific application are worth emphasizing. First, complex aquatic or terrestrial microcosms can be used to assess the transport and accumulation of chemical pollutants in natural ecosystems. Second, the analysis of complex microcosms may reveal significant secondary effects of such chemicals on ecosystem activities. For example, from experiments made with continuous-flow soil column microcosms, continuous low dosing of agricultural soil with the herbicide dalapon (2,2-dichloropropionic acid) was shown to have a marked effect on the microbial turnover of humus: under aerobic conditions the half-life of the total humic fraction was reduced from 430 to 150 days (127).

In this discussion our concern is with homogeneous continuous cultures, especially of mixed populations, and an additional advantage of continuous over batch cultures here is the opportunity for studying chronic effects of pollutants. Most investigations have been made with chemostat populations but, because most pollutant chemicals appear to adversely affect specific growth rate, the turbidostat might be a more appropriate model to consider. Probably the most valuable studies in this field have concerned phytoplankton communities, and of particular note are the series of papers from C. F. Wurster and his colleagues (128, 129, 130) dealing with effects of polychlorinated biphenyls (PCB) and DDT [1,1,1-trichloro-2,2-bis(p-chlorophenyl)ethane] on marine phytoplankton. In one study (129) the effects of PCB on competition in a two-membered community (*Dunaliella tertiolecta* and *Thalassiosira pseudonana*) were examined. PCB concentrations of 0.1 μg/liter (of the order observed in natural waters) had little effect on mono- or mixed-batch cultures of these species. However, in nitrate-limited chemostats where the diatom, *Th. pseudonana,* outgrew but did not completely exclude the chlorophycean alga, the effect of 0.1 μg/liter PCB was to reduce the diatom population drastically such that approximately equal densities of the two species were established. These experiments were made at a constant dilution rate of 0.21 day^{-1}. In subsequent experiments the behavior of natural phytoplankton communities, from which grazers had been removed, was followed in $NO_3{}'$-limited chemostats. Initially the community comprised 15 species and was dominated by *Th. pseudonana, Skeletonema costatum,* and a *Chaetoceros* sp. The species diversity index (131) remained high in control cultures but declined noticeably (1.0 to 0.25) in the presence of 0.1 μg/liter PCB. Of the three dominant phytoplankters, *Th. pseudonana* again proved most susceptible to PCB, while the competitiveness of *Chaetoceros* increased in the presence of the pollutant.

Fisher et al. (129) also provided some slight evidence for an additive effect of dilution rate: the competitiveness of *Th. pseudonana* and *Sk. costatum* fell as D was raised from 0.33 to 0.47 day^{-1}, a result that is consistent with PCB having a species-specific effect on nitrate utilization. Two important conclusions can be drawn from these experiments: 1) PCB appears to have an effect on *Th. pseudonana* that is similar in simple gnotobiotic and complex naturally occurring phytoplankton communities; and 2) the reduced species diversity caused by PCB is likely to have significant repercussions at higher trophic levels due to the selective grazing behavior of zooplankton and marine animals. This in turn would trigger such deleterious developments as eutrophication.

The effect of PCB on nitrate assimilation was studied further by Fisher et al. (130). PCB addition had no effect on the μ_{max} of $NO_3{}'$-limited *Th. pseudonana*, but at a concentration of 1 μg/liter the K_s (nitrate) was increased eightfold to 3.30 μM. Thus PCB acted as a competitive inhibitor of nitrate metabolism. This result has ecological importance in view of the seasonal variations in nitrogen availability and the fact that marine phytoplankton are often nitrogen-limited in nature. One further point needs emphasizing: the toxicity of PCB and other xenobiotics may be modulated by numerous environmental circumstances, such as light intensity, salinity, temperature, and the presence of secondary chemical pollutants. For example, the sensitivity of diatoms to PCB increases markedly as the environmental temperature approaches tolerance limits (131). Such environmental perturbations and cyclic changes can be simulated readily in continuous cultures.

The role of microorganisms in the biodegradation of petroleum and its products has received intensive study, but relatively few data are available on the effects of oil on microbial activities. The problem also is amenable to study by continuous culture techniques and it is well known that exposure to crude and refined oils can have significant and differential effects on ecologically important groups of microorganisms in water and sediments (132). Moreover, it is gradually being appreciated that oil has the capacity to concentrate such environmental pollutants as heavy metals and chlorinated hydrocarbons and in so doing increases its resistance to biodegradation. Modeling of such situations in continuous-flow fermenters is highly desirable but studies of this kind are entirely lacking.

Biodegradation

We remarked earlier that microbiologists generally have adopted monoculture techniques: nowhere are the limitations of this approach more evident than in the study of biodegradation, whether of xenobiotic or natural chemicals. Attempts to isolate *monocultures* of biodegradative microbes may frequently be thwarted when complex long-lived chemicals, such as pesticides, lignins, or melanins, are under investigation. Biodegradation of these latter materials is likely to involve cooperative attack by microbial communities and by such mechanisms as co-metabolism (133).

Similarly, the great majority of biodegradation research has been made with aerobic populations. Evidence has slowly accumulated during the last decade for anaerobic attack on persistent chemical pollutants and for sequential anaerobic-aerobic stages mediated by different microorganisms in different habitats of an ecosystem, e.g., sediment and water column phases of aquatic ecosystems. We argue also that provision of the chemical of interest as sole source of carbon and/or energy may be inadequate to ensure its biodegradation by microorganisms. For example, a readily assimilable carbon source may be required to generate sufficient energy or reducing power to initiate or sustain metabolism of the chemical (see ref. 134 for the effects of glucose on aromatic cleavage and the implications for lignin and pesticide biodegradation).

Two final points requiring consideration are the concept of biodegradation and the possibility of bioconcentration of xenobiotics in microbial biomass. Biodegradation can imply anything from total mineralization to inorganic products to simple chemical modifications of the kind represented by the conversions of DDT to DDD [1,1-dichloro-2,2-bis(p-chlorophenyl)ethane]. The importance of the latter type of metabolism lies in the generation of chemicals, creating increased environmental hazards. A useful discussion of this matter has been made by Stafford and Callely (135). Considerable information has been collected on the accumulation of chemical pollutants in animals, but little study has been made of bioconcentration by microorganisms. Several recent reports, however, emphasize the importance of bacteria and microalgae in introducing toxic compounds into food chains (136, 137, 138) and bioconcentration ratios as high as 56,000 have been observed in heterotrophic bacteria [β-chlordane (exo-1, exo-2,4,5,6,7,8-octachloro-3a,4,7,7a,-tetrahydro-4,7-methanoindan) *Caulobacter* system (136)].

The arguments presented above lead us to the view that many discussions of biodegradation and putative chemical recalcitrance might be too simplistic or invalid due to an adherence to traditional techniques that take no account of community interactions, fluctuating environments, and adaptive responses. The adoption of continuous-flow culture techniques for such work has many advantages and should permit a much more realistic appraisal to be made of biodegradative processes. It is hoped that the few illustrations that follow will substantiate this claim.

Biodegradation of pesticides by microbial communities has rarely been studied, but the biodegradation of such widely different chemicals as trichloroacetic acid (139), dalapon (140), Diazinon (O,O-diethyl O-2-isopropyl-4-methyl-6-pyrimidinyl thiophosphate) (141), parathion (O,O-diethyl O-p-nitrophenyl phosphorothionate) (142), malathion (O,O-dimethyl dithiophosphate of diethyl mercaptosuccinate) (143), and propham (isopropyl phenylcarbamate) (144) has been shown to be susceptible to communities of varying complexity. In some of these cases the nature of the microbial interactions has been elucidated: 1) commensalism is the basis of the interaction between an unidentified bacterium and two *Strepto-myces* spp. in the metabolism of TCA (139), the *Streptomyces* organisms supplying vitamin B_{12} as an essential growth factor for the bacterium; and 2) partial degradation of Diazinon is achieved via the mutualistic action of an *Arthrobacter* sp. and a *Streptomyces* sp. (141) in attacking the pyrimidinyl ring of this insecticide. The use of continuous cultures to study community-pesticide interactions is in its infancy and we are aware of only two such applications—the work of Hsieh on the organophosphorus insecticide parathion and that in one of our laboratories (A. T. Bull) on the chlorinated aliphatic acid herbicide dalapon.

Hsieh and his colleagues have isolated, by continuous-flow enrichment from sewage, a number of microbial communities that utilize parathion. These communities are very complex: one, for example, consisted of at least nine bacteria—five fluorescent pseudomonads, and species of *Azoto-monas, Brevibacterium,* and *Xanthomonas* (145). The physiology of one such community has been studied in detail (142) and has been maintained on parathion in a chemostat for two years. This latter community comprised at least four bacteria, only one of which, *Pseudomonas stutzeri,* could metabolize parathion, with the products being *p*-nitrophenol and ionic diethyl thiophosphate. Cleavage of the parathion molecule occurred extracellularly via a co-metabolic hydrolysis, probably involving a surface-bound phospholipase. A second member of the community, *P. aeruginosa,* utilized the *p*-nitrophenol as a sole source of carbon and energy, but neither species was able to perform the role of the other. Thus these two pseudomonads developed a mutualistic interaction, with *P. stutzeri* presumably growing on the products of metabolism, lysis, or leakage from *P. aeruginosa* and/or the other species. The hydrolysis of parathion was considered to be rate-limiting for the growth of the community and *p*-nitrophenol was considered to be the effective growth-limiting substrate.

Daughton and Hsieh (142) analyzed *p*-nitrophenol–limited chemostat populations and found that this substrate could serve as sole source of carbon or nitrogen. When the phenol was made the limiting carbon source, the μ_{max} and K_s values for the community were determined to be 0.50 h^{-1} and 0.37 mg/liter, respectively; however, there was a considerable scatter of data points on this graphical analysis, a situation that we have also experienced with our dalapon-utilizing community (see below). The steady-state biomass concentration fell by about 15% when the dilution rate was raised above 0.36 h^{-1}, a result that the authors ascribe to the increase of *p*-nitrophenol to concentrations that caused uncoupling of oxidative phosphorylation. Diethyl thiophosphate was produced in theoretical yields from parathion but was not metabolized further, even when it provided sole sources of sulfur or phosphorus. The stability of this and related thiophosphates has obvious environmental implications and is a forceful illustration of the point that we made earlier regarding the distinction between degradation and disappearance of a parent chemical.

The community isolated by Munnecke and Hsieh (145) has been used to explore the variations in parathion metabolism under varying environmental conditions. The primary aerobic pathway was that described above, yielding p-nitrophenol and diethyl thiophosphate. However, a secondary aerobic route involved oxidation to paraoxon (diethyl p-nitrophenyl phosphate) and its subsequent hydrolysis to p-nitrophenol and diethyl phosphate (146). At low oxygen tensions parathion was reduced to p-aminoparathion, which was hydrolyzed to p-aminophenol and diethyl thiophosphate. Munnecke (147, 148) has extended this work by investigating the detoxifying ability of cell-free extracts of the community grown on parathion. This crude enzyme preparation could hydrolyze nine organophosphate insecticides in addition to parathion, at rates significantly faster than chemical hydrolysis, and was stable at temperatures up to 50° C. The preparation has been successfully immobilized by covalent bonding to glass and as such appears to have considerable potential in the treatment of pesticide-containing effluents and cases of pesticide poisoning.

Senior, Bull, and Slater (140, 149, 150) isolated a complex community of microbes growing on the herbicide dalapon (2,2-dichloropropionic acid) that, unlike the parathion community, contained more than one primary species, i.e., species able to catabolize the herbicide as sole carbon source. The community, isolated by continuous-flow enrichment in a dalapon-limited chemostat, originally comprised six species (one filamentous fungus, *Trichoderma harzianum,* a *Rhodotorula*-like yeast and four bacteria). The *Trichoderma* and two of the bacteria could grow as monocultures on dalapon, but the remaining organisms lacked this capacity. Apart from the yeast, which was eliminated from the community at an early stage of chemostat culture, the community was tightly integrated and very stable over a total culture period of > 20,000 h. The composition of the dalapon community is shown in Figure 1.

We have drawn attention to the distinction between stability and steady state in the context of continuous mixed cultures (see "Ecosystem Modeling," above); thus, the dalapon community, while possessing great stability, did not attain true steady states. Undamped oscillations were observed in biomass, residual growth-limiting substrate concentration, and pH and their amplitude was a function of dilution rate: the lower the value of D, the smaller the amplitudes (149). The μ_{max} of the community growing on dalapon was in the range 0.36 h^{-1} (batch data) to 0.40 h^{-1} (determined from washout kinetics) and the apparent K_s was 40 mg/liter. However, because of the nonsteady state and nature of the culture, the reliability of the K_s value is questionable and can only be taken as a guide to the affinity for dalapon. The community had a maximum rate of dalapon degradation of 12 g per liter per day (D = 0.30 h^{-1}, S_R = 1.65 g/liter); the previous highest reported rate of pesticide degradation is 8 g per liter per day for parathion (142). Cooperative effects in dalapon

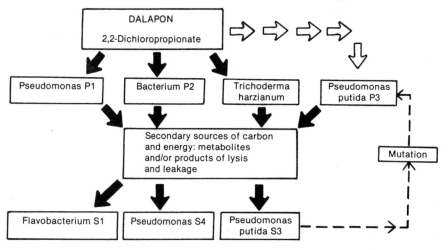

Figure 1. Biodegradation of dalapon by a stable seven-membered microbial community [adapted from Senior et al. (140)].

utilization are most unlikely, but the coexistence of three primary species indicated that interactions other than competition were important in stabilizing the dalapon community; so far these interactions have not been identified.

A major change in the community occurred after approximately 2,900 h of continuous culture with the appearance of a fourth dalapon-utilizing organism. This new primary species arose from a mutation of one of the secondary bacteria and the subject is discussed further in "Continuous Culture and Microbial Evolution," below. The significant feature of this event is the provision within the chemostat of a permissive environment for metabolic evolution of a population that is exposed continually to a substrate that it cannot utilize. All dalapon-utilizing organisms possessed dehalogenase(s) that caused an initial dechlorination of the substrate.

Chemostat cultures have been used by several researchers to examine the biodegradation and utilization of phenol by mono- (151, 152, 153) and mixed (154, 155) cultures, while in one of our laboratories (A. T. Bull) the growth of a three-membered bacterial community on a related compound, orcinol (3-methyl resorcinol), is currently being analyzed (156). This association of orcinol-utilizing organisms was isolated from activated sludge by a chemostat enrichment procedure. Of the three bacteria only one, *P. stutzeri*, metabolized orcinol and could grow as a monoculture on this substrate; the two secondary species, *Brevibacterium linens* and *Curtobacterium* sp., comprised less than 20% of the community by number and were unable to cleave the aromatic ring (Figure 2). The secondary species grew at the expense of products of *Pseudomonas* metabolism and, probably more important, on products of cell lysis and on leakage caused

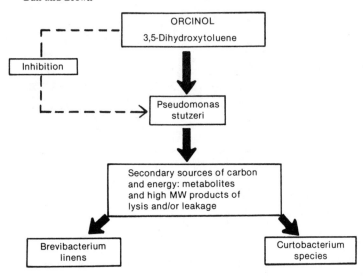

Figure 2. A stable three-membered bacterial commmunity growing on orcinol (unpublished data of A. Osman, J. H. Slater, and A. T. Bull).

by the phenolic substrate. *Br. linens*, for example, produces extracellular protease and nucleases that enable high molecular weight lysates to be utilized. However, the selective advantage for the primary organism of such a tight and stable association appears to have a kinetic rather than a nutritional basis. Values for maximum specific growth rate and apparent half-saturation constant were determined from batch and chemostat data for the community, for *P. stutzeri,* and for combinations of *P. stutzeri* with the secondary species (Table 1).

A particularly interesting, but as yet unexplained, observation was the increased affinity for orcinol shown by *P. stutzeri* when grown with the

Table 1.   Kinetic data obtained for a bacterial community growing on orcinol (previously unpublished data of A. Osman, J. H. Slater, and A. T. Bull).

	μ_{max} $(h^{-1})^a$	K_s (g/l)	K_i (g/l)	$\sqrt{K_s \cdot K_i}$ $(g/l)^b$	S_{max} (observed) (g/l)
Community	0.28	72	1750	354	600
Pseudomonas stutzeri	0.29	100	690	266	450
Pseudomonas stutzeri +					
Brevibacterium linens	0.30	83	570	217	400
Pseudomonas stutzeri +					
Curtobacterium	0.35	59	790	216	290

[a] $S_R = 1.0$ g/liter.
[b] Theoretical concentration of orcinol producing maximum value of μ.

secondary species. Construction of μ versus s plots for each organism combination described in Table 1 indicates that the *Pseudomonas-Curtobacterium* dual culture should be the most competitive at any imposed dilution rate in the chemostat. Nevertheless, *Br. linens* was not excluded and this simpler community was not selected. Clearly, factors other than μ_{max} and K_s were significant in the selection of this stable community.

Batch cultures over a range of orcinol concentrations revealed typical substrate-inhibited growth and, assuming that Haldane inhibition kinetics applied to this system, values for the inhibition constant (K_i) were calculated. The orcinol concentration (S_{max}) giving a maximum value of μ is defined by the Haldane equation as $\sqrt{K_s K_i}$, and the data in Table 1 now -indicate the selective advantage possessed by the whole community as opposed to any other combination of organisms. Values for S_{max} were also obtained by experiment but agreement with the calculated values was poor. Similar differences among calculated and observed values of S_{max} were reported by Sato and Akiyama (157) for the oxidation of organic acids by activated sludges. Both sets of results suggest that the mathematical modeling of these complex microbiological systems is currently inadequate. Tyler and Finn (158) grew *Pseudomonas* sp. NCIB 9340 on 2,4-dichlorophenol in chemostat and in batch culture and found that growth was strongly inhibited at substrate concentrations above 25 mg/liter. In this case also the Haldane function provided an unsatisfactory fit to the data.

A final category of xenobiotic pollutants that will be mentioned here are surfactants. The most extensively used surfactants are alkyl-benzene sulfonates (ABS), and their biodegradation has been widely studied. Some recent results of Johanides and Hrsak (159) are of considerable interest and again focus on the importance of cooperative microbial action in biodegradation. A community of bacteria growing on a mixture of linear ABS was obtained from a detergent plant effluent by continuous enrichment culture. The community was dominated by a *Pseudomonas* sp. and and an *Alcaligenes* sp. and their proportions were dependent on the dilution rate imposed. Neither the *Pseudomonas* nor the *Alcaligenes* (nor other members of the community) had the capacity to degrade the ABS when grown as monocultures. The biochemical basis of this interaction has not yet been described.

Apart from their intrinsic interest, one can speculate on the possible practical significance of community biodegradation studies. The most obvious areas of relevance are microbial seeding and ecosystem modification in relation to chemical pollution of the environment. Such options have been advocated for the cleanup of oil spills and, in this case, seeding would have to be made with a mixture of microbes because single species do not have the capacity to degrade the plethora of compounds present. Alternatively, the introduction of appropriate catabolic plasmids into the contaminated environment has been suggested (103, 160). In either instance virtually nothing is known about the competitiveness or activities (including secondary effects on, for example, the existing biota) of the organisms introduced; the behavior of chemostat communities could provide rapid and helpful evaluation data. Similarly, the effects of environmental modification (e.g., addition of nutrients or forced aeration) on ecosystem functioning have received little experimental analysis and continuous cultures may serve as useful models in this context. Readers interested in these problems should refer to the recent comprehensive reviews of Colwell and Walker (132) and Atlas (161).

Microbial Technology

Before concluding this section we wish to make brief reference to the technological importance of mixed cultures, especially in continuous-flow operations. Defined mixed culture fermentations have been developed for a number of microbial products ranging from metabolites (e.g., the Amylo process and β-carotene) and biomass (e.g., Symba process) to foods (e.g., cheese, sauerkraut), while such diverse activities as the biological treatment of wastes and effluents and metal leaching (162, 163) also may involve the cooperative actions of microbial communities. Excellent discussions of much of this material can be found in the text by Bailey and Ollis (111).

In recent years very interesting developments have occurred in the understanding of methanogenesis and methane- (and methanol-) based fermentations and the mixed microbial populations involved. Methane synthesis occurs widely under anaerobic conditions, e.g., sediments and the rumen, and various processes are being developed for commercial fuel production. In anaerobic sludge digesters, for example, a trophic hierarchy of organisms converts insoluble substrates to methane: 1) primary biodegraders convert polysaccharides, such as cellulose, proteins, and lipids, to soluble substrates; 2) fermentative bacteria metabolize the solubilized substrates largely to acetate (and other volatile acids), CO_2, and H_2; and 3) methanogenic bacteria utilize H_2-CO_2 or acetate as major precursors in CH_4 synthesis.

Within the last five or six years considerable headway has been made in defining the critical metabolic interactions between these various groups of bacteria, and the unifying concept of "interspecies hydrogen transfer" has gained wide acceptance (164). Fermentative bacteria dispose of electrons generated in glycolysis by producing reduced end products like ethanol, lactate, formate, succinate, propionate, and H_2. However, in the presence of methanogenic bacteria there is an altered flow of electrons and hydrogen with the reduction of CO_2 to methane. Methanogenic bacteria can scavenge very low concentrations of H_2 and it may be difficult to detect the latter in methanogenic ecosystems.

Methane production in sediments has been modeled in chemostats by Cappenberg (165). He grew a *Methanobacterium* sp. and *Desulfovibrio desulfuricans* in a chemostat, the latter generating acetate for the methanogen. This commensal interaction was stable only when the hydrogen sulfide concentration was less than about 0.1 mM; at higher concentrations the *Methanobacterium* was excluded, in Cappenberg's opinion because of sulfide toxicity. Recently this system has been reexamined in Bryant's laboratory (166) and the conclusion reached that the elimination of *Methanobacterium* on addition of sulfate reflected the preference of sulfate as the electron acceptor. Sineriz and Pirt (167) isolated a methanogenic community in a formate-limited chemostat. The community was dominated by two species of *Methanobacterium*, but when it was fed glucose the small *Citrobacter* population originally present became established as a major component and the products of its fermentative catabolism of glucose were metabolized by the methanogens in the now predictable manner.

The conversion of methane into single cell protein (SCP) has excited much interest and research in recent times and, because mixed cultures growing on CH_4 as the sole source of carbon and energy generally are more stable than monocultures of methane utilizers, continuous mixed culture fermentation technology has been given a big boost. The most intensive studies in this field have been made by researchers at Shell Research in England (168, 169, 170). One methane-utilizing community isolated by the Shell group was comprised of four bacteria: a CH_4-utilizing *Pseudomonas* sp., a methanol-utilizing *Hyphomicrobium* sp., and an *Acinetobacter* sp. and a *Flavobacterium* sp. that did not metabolize C_1 compounds (168). For sustained growth on CH_4 the pseudomonad required the presence of *Hyphomicrobium* and the two secondary species for maximum-biomass yields. The CH_4 utilizer always constituted more than 90% of the community numbers and the relative proportions of the four organisms were independent of dilution rate. Methane is initially oxidized to methanol and it was postulated that the *Hyphomicrobium* utilized the alcohol, which, in concentrations above about 0.16 mM, caused inhibition of CH_4 oxidation. Therefore, it methanol is added to a steady-state chemostat culture of the community, an immediate inhibition of CH_4 utilization occurs, followed by predicted population changes. As methanol is gradually removed, CH_4 oxidation recommences. Substrate affinities of the community for CH_4 and methanol were found to be 1.9 $\times$ 10^{-5}M and 2.9 $\times$ 10^{-2}M, respectively; the K_s of *Hyphomicrobium* for methanol was 8.3 $\times$ 10^{-6}M (170). Thus, the *Hyphomicrobium* has a high scavenging capacity for methanol produced in the mixed culture. Although exact metabolic roles were not assigned to the two secondary species, it was suggested (169) that they removed potentially inhibitory organic molecules derived from growth and lysis of the community. The methane utilizing community is depicted in Figure 3.

More recently the role of secondary (heterotrophic) organisms in methane-utilizing communities has been investigated by Linton and Buckee (171). These authors studied a five-membered bacterial community comprising a single CH_4-oxidizer, *Methylococcus* sp., and four secondary

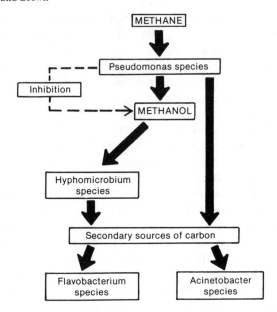

Figure 3. Methane-utilizing bacterial community [adapted from Wilkinson et al. (168)].

species unable to grow on CH_4 (two *Pseudomonas* spp., a *Mycobacterium/ Nocardia* sp., and a *Moraxella* sp.). When *Methylococcus* was grown alone on CH_4, organic carbon (as protein and nucleic acids) accumulated in the medium. Such accumulation did not occur in mixed cultures. The four secondary species produced a battery of extracellular depolymerases and the dominant organism, *Pseudomonas* sp. NCIB 11310, excreted a neutral protease that degraded the proteins in the culture liquor of the *Methylococcus*.

Microbial communities growing on methanol as sole carbon and energy source have also been analyzed (169). One such community contained a single obligate methanol utilizer and four secondary bacterial species. A mutualistic interaction between the primary and secondary species was evident from chemostat experiments: when the whole community was cultured the D_{crit} was approximately 0.64 h^{-1} and the maximum $Y_{methanol}$ was 0.52 g/g. In contrast, D_{crit} and $Y_{methanol}$ values for the methanol utilizer in monoculture were 0.20 h^{-1} and 0.30 g/g, respectively. The benefits of using mixed cultures to produce SCP from CH_4 and methanol include enhanced growth rate and biomass yield, greater culture stability, and, according to Harrison and his colleagues (169), resistance to contamination and reduced foaming of the culture.

HETEROGENEOUS CONTINUOUS CULTURE SYSTEMS

Implicit in our discussions so far has been the acceptance of continuous cultures as completely mixed stirred tank reactors. However, many experimental deviations from chemostat theory have been documented (7) and the growth of organisms on walls, baffles, probes, etc. is known to be a contributory factor. In this section we make brief reference to this phenomenon and consider its significance, in metabolic terms, for fermentation technology and ecophysiological studies.

A useful kinetic analysis of chemostat cultures having wall growth has been made by Topiwala and Hamer (172). These authors included a term (K) in the steady-state chemostat equations that defined the total adherent biomass. For monolayers of bacteria the values of K can be expected to fall in the range 0.02–0.05. The effect of K is to extend the operational range of the chemostat so that dilution rates can be realized that may be considerably greater than the theoretical critical dilution rate. For example, a thirtyfold increase in *Serratia marcescens* productivity was achieved on adding glass wool as a surface extender to a chemostat vessel (173). An early demonstration of the benefits of surface growth was made by Breden and Buswell (174), who added shredded asbestos to methane fermentations. More recently, the effects of wall growth in mixed cultures growing on methane have been examined by the Shell group (175) and experimental results closely matched predictions of the model (172). The ability of an organism to attach to a surface (176), or of organisms to attach to each other ("clumping"; see next section for an example), has important repercussions for competition in continuous cultures and may confer considerable selective advantage. On the other hand, with stable mixed cultures the whole community may be maintained in the chemostat at $D > D_{crit}$ as a result of wall growth, despite the fact that such growth may be initiated by one component of the mixture [probably the *Hyphomicrobium* sp. of the methane-utilizing system and *Tr. harzianum* of the dalapon community (149)].

This fortuitous adhesion capacity of microorganisms can be exploited in new fermenter designs; indeed, such long-established processes as waste water treatment by trickle filtration and the "quick" vinegar fermentation are based on precisely this property. The development of microbial film fermenters has been strongly advocated by Atkinson and his colleagues (177, 178). Such films may be mixed or nonmixed and for the mixed type conventional stirred fermenters or fermenters based on the fluidized bed and air-lift principles are feasible. Various laboratory-scale designs have been published (178, 179, 180) and kinetic analyses of growth in such fermenters have been attempted (178, 181, 182). So far the performance of continuous-flow film fermenters has been restricted to laboratory study and even here the potential applications have attracted only slight attention. Most effort has been directed toward effluent, especially toxic effluent, treatment, and preliminary reports are encouraging [e.g., a tenfold increased efficiency in phenol degradation from fluidized bed versus conventional stirred fermenter (183)].

One area of microbiology where heterogeneous continuous culture methods are being effectively applied is ecophysiology (109). Obviously the effect(s) that particulate matter has on microbial activity is dependent on numerous considerations—nature of particle, organism, and environmental factors, such as pH and salinity; concentration of particles and nutrients; and complexity of the microflora. However, this bewildering array of variables notwithstanding, it is possible to mimic many situations that exist in natural ecosystems, and the fact that the majority of microorganisms probably grow attached to surfaces makes this an important field of research. ZoBell's early observations on the stimulative effects of surfaces on bacterial growth (184) have been followed up by several continuous culture studies on the interaction of nutrients with clay and organic colloids (185, 186, 187), while some preliminary information is appearing on the effects of clay particles on interspecies competition (188).

Other kinds of interaction can also be modified by the presence of particles. An interesting illustration is provided by Roper and Marshall's study of host-bacteriophage interaction in the presence of saline sediments and clay minerals (189). Both bacteria (*E. coli*) and phage adsorbed strongly to such materials but were desorbed as the salinity was reduced. However, the bacteria were protected from phage infection by the sediments at both high and low salinities: at high salinities interaction was prevented by adsorption of host and parasite to particles, and at low salinities it was prevented by the adsorption of colloidal matter (released from the sediment) around the bacteria. Finally, in addition to examining specific interactions, continuous cultures and related flow systems can be used effectively to model ecosystems like rivers (190, 191) and soil (192).

CONTINUOUS CULTURE AND MICROBIAL EVOLUTION

The theoretical and experimental basis of the use of the chemostat in studies of microbial evolution was set out by Novick and Szilard (193). Kubitschek (3, 194) has provided a useful summary of early work in this area and points out that chemostats are less useful than batch liquid or plate methods in many instances, because the use of traditional mutagens would most probably lead to culture washout. Turbidostats, however, are useful in this context and Bryson and Szybalski (195) described a 'turbidostat selector' especially for this purpose. The chemostat, however, is of particular significance in situations where a mutation leads to a variation in the ability to utilize the limiting substrate. If, for example, the mutant uses the substrate more efficiently (with higher μ_{max}), then it will be selected for and will rapidly take over the culture. A mutant with a less efficient use of the limiting substrate will be washed out. Between these two extremes fall mutants that use the substrate in a manner equivalent to the parent—such neutral mutants will increase in number in a linear manner at a rate determined by the mutation rate.

Novick and Szilard exploited this fact to measure spontaneous mutation rates in *E. coli* during tryptophan-limited growth. Ten days cultivation of *E. coli* B/1 at D 0.5 h^{-1} produced a variant B/1/f that grew five times faster than the parent at low tryptophan concentrations and took over the culture. This organism was used in a study of the spontaneous mutation to phage T5 resistance. Such a mutation was neutral in the sense outlined above and mutant numbers increased linearly with time. The mutation rate at 37°C was found to be about double that observed at 25°C. Periodically, however, the culture mutated to more efficient use of tryptophan, at which point parent and T5 resistant mutants were eliminated. Subsequently T5 resistance in this new population increased linearly until a further mutation affecting tryptophan uptake and metabolism occurred. Thus the number of T5 resistant mutants rose and fell in a 'saw-tooth' manner (3). In extending these observations to carbon-limited cultures, Kubitschek (3) reported that the spontaneous mutation rate varied directly with μ and therefore with the rate of DNA replication.

The experiments with tryptophan limitation illustrate the use of the chemostat in selection for organisms better fitted to the prevailing environmental conditions. Kubitschek and Bendigkeit (196, 197) used caffeine, ultraviolet irradiation, and 2-amino purine to increase the mutation rate in chemostat cultures of *E. coli*, and reported that while ultraviolet light-induced mutation was not μ dependent there was a μ dependence of mutation rate with caffeine and 2-amino purine.

Another way to increase the rate of mutation is to use "mutator mutants." Mutator (or antimutator) genes increase (or decrease) the rates of mutation in other genes. For example, the Triffers mutator gene of *E. coli* increases the mutation rate more than 100 times at most if not all chromosomal loci (198). This mutator gene (*mut* T1) preferentially increases the number of transversions of AT to GC, with DNA exhibiting a growth-dependent increase in buoyant density as a result of an increased GC content. In competition experiments a *mut* T1 *lac*$^-$ (no lactose transport) was used because this was useful in screening experiments and in preliminary experiments a *mut* T *lac*$^+$ strain in a non-lactose–containing medium was at a slight disadvantage. Thus it was unlikely that the *lac*$^-$ genotype would confer an advantage in competition with *mut*$^+$ organisms. In competition between *mut* T1 *lac*$^-$ and *mut*$^+$ *lac*$^+$, the mutator strain consistently outgrew the wild type in glucose-limited cultures, indicating a selective advantage of a high mutation rate in a population under a selection pressure. These results illustrate experimentally the importance of mutation in evolution.

In a further series of experiments it was shown that, while the μ_{max} and K_s values of 'uncompeted' *mut*$^+$ and *mut* T1 were the same, the *mut*$^+$ organisms selected from chemostat competition experiments were more 'fit' than the starting cultures. For example, some mutants isolated were more resistant to glucose starvation while others were able to stick to glass, thus conferring a selective advantage in a well-stirred system. The growth medium used contained citrate as chelator and another class of mutant was able to utilize this as a carbon source in addition to glucose. Genetic evidence was presented that these results were due to the increased rate of mutation in the *mut* T₁ populations and loss of *mut* T genotype of mutants did not result in loss of mutant characteristics (199).

Nestermann and Hill (200) reported results from experiments using another mutator strain of *E. coli* (*mut* H₁). They studied the mutation from sensitivity to resistance to T₅ phage (discussed above) in carbon-limited chemostats at D 0.46 h⁻¹ and reported that the periodic selection of organisms better fitted to the cultural conditions occurred at a higher rate for *mut* H₁ than *mut*⁺ populations. In competition experiments *mut* H₁ and *mut*⁺ were distinguished by a *Gal⁻* marker. With a starting mixture of *Gal⁺ mut* H₁ and *Gal⁻ mut*⁺, the *mut* H₁: *mut*⁺ ratio increased with time in glucose-limited cultures; i.e., numbers of galactose utilizers increased. With a starting mixture of *Gal⁺ mut*⁺ and *Gal⁻ mut* H₁, the *mut* H₁: *mut*⁺ ratio again increased, reflected by a decrease in the numbers of galactose utilizers. These results indicated that (as with *mut* T₁) mutation to increased fitness occurred with greater frequency in mutator populations. While *mut*⁺ and *mut* H₁ strains exhibited identical growth rates on glucose in batch cultures, the growth rate of *mut* H₁-derived organisms that had competed successfully in chemostats was increased by up to 50%.

Novick and Szilard's spontaneous mutations to more efficient tryptophan utilization (193) was an early example of the chemostat selection of organisms better able to utilize the limiting substrate. Little is known of the biochemical basis of these "tryptophan mutants," however, and more attention has been given to catabolic systems in which the activity of some "substrate capturing enzymes" (201) might be considered rate limiting. Perhaps the first example of this was the work of Horiuchi et al. (202) on the β-galactosidase levels of *E. coli* grown in lactose-limited chemostats. For the first 10 generations the level of β-galactosidase was low and inducible, but then the level rose until after 20 generations a 14-fold increase was noted. This rise was associated with the spontaneous production of mutants constitutive for β-galactosidase. Their selection may be explained by the fact that in the presence of low and limiting lactose concentrations the enzyme of the inducible strain was not fully expressed. Therefore the constitutive mutant with a higher enzyme level has a distinct growth advantage.

After several hundred generations organisms exhibiting much higher ("hyper") levels of β-galactosidase appeared (with up to five times more enzyme than the constitutive mutants). Some of these hyper mutants were inducible. The β-galactosidase of the hyper mutants accounted for up to 25% cell protein content, but was identical to wild-type organisms with respect to K_m for lactose, heat sensitivity, sedimentation constant, and antigenic properties. When subcultured in batch, hyper strains grew 20% more slowly than wild type under conditions when high β-galactosidase activity was not an advantage. Continued subculture led to loss of hyper-activity, although this was stable in lactose-limited continuous culture or when grown on glucose (β-galactosidase partly repressed). It was interesting to note that mutants with hyper β-galactosidase activity showed no increase in lactose uptake rate, i.e., the "permease" was not affected. The acetylase was not studied. Hyper characteristics were associated with *lac* DNA and could be transferred by conjugation. Hyper production may be the result of multiple gene copies for β-galactosidase or of a promotor mutation leading to a higher transcription rate (203). The former seems most likely because promotor mutants should have higher levels of both β-galactosidase and permease.

An example of hyper enzyme levels that is better understood at a biochemical level is the formation of ribitol dehydrogenase in *K. aerogenes*. Mortlock et al. (204) and Mortlock and Wood (205) reported that wild-type *K. aerogenes* would grow on ribitol and D-arabitol as sole carbon source because the organism contained an inducible pentitol uptake system, ribitol dehydrogenase (RDH), and D-arabitol dehydrogenase. Two further pentitols, xylitol, and L-arabitol, however, did not support growth of the wild-type organism. Mutants constitutive for RDH did grow slowly on these unnatural pentitols due to a side specificity of RDH. The K_m of RDH for ribitol was about 1 mM, while for xylitol a value approaching 1 M was reported. It was likely that the RDH content limited growth on xylitol, because this substrate was transported into the cell by the pentitol uptake system. Mortlock and Wood (205) quoted data obtained by Bisson (1968) in which an RDH constitutive mutant was grown under xylitol limitation in a chemostat, and a second mutation occurred spontaneously that increased the batch culture growth rate on xylitol from 0.26 to 0.5 h^{-1}. This mutant was of the hyper variety, with no alteration in the specificity of RDH for ribitol of xylitol but with a RDH level some fourfold higher than that of the original constitutive mutant.

Similar results were reported by Hartley's group (206, 207). During xylitol-limited growth of the constitutive parent A, a spontaneous mutation to more efficient growth on xylitol occurred after about 50 generations (to produce A1). A further 50 or so generations later A1 was succeeded by A11. The μ_{max} of A1 on xylitol was 20% greater than A and the μ_{max} of A11 some 28% greater than A1, with enzyme specific activity increasing 4.5 times from A to A1 and 3.6 times from A1 to A11. Neither the K_m for xylitol nor the xylitol/ribitol activity ratio of the purified RDHs varied significantly, but it was apparent that A1 made five times and A11 fifteen times more enzyme than A. Mutagenesis with ultraviolet radiation was possible by building an ultraviolet lamp into the chemostat and irradiating at a low level. Several more hyper mutants were selected by this method, with A6111 containing some 20% total protein as RDH. Thus mutants produced spontaneously or with UV and with nitrosoguanidine in the chemostat were of the hyper variety with no increased specificity for xylitol.

Evidence based on the frequency of mutation to absence of RDH suggested that A11 contained multiple copies of the RDH gene, and this was confirmed by segregation analysis. [Hartley (208) has discussed the wider implications of multiple gene copies in microbial evolution.] Batch treatment with nitrosoguanidine produced a mutant B with an increased specificity for xylitol, and the authors concluded that the frequency of mutation of hyper enzyme production was much greater than that leading to improved enzyme specificity. This suggested that a change in enzyme specificity might be due to multiple mutations and that evolution toward new specificity to xylitol by single step mutations would not be possible in a chemostat under continuous selection pressure. Inderlied and Mortlock (209) confirmed the gene duplication results outlined above and further reported that while hyper RDH mutants contained up to 60 times more enzyme than constitutive organisms the levels of the closely linked D-xylulokinase, D-arabitol dehydrogenase, and D-ribulokinase were not increased. It was suggested that some intermediate hyper mutants existed with increased levels of all these enzymes but those not required were diluted out. Perhaps the same is true for the *lac* proteins.

During studies on the mixed culture–degrading dalapon (see "Biodegradation," above) Senior et al. (140) reported that after 2,900 h of growth an additional primary dalapon utilizer was formed. This organism was identified as *Pseudomonas putida* and was similar in all other respects to a secondary organism in the original culture. Carbon-limited chemostat culture of the non-dalapon–utilizing *P. putida* in monoculture on a mixture of propionate and dalapon showed the development of a dalapon utilizer by spontaneous mutation. Further studies (210, 211) showed that dalapon-degrading halogenase from the mutant most likely arose from preexisting halogenases, with activity toward 2-monochloropropionate and dichloroacetate.

Studies on enzyme evolution in the chemostat have not been confined to bacteria. Francis and Hansche (212) reported the modification of acid phosphatase in cultures of *Sa. cerevisiae* limited by β-glycerophosphate. The pH optimum of the yeast acid phosphatase was between 3 and 4, but cultures were maintained at pH 6.0. After 180 generations a spontaneous mutation resulting in an increased population density was found to be due to an increased efficiency of uptake of inorganic phosphate released on the hydrolysis of β-glycerophosphate. After 400 generations a further mutation resulted in a 60% increase in acid phosphatase level together with a shift in the pH optimum of the enzyme from 4.2 to 4.8. A further mutation after 800 generations led to cells clumping and therefore increasing in density due to incomplete mixing (a phenomenon equivalent to the advantage of the mutant capable of sticking to glass mentioned above).

Zamenhof and Eichhorn (213) used chemostats in studies of the selective advantage of auxotrophic mutants of *B. subtilis* over prototrophs under conditions where the auxotrophic requirements were satisfied. For example, a histidine-requiring mutant outgrew a non-histidine–requiring back mutant in the presence of 0.3 mM histidine and an indole-requiring mutant outgrew a non-indole–requiring back mutant in the presence of 0.25 mM tryptophan. These results were interpreted in terms of the economy of not synthesizing unnecessary enzyme(s) and intermediates. This economy is greater when the metabolic block is relatively early in a biosynthetic sequence; thus an anthranilate-requiring mutant outgrew a tryptophan-requiring strain lacking tryptophan synthetase in a medium containing 0.25 mM tryptophan. Wild-type B. subtilis grown in the presence of tryptophan partly repressed the synthesis of tryptophan synthetase. This organism outgrew a derepressed mutant (which synthesized the amino acid constitutively) when both organisms were grown in the presence of tryptophan.

The genotype of an organism may vary by mutation and by gene transfer, and most chemostat studies have been concerned with mutation. To the authors' knowledge there is no report of gene transfer in continuous culture; indeed, the rapid mixing used in this apparatus may preclude such events occurring. There are, however, a few reports of related studies on microbial evolution in which continuous culture has played an important role and examples of these are given below.

Bull and Slater (9) described the evolution of E. coli containing plasmids that carried information essential for the growth of the host. The organism used had chromosomal DNA with the lac Z gene deleted so that no revertants were possible. The Z gene was carried on a plasmid so that the merozygote was able to grow on lactose as carbon and energy source. The precise genotype was $F'Z^+ tra^-/Z^-$ with tra^- referring to transfer deficient, i.e., the plasmid could not reinfect an organism that had become plasmid$^-$. When grown at D 0.4 h^{-1} for 5 days in a lactose-limited chemostat some organisms were present that were plasmid$^-$ (determined by testing colonies for ability to inhibit replication of T$_3$ phage). After 10 days over 90% of the population was plasmid$^-$, and this loss was associated with a rise in the culture β-galactosidase activity. These results indicated that recombination of Z$^+$ from plasmid had occurred prior to the loss of the plasmid. The reason for the higher enzyme activity of the recombinants is unknown but clearly this is a selective advantage in the chemostat.

It is also of interest to study plasmid stability in the absence of selection pressure, and this was the subject of a report by Melling et al. (214). The presence of plasmid RP1 confers resistance towards ampicillin, carbenicillin, tetracycline, kanomycin, and neomycin on cultures of E. coli W3110. The RP1$^+$ phenotype was retained during growth in carbon-, Mg^{2+}-, and phosphate-limited cultures at growth rates in the range 0.05–1.0 h^{-1}. This was found to be due to the retention of the plasmid (tested by mating with strain J53) and there was no evidence for recombination. These results could be explained if the RP1$^-$ phenotype conferred a disadvantage, but competition experiments showed that the RP1$^-$ strain (wild type) could outgrow

RP1 $^+$ under phosphate limitation. There was evidence of the outcome of competition experiments in Mg^{2+}- and carbon-limited cultures being dependent upon the initial numbers of the two strains used. Thus the RP1 $^+$ antibiotic resistant phenotype was able to compete with RP1 $^-$ except under phosphate limitation, and the stability of antibiotic resistant character in the absence of antibiotics is of obvious practical significance.

B. subtilis may be transformed by bacterial DNA and undergo transfection with phage DNA. The ability to act as DNA recipient is environmentally determined and is defined as the "competence" of the culture. Portoles et al. (215) and Espinosa et al. (216) have attempted to define the state of competence by growing recipient cultures in chemostats. The frequency of both transformation and transfection was found to be determined by recipient growth rate, with a main peak at a D of 0.4 h^{-1} and a smaller peak at D 0.1 h^{-1}. The addition of such amino acids as alanine, arginine, and glutamate had a marked effect on both transformation and transfection frequency. This approach promises to yield further results in the future.

ACKNOWLEDGMENTS

We are grateful to our colleagues for permission to cite unpublished work and to Professor D. C. Ellwood and Dr. J. H. Slater for many stimulating discussions.

REFERENCES

1. Dean, A. C. R., Ellwood, D. C., Evans, C. G. T., and Melling, J. (1976). Continuous Culture 6. Applications and New Fields. Ellis Horwood, Chichester, England.
2. Malek, I., and Fencl, Z. (1966). Theoretical and Methodological Basis of Continuous Culture of Microorganisms. Academic Press, New York and London.
3. Kubitschek, H. E. (1970). Introduction to Research with Continuous Cultures. Prentice-Hall, Inc., Englewood Cliffs, New Jersey.
4. Pirt, S. J. (1975). Principles of Microbe and Cell Cultivation. Blackwell Scientific Publications, Oxford.
5. Evans, C. G. T., Herbert, D., and Tempest, D. W. (1970). *In* J. R. Norris and D. W. Ribbons (eds.), Methods in Microbiology, Vol. 2, p. 277. Academic Press, London.
6. Tempest, D. W. (1970). Adv. Microb. Physiol. 4:223.
7. Bull, A. T. (1974). *In* A. T. Bull, J. R. Lagnado, J. O. Thomas, and K. F. Tipton (eds.), Companion to Biochemistry, p. 415. Longman Group Ltd., London.
8. Veldkamp, H. (1976). Continuous Culture in Microbial Physiology and Ecology. Meadowfield Press Ltd., Shildon, Co. Durham, England.
9. Bull, A. T., and Slater, J. H. (1976). *In* A. C. R. Dean, D. C. Ellwood, C. G. T. Evans, and J. Melling (eds.), Continuous Culture 6. Applications and New Fields, p. 49. Ellis Horwood, Chichester, England.
10. Tempest, D. W., and Neijssel, O. M. (1970). *In* A. C. R. Dean, D. C. Ellwood, C. G. T. Evans, and J. Melling (eds.), Continuous Culture 6. Applications and New Fields, p. 283. Ellis Horwood, Chichester, England.
11. Herbert, D. (1961). *In* G. G. Meynell and H. Gooder (eds.), Microbial Reaction to Environment, p. 391. Cambridge University Press, Cambridge, England.

12. Herbert, D. (1976). *In* A. C. R. Dean, D. C. Ellwood, C. G. T. Evans, and J. Melling (eds.), Continuous Culture 6. Applications and New Field, p. 1. Ellis Horwood, Chichester, England.
13. Schaechter, M., Maaløe, O., and Kjeldgaard, N. O. (1958). J. Gen. Microbiol. 19:592.
14. Tempest, D. W., and Hunter, J. R. (1965). J. Gen Microbiol. 41:267.
15. Evans, C. G. T. (1976). *In* A. C. R. Dean, D. C. Ellwood, C. G. T. Evans, and J. Melling (eds.), Continuous Culture 6. Applications and New Fields, p. 346. Ellis Horwood, Chischester, England.
16. Tempest, D. W. (1976). *In* A. C. R. Dean, D. C. Ellwood, C. G. T. Evans, and J. Melling (eds.), Continuous Culture 6. Applications and New Fields, p. 349. Ellis Horwood, Chichester, England.
17. Brown, C. M., and Rose, A. M. (1969). J. Bacteriol. 97:261.
18. Rose, A. H. (1972). J. Appl. Chem. Biotech. 22:527.
19. Tempest, D. W. (1969). *In* P. M. Meadow and S. J. Pirt (eds.), Microbial Growth, p. 87. Cambridge University Press, Cambridge, England.
20. Ellwood, D. C., and Tempest, D. W. (1969). Biochem. J. 111:1.
21. Archibald, A. R., and Coapes, H. E. (1971). Biochem. J. 125:667.
22. Archibald, A. R., and Coapes, H. E. (1976). J. Bacteriol. 125:1195.
23. Archibald, A. R. (1976). *In* A. C. R. Dean, D. C. Ellwood, C. G. T. Evans, and J. Melling (eds.), Continuous Culture 6. Applications and New Fields, p. 262. Ellis Horwood, Chichester, England.
24. Archibald, A. R. (1976). J. Bacteriol. 127:956.
25. Melling, J., and Brown, M. R. W. (1975). *In* M. R. W. Brown (ed.), Resistance of *Pseudomonas aeruginosa*, p. 35. John Wiley and Sons, London.
26. Melling, J., Robinson, A., and Ellwood, D. C. (1974). Proc. Soc. Gen. Microbiol. 1:61.
27. Finch, J. E., and Brown, M. R. W. (1975). J. Antimicrob. Chemother. 1:379.
28. Dean, A. C. R., Ellwood, D. C., Melling, J., and Robinson, A. (1976). *In* A. C. R. Dean, D. C. Ellwood, C. G. T. Evans, and J. Melling (eds.). Continuous Culture 6. Applications and New Fields, p. 251. Ellis Horwood, Chichester, England.
29. Gilbert, P., and Brown, M. R. W. (1977). Proc. Soc. Gen. Microbiol. 4:79.
30. Brown, M. R. W. (1975). Pharm. J. 215:239.
31. Brown, M. R. W. (1977). J. Antimicrob. Chemother. 3:198.
32. Ellar, D. J. (1978). *In* A. T. Bull and P. M. Meadow (eds.), Companion to Microbiology, p. 265. Longman Group Ltd., London.
33. Minnikin, D. E., Abdolrahimzadeh, H., and Baddiley, J. (1972). FEMS Microbiol. Lett. 27:16.
34. Minnikin, D. E., and Abdolrahimzadeh, H. (1974). J. Bacteriol. 120:999.
35. Minnikin, D. E., Abdolrahimzadeh, H., and Baddiley, J. (1974). Nature 249:268.
36. Minnikin, D. E., and Abdolrahimzadeh, H. (1974). FEMS Microbiol. Lett. 43:257.
37. Johnson, B., Brown, C. M., and Minnikin, D. E. (1973). J. Gen. Microbiol. 75:*x*.
38. Brown, C. M., and Hough, J. S. (1965). Nature 206:676.
39. Johnson, B. (1975). Proc. Soc. Gen. Microbiol. 3:55.
40. Luscombe, B. M., and Gray, T. R. G. (1971). J. Gen. Microbiol. 69:433.
41. Luscombe, B. M., and Gray, T. R. G. (1974). J. Gen. Microbiol. 82:213.
42. Duxbury, T., Gray, T. R. G., and Sharples, G. P. (1977). J. Gen. Microbiol. 103:91.
43. Bull, A. T., and Trinci, A. P. J. (1977). Adv. Microb. Physiol. 15:1.
44. Lin, E. C., Levin, A. P., and Magasanik, B. (1960). J. Biol. Chem. 235:1824.

45. Neijssel, O. M., Hueting, S., Crabbendam, K. J., and Tempest, D. W. (1975). Arch. Microbiol. 104:83.
46. Calcott, P. H., and Postgate, J. R. (1974). J. Gen. Microbiol. 85:85.
47. Dawes, E. A., Midgley, M., and Whiting, P. H. (1976). In A. C. R. Dean, D. C. Ellwood, C. G. T. Evans, and J. Melling (eds.), Continuous Culture 6. Applications and New Fields, p. 195. Ellis Horwood, Chichester, England.
48. Whiting, P. H., Midgley, M., and Dawes, E. A. (1976). J. Gen. Microbiol. 92:304.
49. Tempest, D. W., Meers, J. L., and Brown, C. M. (1970). Biochem. J. 117:405.
50. Tempest, D. W., Meers, J. L., and Brown, C. M. (1973). In S. Prusiner and E. R. Stadtman (eds.), The Enzymes of Glutamine Metabolism, p. 167. Academic Press, New York.
51. Brown, C. M., Macdonald-Brown, D. S., and Meers, J. L. (1974). Adv. Microb. Physiol. 11:1.
52. Brown, C. M. (1976). In A. C. R. Dean, D. C. Ellwood, C. G. T. Evans, and J. Melling (eds.), Continuous Culture 6. Applications and New Fields, p. 170. Ellis Horwood, Chichester, England.
53. Miflin, B. J., and Lea, P. J. (1976). Phytol. Chem. 15:873.
54. Sims, A. P., and Folkes, B. F. (1964). Proc. Roy. Soc. Lond. (Biol.) 159:479.
55. Ferguson, A. R., and Sims, A. P. (1974). J. Gen. Microbiol. 80:159.
56. Ferguson, A. R., and Sims, A. P. (1974), J. Gen. Microbiol. 80:173.
57. Folkes, B. F., and Sims, A. P. (1974). J. Gen. Microbiol. 82:77.
57a. Burn, V. J., Turner, P. R., and Brown, C. M. (1974). Antonie van Leeuwenhoek 40:93.
58. van Andel, J. G., and Brown, C. M. (1977). Arch. Microbiol. 111:265.
59. Clark, B., and Holms, W. H. (1976). J. Gen. Microbiol. 95:191.
60. Herbert, D., and Kornberg, H. L. (1976). Biochem. J. 156:477
61. Carter, I. S., and Dean, A. C. R. (1977). Biochem. J. 166:643
62. Alim, S., and Ring, K. (1976). Arch. Microbiol. 111:105.
63. Dean, A. C. R. (1972). J. Appl. Chem. Biotech. 22:245
64. Herbert, D., and Phipps, P. J. (1974). Proc. Soc. Gen. Microbiol. 1:70.
65. Veenhuis, M., van Dijken, J. P., and Harder, W. (1976). Arch. Microbiol. 111:123.
66. van Dijken, J. P., Otto, R., and Harder, W. (1976). Arch. Microbiol. 111:137.
67. Harrison, D. E. F. (1972). J. Appl. Chem. Biotech. 22:417.
68. Harrison, D. E. F. (1976). Adv. Microb. Physiol. 14:243.
69. MacLennan, D. G., and Pirt, S. J. (1966). J. Gen. Microbiol. 45:289.
70. Senior, P. J., Beech, G. A., Ritchie, G. A. F., and Dawes, E. A. (1972). Biochem. J. 128:1193.
71. Jackson, F. A., and Dawes, E. A. (1976). J. Gen. Microbiol. 97:303.
72. Ward, A. C., Rowley, B. I., and Dawes, E. A. (1977). J. Gen. Microbiol. 102:61.
73. Light, P. A., and Garland, P. B. (1971). Biochem. J. 124:123.
74. Light, P. A. (1972). J. Appl. Chem. Biotech. 22:509.
75. Aiking, H., and Tempest, D. W. (1976). Arch. Microbiol. 108:117.
76. Aiking, H., Sterkenberg, A., and Tempest, D. W. (1977). Arch. Microbiol. 113:65.
77. Harder, W., and Dijkshuizen, L. (1976). In A. C. R. Dean, D. C. Ellwood, C. G. T. Evans, and J. Melling (eds.), Continuous Culture 6. Applications and New Fields, p. 297. Ellis Horwood, Chichester, England.
78. Monod, J. (1950). Ann. Inst. Pasteur 79:390.
79. Standing, C. N., Fredrickson, A. G., and Tsuchiya, H. M. (1972). Appl. Microbiol. 23:354.
80. Smith, M. E., and Bull, A. T. (1976). J. Appl. Bacteriol. 41:81.

81. Mateles, R. I., Chian, S. K., and Silver, R. (1967). *In* E. O. Powell, C. G. T. Evans, R. E. Strange, and D. W. Tempest (eds.), Microbial Physiology and Continuous Culture, p. 232. H.M.S.O., London.
82. Chian, S. K., and Mateles, R. I. (1968). Appl. Microbiol 16:1337.
83. Silver, R., and Mateles, R. I. (1969). J. Bacteriol. 97:535.
84. Edwards, V. H. (1969). Biotech. Bioeng. 11:99.
85. Hamlin, B. T., Ng, F. M. W., and Dawes, E. A. (1967). *In* E. O. Powell, C. G. T. Evans, R. E. Strange, and D. W. Tempest (eds.), Microbial Physiology and Continuous Culture, p. 211. H.M.S.O., London.
86. Hamilton, W. A., and Dawes, E. A. (1959). Biochem. J. 71:25.
87. Sariaslani, F. S., Westwood, A. W., and Higgins, I. J. (1975). J. Gen. Microbiol. 91:315
88. Hough, J. S., Keevil, C. W., Maric, V., Philliskirk, G., and Young, T. W. (1976). *In* A. C. R. Dean, D. C. Ellwood, C. G. T. Evans, and J. Melling (eds.), Continuous Culture 6. Applications and New Fields, p. 226. Ellis Horwood, Chichester, England.
89. Macquillan, A. M., and Halvorson, H. O. (1962). J. Bacteriol. 84:23.
90. Sutton, D. D., and Lampen, J. O. (1962). Biochim. Biophys. Acta 56:303.
91. Millin, D. J. (1963). J. Inst. Brew. 69:389.
92. Harris, G., and Millin, D. J. (1963). Biochem. J. 88:89.
93. Portno, A. D. (1968). J. Inst. Brew. 74:448.
94. Portno, A. D. (1969). J. Inst. Brew. 75:468.
95. Pohland, D., Ringpfeil, M., and Behrends, U. (1966). Z. Allg. Mikrobiol. 6:386.
96. Baidya, T. K. N., Webb, F. C., and Lilly, M. D. (1967). Biotech. Bioeng. 9:195.
97. Gaudy, A. F., Gaudy, E. T., and Komolrit, R. (1963). Appl. Microbiol. 11:157.
98. Gaudy, A. F. (1962). Appl. Microbiol. 10:264
99. Postgate, J. R. (1973). *In* T. Rosswall (ed.), Modern Methods in the Study of Microbial Ecology, p. 287. Swedish Natural Science Research Council, Bulletins from the Ecological Research Committee 17, Stockholm.
100. Gardner, M. R., and Ashley, W. R. (1970). Nature 228:784.
101. May, R. M. (1973). Stability and Complexity in Model Ecosystems. Princeton University Press, Princeton, New Jersey.
102. Saunders, P. T., and Bazin, M. J. (1975). Nature 256:120.
103. Friello, D. A., Mylroie, J. R., and Chakrabarty, A. M. (1976). Proceedings of the 3rd International Biodegradation Symposium, p. 205. Applied Science Publications, London.
104. Slater, J. H., and Bull, A. T. (1978). *In* A. T. Bull and P. M. Meadow (eds.), Companion to Microbiology, p. 181. Longman Group Ltd., London.
105. Bungay, H. R., and Bungay, M. L. (1968). Adv. Appl. Microbiol. 10:269.
106. Hobson, P. N. (1969). *In* S. J. Pirt and P. M. Meadow (eds.), Microbial Growth, p. 43. Cambridge University Press, Cambridge, England.
107. Veldkamp, H., and Jannasch, H. W. (1972). J. Appl. Chem. Biotech. 22:105.
108. Meers, J. L. (1973). C. R. C. Crit. Rev. Microbiol. 2:139.
109. Jannasch, H. W., and Mateles, R. I. (1974). Adv. Microb. Physiol. 11:165.
110. Veldkamp, H. (1976). *In* A. C. R. Dean, D. C. Ellwood, C. G. T. Evans, and J. Melling (eds.), Continuous Culture 6. Applications and New Fields, p. 315. Ellis Horwood, Chichester, England.
111. Bailey, J. E., and Ollis, D. F. (1977). Biochemical Engineering Fundamentals. McGraw-Hill Book Co., New York.
112. Veldkamp, H. (1970). *In* J. R. Norris and D. W. Ribbons (eds.), Methods in Microbiology, Vol. 3A, p. 305. Academic Press, London and New York.

113. Campacci, E. F., New, P. R., and Tchan, Y. T. (1977). Nature 266:164.
114. Jost, J. L., Drake, J. F., Fredrickson, A. G., and Tsuchiya, H. M. (1973). J. Bacteriol. 113:834.
115. Lee, I. H., Fredrickson, A. G., and Tsuchiya, H. M. (1976). Biotech. Bioeng. 18:513.
116. Bryant, M. P., Wolin, E. A., Wolin, M. J., and Wolfe, R. S. (1967). Arch. Mikrobiol. 59:20.
117. Gray, B. H., Foudler, C. F., Nogent, N. A., Rigopoulos, N., and Fuller, R. C. (1973). Int. J. Syst. Bacteriol. 23:256.
118. Coulter, W. A., and Russell, C. (1976). J. Appl. Bacteriol. 40:73.
119. Dibdin, G. H., Shellis, R. P., and Wilson, C. M. (1976). J. Appl. Bacteriol. 40:261.
120. Sudo, S. Z. (1977). Appl. Environ. Microbiol. 33:450.
121. Ellwood, D. C., Hunter, J. R., and Longyear, V. M. C. (1974). Arch. Oral Biol. 19:659.
122. Ellwood, D. C., and Hunter, J. R. (1976). In A. C. R. Dean, D. C. Ellwood, C. G. T. Evans, and J. Melling (eds.), Continuous Culture 6. Applications and New Fields, p. 270. Ellis Horwood, Chichester, England.
123. Mikx, F. H. M., and van der Honen, J. S. (1975). Arch. Oral Microbiol. 20:407.
124. Hunter, J. R., Baird, J. K., and Ellwood, D. C. (1974). J. Dent. Res. 5:954.
125. Draggan, S. (1976). Int. J. Environ. Stud. 10:65.
126. Metcalf, R. L., Sangha, G. K., and Kapoor, I. P. (1971). Environ. Sci. Technol. 5:709.
127. Hussain, H. S. N., and Bull, A. T. (1978). In preparation.
128. Mosser, J. L., Fisher, N. S., and Wurster, C. F. (1972). Science 176:533.
129. Fisher, N. S., Carpenter, E. J., Remson, C. C., and Wurster, C. F. (1974). Microb. Ecol. 1:39.
130. Fisher, N. S., Guillard, R. R. L., and Wurster, C. F. (1976). In R. P. Canale (ed.), Modeling Biochemical Processes in Aquatic Ecosystems, p. 305. Ann Arbor Science Publications, Inc., Ann Arbor, Michigan.
131. Fisher, N. S., and Wurster, C. F. (1973). Environ. Pollut. 5:205.
132. Colwell, R. R., and Walker, J. D. (1977). C. R. C. Crit. Rev. Microbiol. 6:423.
133. Horvath, R. S. (1972). Bacteriol. Rev. 36:146.
134. Huber, T. J., Street, J. R., Bull, A. T., Cook, K. A., and Cain, R. B. (1975). Arch Microbiol. 102:139.
135. Stafford, D. A., and Callely, A. G. (1977). In A. G. Callely, C. F. Forster, and D. A. Stafford (eds.), Treatment of Industrial Effluents, p. 129. Hodder and Stoughton, London.
136. Grimes, D. J., and Morrison, S. M. (1975). Microb. Ecol. 2:43.
137. Isensee, A. R. (1976). Int. J. Environ. Stud. 10:35.
138. Scura, E. D., and Theilacker, G. H. (1977). Marine Biol. 40:317.
139. Jensen, H. L. (1957). Can. J. Microbiol. 3:151.
140. Senior, E., Bull, A. T., and Slater, J. H. (1976). Nature 263:476.
141. Gunner, H. B., and Zuckerman, B. M. (1968). Nature 217:1183.
142. Daughton, C. G., and Hsieh, D. P. H. (1977). Appl. Env. Microbiol. 34:175.
143. Paris, D. F., Lewis, D. L., and Wolfe, N. L. (1975). Env. Sci. Technol. 9:135.
144. McClure, G. W. (1973). Water Res. 7:1683.
145. Munnecke, D. M., and Hsieh, D. P. H. (1975). Appl. Microbiol. 30:575.
146. Munnecke, D. M., and Hsieh, D. P. H. (1976). Appl. Environ. Microbiol. 31:63.
147. Munnecke, D. M. (1976). Appl. Environ. Microbiol. 32:7.
148. Munnecke, D. M. (1977). Appl. Environ. Microbiol. 33:503.
149. Senior, E. (1977). Ph.D. thesis, University of Kent at Canterbury, England.

150. Senior, E., Slater, J. H., and Bull, A. T. (1975). J. Appl. Chem. Biotech. 25:329.
151. Evans, C. G. T., and Kite, S. (1961). *In* Continuous Culture of Microorganisms, S.C.I. Monograph No. 12, p. 175. Society of Chemical Industry, London.
152. Wase, D. A. J., and Hough, J. S. (1966). J Gen. Microbiol. 42:13.
153. Jones, G. L., Jansen, F., and McKay, A. J. (1973). J. Gen. Microbiol. 74:139.
154. Yang, R. D., and Humphrey, A. E. (1975). Biotech. Bioeng. 17:1211.
155. Sauer, M., and Johanides, V. (1976). *In* H. Dellweg (ed.), Abstracts of Papers, 5th International Fermentation Symposium, p. 427. Verlag Versuchs—und Lehranstalt für Spiritusfabrikation und Fermentationstechnologie im Institut für Garungsgewerbe und Biotechnologie, Berlin.
156. Osman, A., Bull, A. T., and Slater, J. H. (1976). *In* H. Dellweg (ed.), Abstracts of Papers, 5th International Fermentation Symposium, p. 124. Verlag Versuchs—und Lehranstalt für Spiritusfabrikation und Fermentationstechnologie im Institut für Garungsgewerbe und Biotechnologie, Berlin.
157. Sato, T., and Akiyama, T. (1972). Water Res. 6:1059.
158. Tyler, J. E., and Finn, R. K. (1974). Appl. Microbiol. 28:181.
159. Johanides, V., and Hrsak, D. (1976). *In* H. Dellweg (ed.), Abstracts of Papers, 5th International Fermentation Symposium, p. 426. Verlag Versuchs—und Lehranstalt für Spiritusfabrikation und Fermentationstechnologie im Institut für Garungsgewerbe und Biotechnologie, Berlin.
160. Cain, R. B. (1977). *In* A. G. Callely, C. F. Forster, and D. A. Stafford (eds.), Treatment of Industrial Effluents, p. 283. Hodder and Stoughton, London.
161. Atlas, R. M. (1977). C. R. C. Crit. Rev. Microbiol. 6:371.
162. Balashova, V. V., Vedenina, I. Y., Markosyan, C. E., and Zavarzin, G. A. (1973). Mikrobiologiia 43:581.
163. Markosyan, C. E. (1973). Dokl. Acad. Nauk SSSR. 211:1205.
164. Zeikus, J. G. (1977). Bacteriol. Rev. 41:514.
165. Cappenberg, T. E. (1975). Microb. Ecol. 2:60.
166. Bryant, M. P., Campbell, L. L., Reddy, C. A., and Crabill, M. R. (1977). Appl. Environ. Microbiol. 33:1162.
167. Sineriz, F., and Pirt, S. J. (1977). J. Gen. Microbiol. 101:57.
168. Wilkinson, T. G., Topiwala, H. H., and Hamer, G. (1974). Biotech. Bioeng. 16:41.
169. Harrison, D. E. F., Wilkinson, T. G., Wren, S. J., and Harwood, J. H. (1976). *In* A. C. R. Dean, D. C. Ellwood, C. G. T. Evans, and J. Melling (eds.), Continuous Culture 6. Applications and New Fields, p. 122. Ellis Horwood, Chichester, England.
170. Wilkinson, T. G., and Harrison, D. E. F. (1973). J. Appl. Bacteriol. 36:309.
171. Linton, J. D., and Buckee, J. C. (1977). J. Gen. Microbiol. 101:219.
172. Topiwala, H. H., and Hamer, G. (1971). Biotech. Bioeng. 13:919.
173. Larsen, D. H., and Dimmock, R. L. (1964). J. Bacteriol. 88:1380.
174. Bredan, C. R., and Buswell, A. M. (1933). J. Bacteriol. 26:379.
175. Wilkinson, T. G., and Hamer, G. (1974). Biotech. Bioeng. 16:251.
176. Munson, R. J., and Bridges, B. A. (1964). J. Gen. Microbiol. 37:411.
177. Atkinson, B., and Knights, A. J. (1975). Biotech. Bioeng. 17:1245.
178. Atkinson, B. (1974). Biochemical Reactors. Pion Press, London.
179. Scott, C. D., and Haucher, C. W. (1976). Biotech. Bioeng. 18:1393.
180. Ashby, R. E., and Bull, A. T. (1977). Lab. Pract. 26:327.
181. Hattori, R. (1972). J. Gen. Appl. Mircrobiol. 18:319.
182. Lamotta, E. J. (1976). Appl. Environ. Microbiol. 31:286.
183. Scott, C. D., Haucher, C. W., Holladay, D. W., and Dinsmore, G. B. (1975). *In* Proceedings of the Symposium on Environmental Aspects of Fuel Conver-

sion Technology II, p. 233. Environmental Protection Agency, Washington, D.C.
184. ZoBell, C. E. (1943). J. Bacteriol. 46:39.
185. Button, D. K. (1969). Limnol. Oceanog. 14:95.
186. Jannasch, H. W., and Pritchard, P. H. (1972). Mem. Inst. Ital. Idrobiol. Suppl. 29:289.
187. Martin, J. P., and Filip, Z., and Haider, K. (1976). Soil Biol. Biochem. 8:409.
188. Maigetter, R. Z., and Pfister, R. M. (1975). Can. J. Microbiol. 21:173.
189. Roper, M. M., and Marshall, K. C. (1974). Microb. Ecol. 1:1.
190. Hendricks, C. W. (1974). Appl. Microbiol. 28:572.
191. Legner, M., Puncochar, P., and Straskrabova, V. (1976). In A. C. R. Dean, D. C. Ellwood, C. G. T. Evans, and J. Melling (eds.), Continuous Culture 6. Applications and New Fields, p. 329. Ellis Horwood, Chichester, England.
192. Bazin, M. J., Saunders, P. T., and Prosser, J. I. (1976). C. R. C. Crit. Rev. Microbiol. 4:413.
193. Novick, A., and Szilard, L. (1950). Proc. Natl. Acad. Sci. USA 36:708.
194. Kubitschek, H. E. (1974). In M. J. Carlile and J. J. Skehel (eds.), Evolution in the Microbial World, p. 105. Cambridge University Press, London.
195. Bryson, V., and Szybalski, W. (1952). Science 116:45.
196. Kubitschek, H. E., and Bendigkeit, H. E. (1964). Mutat. Res. 1:113.
197. Kubitschek, H. E., and Bendigkeit, H. E. (1964). Mutat. Res. 1:209.
198. Gibson, T. C., Scheppe, M. L., and Cox, E. C. (1970). Science 169:686.
199. Cox, E. C., and Gibson, T. C. (1974). Genetics 77:169.
200. Nestermann, E. R., and Hill, R. F. (1973). Genetics 73:41.
201. Harder, W., Kuenen, J. G., and Matin, A. (1977). J. Appl. Microbiol. 43:1.
202. Horiuchi, T., Tomzawa, J., and Novick, A. (1962). Biochim. Biophys. Acta 55:152.
203. Clarke, P. H. (1974). In M. J. Carlile and J. J. Skehel (eds.), Evolution in the Microbial World, p. 183. Cambridge University Press, London.
204. Mortlock, R. P., Fossitt, D. D., and Wood, W. A. (1965). Proc. Natl. Acad. Sci. USA 54:572.
205. Mortlock, R. P., and Wood, W. A. (1971). In I. A. Bernstein (ed.), Biochemical Responses to Environmental Stress, p. 1. Plenum Press, London.
206. Hartley, B. S., Burleigh, B. D., Midwinter, G. G., Moore, C. H., Morris, H. R., Rigby, P. W. J., Smith, M. J., and Taylor, S. S. (1972). In Enzymes: Structure and Function, p. 151. North Holland, Amsterdam.
207. Rigby, P. W. J., Burleigh, B. D., and Hartley, B. S. (1974). Nature 251:200.
208. Hartley, B. S. (1974). In M. J. Carlile and J. J. Skehel (eds.), Evolution in the Microbial World, p. 151. Cambridge University Press, London.
209. Inderlied, C. B., and Mortlock, R. P. (1977). J. Molec. Evol. 9:181.
210. Bull, A. T., Senior, E., and Slater, J. H. (1976). Proc. Soc. Gen. Microbiol. 4:39.
211. Slater, J. H., Senior, E., and Bull, A. T. (1976). Proc. Soc. Gen. Microbiol. 4:39.
212. Francis, J. C., and Hansche, P. E. (1972). Genetics 70:59.
213. Zamenhof, S., and Eichhorn, H. H. (1967). Nature 216:456.
214. Melling, J., Ellwood, D. C., and Robinson, A. (1977). FEMS Microbiol. Lett. 2:87.
215. Portoles, A., Lopez, R., and Fapia, A. (1973). In L. J. Archer (ed.), Bacterial Transformation, p. 65. Academic Press, London.
216. Espinosa, M., Lopez, R., Perez-Urina, M. T., Garcia, E., and Portoles, A. (1976). J. Gen. Microbiol. 97:297.

International Review of Biochemistry
Microbial Biochemistry, Volume 21
Edited by J. R. Quayle
Copyright 1979   University Park Press   Baltimore

6
Utilization of Inorganic Nitrogen by Microbial Cells

H. DALTON
Department of Biological Sciences,
University of Warwick,
Conventry, England

The vast majority of organisms on this planet, apart from most animal cells, obtain their nitrogen from an inorganic source. This ranges from the most oxidized form, the nitrate anion, through the uncharged dinitrogen molecule to the most reduced form, the ammonium ion. No matter what the primary source of nitrogen is for growth, however, it is transformed to ammonia before it is assimilated into the amino acids and hence into proteins and nucleic acids in the cell.

Although I only intend to cover the assimilation of inorganic nitrogen compounds, it should be realized that many organic nitrogen sources can be broken down to yield ammonia, which can then be assimilated by one or more of the mechanisms outlined below.

Several inorganic nitrogen compounds, e.g., NO^-_2, NO^-_3, and NH_3, are metabolized by microbes in a variety of energy-yielding reactions in which there is no net incorporation of nitrogen into biomass. These reactions, for the purposes of this chapter, are referred to as dissimilatory, to distinguish them from reactions in which nitrogen is incorporated into biomass, which are termed assimilatory (Figure 1). Only the assimilatory reactions are considered here.

Because all inorganic nitrogen sources must be converted to ammonia before true assimilation of the nitrogen can be undertaken, I propose to consider the incorporation of the ammonium ion into amino acids first. The subsequent fate of the amino acid so formed is not considered, and the reader is referred to a number of articles on this subject. The trans-

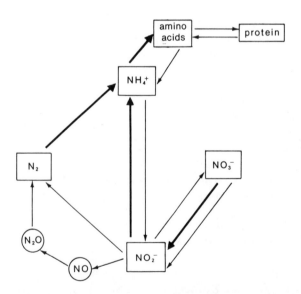

Figure 1. Mobilization of inorganic nitrogen compounds by microorganisms: →, assimilatory reactions leading to a net incorporation into biomass (considered in this chapter); ⇀, reactions involving intermediates or dissimilation of inorganic nitrogen (not considered here).

formation of inorganic nitrogen compounds more oxidized than ammonia is reviewed, as well as current ideas on the actual uptake and entry of nitrogen compounds across cell membranes. By adopting this approach, working from the inside of the cell to the outside, I hope that the reader will be able to view one of the more intriguing aspects of microbial metabolism—that reactions of fundamental importance are basically the same, no matter what the organism: it is only in the initial transformation that most organisms differ.

ASSIMILATION OF AMMONIA

Discovery of Glutamate Synthase

Because ammonium ions readily dissociate at alkaline pH values to form ammonia, it is not always clear exactly what is the substrate for the ammonia-assimilating enzymes. I have used the word "ammonia" in this context, to denote either NH_4^+ or NH_3.

The mechanism of ammonia assimilation by microorganisms had aroused only polite interest until 1970, probably for a number of reasons. First, it was commonly accepted that the dehydrogenases for alanine and glutamate to a greater extent, and leucine and valine to a lesser extent, were responsible for the assimilation of ammonia in microbes. Aspartate ammonia lyase, in reverse, was also assumed to play a role in microbial ammonia assimilation. Second, organisms had been grown, for the most part, in cultures in which the ammonia concentration in the medium was high and generally largely in excess of the K_m for these enzymes to operate in a biosynthetic role (between 20 and 300 mM), and so there was not any pressure to look for enzymes capable of assimilating ammonia at low substrate concentrations.

With the increasing use of chemostat cultures, in which it was possible to maintain very low concentrations of a substrate to an actively growing population of organisms, an investigation into the factors affording growth at low ammonia concentrations was made. Meers, Tempest, and Brown (1,2) observed that at high ammonia concentrations with either glucose or phosphate as the growth-limiting nutrient the main pathway of ammonia assimilation in a number of bacteria was via glutamate dehydrogenase (GDH). The K_m of GDH for ammonia in *Aerobacter aerogenes* in particular was about 10mM, and so this enzyme would function in a biosynthetic role under these circumstances (Figure 2). When grown under ammonia-limiting conditions in which the intracellular concentration of ammonia was less than 0.5mM, the level of glutamate dehydrogenase in all the bacteria tested was very small and it could not function as an ammonia-assimilating enzyme. Under these circumstances the organisms assimilated ammonia in a two-step process involving glutamine synthetase (GS) and glutamate synthase (GOGAT). The reactions involved first the amidation of endogenous glu-

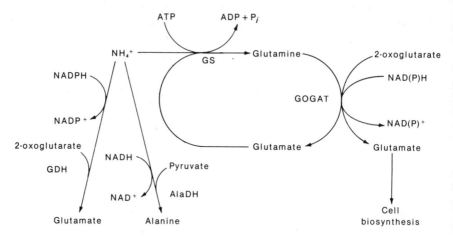

Figure 2. Pathways involved in the assimilation of ammonia: GS, glutamine synthetase; GDH, glutamate dehydrogenase; GOGAT, glutamine(amide):2-oxoglutarate amidotransferase oxidoreductase (glutamate synthase); AlaDH, alanine dehydrogenase.

tamate to glutamine and then a reductive transfer of the amide nitrogen of glutamine to the 2 position of 2-oxoglutarate, giving rise to the net synthesis of 1 mol of glutamate from 1 mol of ammonia (Figure 2).

The real significance of their findings lies in the identification of GOGAT. GS had been observed in bacteria before (3) and indeed it was postulated as an ATP-driven "pump" that could effectively scavenge trace amounts of ammonia from the environment (4), but this only led to the synthesis of glutamine in which incorporation of the nitrogen from ammonia was into the amide-N of glutamine. This nitrogen could then be donated to carbamyl phosphate and from this to purines. The nitrogen was only considered to contribute to the non-α-amino-N of tryptophan, histidine, and arginine, (5) and glutamate was presumed to be synthesized by GDH alone. The K_m for ammonia by GS was less than 1 mM (6) and the values for glutamine by GOGAT from several bacteria ranged from 0.1 to 2 mM (2). Clearly such a system would be able to operate at ammonia concentrations well below those required for assimilation by the GDH system.

Distribution of GS/GOGAT and GDH in Microbes

Since these original observations were reported there have been a large number of reports on the incidence of the GS-GOGAT system for ammonia assimilation in microorganisms. In some cases, the biosynthetic role of the system for glutamate synthesis from ammonia is now clearly established; in others, the evidence is not so clear. The mere presence of the enzymes glutamine synthetase and glutamate synthase is not sufficient evidence to indicate that the GS/GOGAT system is operating in its biosynthetic role, because there are many factors responsible for controlling their activities (see below).

Bacteria   The incidence of the GS/GOGAT system for assimilation of low concentrations of ammonia is now well established in many species of bacteria. In general terms, it is fair to say that ammonia is assimilated by GDH or other amino acid dehydrogenases when the environmental concentration of ammonia is high and via GS/GOGAT when it is low. The GS/GOGAT system is ATP-dependent and physiologically irreversible, whereas the GDH system does not require ATP and is reversible.

Based on the identification of the GS/GOGAT pathway in bacteria, it is fair to conclude that the importance of the amino acid dehydrogenases in biosynthesis of amino acids from ammonia only assumes significance when the ammonia concentration in the environment is high (because of their high K_m values for ammonia). Because this is generally an unlikely environmental situation, the main physiological role of these enzymes appears to be catabolic. Indeed, Meers and Pedersen (7) found that the NADP-linked GDH in *Bacillus licheniformis* not only served its biosynthetic role at high NH_3 concentrations, but also must play an important catabolic role, because there was an increase in enzyme activity when carbon-limited cultures were pulsed with 30 mM glutamate. This increase in GDH corresponded to an increase in ammonia production and, therefore, the enzyme could also act, in *B. licheniformis* at least, in its deaminating role and therefore prevent an excessive accumulation of the pool glutamate level. There is some evidence also to suggest that there are two distinct GDH activities in bacteria, one NAD-linked, the other NADP-linked. For example, in *Thiobacillus novellus* the NADPH-linked enzyme appears to be more important during early phases of growth in which a biosynthetic activity could presumably be ascribed to the enzyme; the NAD-linked enzyme assumed greater importance during the late logarithmic phase of growth and when glutamate was the nitrogen source (8). Further observations that the NADP-linked enzyme may be biosynthetically active has come from studies on *Hydrogenomonas* (9) and *Pseudomonas* sp. (10). Under conditions in which the environmental ammonia concentration was in excess (and hence the GS/GOGAT system was repressed) these organisms synthesized the NADP-linked GDH, suggesting a biosynthetic role for the enzyme. The NAD-linked enzyme, in the case of *Hydrogenomonas*, was found in high concentrations only when ammonia was limiting, suggesting that this enzyme was acting catabolically.

Most bacteria, however, possess only one type of GDH activity, which may require either NADH or NADPH. The enzyme presumably adopts a biosynthetic role when the cells are grown at high ammonia concentrations and a catabolic role when the cells are grown at low ammonia concentrations or in the presence of glutamate. When the ammonia concentration is limiting growth, then the level of GDH in *Aerobacter aerogenes* was found to have dropped to 3% of its level in glucose-limited cultures (2, 11). Furthermore, the level of GOGAT was observed to increase about 70-fold over its level in the carbon-limited cultures.

Yeasts and Fungi   Until 1973 the biosynthetic glutamate dehydrogenase (NADP⁺-linked) was assumed to be responsible for ammonia assimilation in yeasts and attempts to find either GS or GOGAT had been unsuccessful. In order to permit assimilation at low concentrations of ammonia, it was considered that the ability of some yeasts to concentrate ammonia in the cell (see "Transport of Ammonia into Cells," below) or to derepress GDH at low external concentrations of ammonia was responsible for assimilation under these conditions (12).

Investigations into the assimilation of ammonia by the genus *Schizosaccharomyces* (fission yeasts) showed that there was no derepression of GDH under ammonia-limiting conditions (13). Because the K_m for the enzyme was 25mM and there was no significant intracellular accumulation of ammonia, it was clear that some other mechanism was operative. Indeed, the presence of a GS/GOGAT system was observed in which GOGAT had a low K_m for its substrates (about 0.8 mM for glutamine and 2-oxoglutarate) and a pH optimum of 6.4, which was about one pH unit lower than that observed for the enzyme from bacterial systems. The GS/GOGAT system has also been found in appreciable quantities in *Schiz. malidevorans* and *Saccharomycodes ludwigii* (14). Although the presence of GOGAT has been observed in the budding yeast *Saccharomyces cerevisiae* (15), it is unlikely to play any significant role in the assimilation of ammonia at low concentrations because it is only present in very low concentrations under these conditions and GDH-NADP is derepressed. Furthermore, GS, which is required for the biosynthetic functioning of GOGAT, is not present when GOGAT is synthesized (16).

The assimilation of ammonia by most yeasts and fungi appears to occur through NADP-linked GDH (the NAD⁺-linked form is considered as a catabolic enzyme) in which a substantial derepression of the enzyme occurs when grown on limiting ammonia (17). Under these conditions it is assumed that there is sufficient enzyme present to assimilate low concentrations of ammonia even though the K_m is high. It has been calculated that, if all enzyme sites were fully saturated, only about 30 units of GDH enzyme would be required to satisfy growth of yeasts at a specific growth rate of 0.1 h⁻¹. In practice, values of between 1,600 and 2,100 units of activity are found (12), which would adequately compensate for the high K_m of the enzyme.

Regulation of Ammonia Metabolism

Modification of Glutamine Synthetase in Vitro   We have already seen that many microbes can switch their pathways of ammonia assimilation depending on the environmental concentration of ammonia. Even before Tempest and co-workers discovered that glutamate synthase was essential in cyclical assimilation of ammonia into glutamate, Holzer and colleagues (18) and Stadtman and colleagues (19) had established that glutamine synthetase in *Escherichia coli* would respond to changes in the concentration of ammonia. When the external concentration of ammonia was high, glutamine synthetase rapidly lost its ability to form glutamine (equation 1):

$$\text{L-glutamate} + NH_3 + ATP \underset{}{\overset{Mg^{2+}}{\rightleftharpoons}} \text{L-glutamine} + ADP + P_i \qquad (1)$$

although the enzyme still retained glutamyl transferase activity (equation 2):

$$\text{L-glutamine} + NH_2OH \underset{Mg^{2+} \text{ or } Mn^{2+}}{\overset{ADP,\ P_i,\ \text{or arsenate}}{\rightleftharpoons}} \gamma\text{-glutamyl hydroxamate} + NH_3 \qquad (2)$$

Upon removal of the excess ammonia from the growth medium there was a rapid conversion of the enzyme from the inactive to the biosynthetically active form. Furthermore, the ability of GS to form glutamine in the biosynthetic reaction is rapidly decreased by the attachment of adenylyl groups to the enzyme. Removal of the adenylyl groups results in a rapid increase in the biosynthetic activity of GS. Both forms of the enzyme, i.e., adenylylated or unadenylylated, have glutamyl transferase activity.

Glutamine synthetase is composed of 12 identical subunits, each with a molecular weight of 50,000, so that complete inactivation of the enzyme occurs when an adenylyl group has been attached to each tryrosine residue on each of the subunits. The adenylylation reaction requires ATP and a specific enzyme, adenylyl transferase (ATase), the activity of which is stimulated by glutamine. Furthermore, this enzyme is also under the control of a regulator protein (P_{II}). The ATase and P_{II} protein are also responsible for the de-adenylylation reaction, thereby restoring the biosynthetic activity of glutamine synthetase. Whether GS is adenylylated or de-adenylylated depends on the form of the P_{II} protein. The unmodified form, P_{IIA}, stimulates adenylylation, whereas the modified form, P_{IID}, in which uridylyl groups are attached, is necessary for de-adenylylation. The uridylylation is accomplished by the enzyme uridylyl transferase (UTase) (20, 21) and another enzyme (UR) catalyzes the removal of the uridylyl groups from P_{IID} (21).

Investigations into the activities of the two enzymes UTase and ATase in cell extracts has revealed that their activities are closely controlled by the levels of two metabolites in the cell. 2-Oxoglutarate stimulates and glutamine inhibits UTase, whereas the opposite is true for the ATase P_{IIA} complex, in which 2-oxoglutarate inhibits and glutamine stimulates the enzyme (Figure 3). Thus the state of adenylylation of glutamine synthetase, and hence its biosynthetic activity, can be readily controlled by the relative concentration of these two metabolites through their effects on adenylylation and uridylylation. Finally, to complete the cascade-type regulation, the controlling element over the levels of 2-oxoglutarate and glutamine in the cell is the level of ammonia.

The operation of such a control system is envisaged as follows. Cells

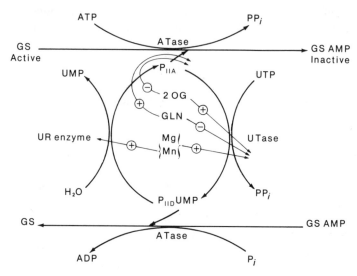

Figure 3. Scheme for the adenylylation and deadenylylation of glutamine synthetase: 2 OG, 2-oxoglutarate; GLN, glutamine; ⊕, reaction stimulated; ⊖, reaction inhibited. See text for details of operation. (Adapted from refs. 19, 20, 21.)

growing with ammonia as the growth-limiting nutrient would possess a high endogenous level of 2-oxoglutarate, due to the inactivity of GDH, whose K_m for ammonia is high, and a low level of glutamine. This high 2-oxoglutarate:glutamine ratio would stimulate UTase activity, resulting in uridylylation of the P_{IIA} protein to the P_{IID} UMP form (20). In this form the P_{II} protein will interact with ATase to de-adenylylate any endogenous adenylylated GS present. Because the P_{II} protein is uridylylated under these conditions it only stimulates the de-adenylylation reaction catalyzed by ATase. ATase will also catalyze the adenylylation reaction but only in the presence of the non-uridylylated P_{II} protein (P_{IIA}). Because the ATase appears to be a constitutive enzyme, its activity can only be regulated by the P_{II} protein.

When cells growing under limiting ammonia are given an excess of ammonia the level of 2-oxoglutarate would be reduced, due to its conversion to glutamate, and the level of glutamine would rise. It is envisaged that the P_{IID} UMP protein would then lose its uridylyl groups through the activity of the uridylyl-removing (UR) enzyme, converting it to P_{IIA}[1]. The low 2-oxoglutarate:glutamine ratio stimulates adenylylation by the ATase

---

[1] So far, the UR enzyme and the UTase have not been separated from one another in vitro and may be different manifestations of the same enzyme. The UTase activity is inhibited by glutamine and stimulated by 2-oxoglutarate. These metabolites do not appear to affect the UR enzyme but it is assumed that some metabolites, as yet unidentified, will affect its activity; otherwise it would couple with the UTase to form UMP and PP_i from UTP in a futile cycle involving P_{II}.

P_{IIA} complex, resulting in a conversion of the GS to the inactive adenylylated form.

Control of Glutamate Dehydrogenase   Studies in Magasanik's laboratory of mutants of *Klebsiella aerogenes* revealed that glutamine synthetase played an important role in the control of glutamate dehydrogenase (GDH) (22). In mutants in which GS was absent, GDH levels were high whether ammonia was limiting or in excess; mutants that constitutively produced GS under all conditions had depressed levels of GDH. Thus there is an inverse relationship between GS and GDH: when the ammonia supply is limiting the GS level is high and there is a low level of GDH, while with excess ammonia the level of GDH is high and that of GS low.

This switch between GS/GOGAT and GDH assimilation pathways does not appear to operate in all microbes. Some organisms, e.g., *Erwinia carotovora* (2), *Bacillus subtilis* W.23 (2), *Bacillus megaterium* (2), and *Clostridium pasteurianum* (23), lack GDH and therefore assimilate ammonia exclusively via the GS/GOGAT system. Others, e.g., *E. coli* (24, 25), *Salmonella typhimurium* (26), and *Schizosaccharomyces pombe* (13), possess both GDH and the GS/GOGAT enzyme systems, but the levels of GDH and GOGAT vary little over the range of ammonia concentrations that produce a considerable variation in the level of GS.

From their studies of chemostat cultures of *E. coli*, Cole et al. (25) have suggested that under ammonia-limiting conditions GS is derepressed (and therefore similar to *K. aerogenes*); the high levels of GDH also observed under these conditions (and therefore unlike *K. aerogenes*) are due to stimulation of GDH synthesis by high cellular levels of 2-oxoglutarate and low levels of glutamate. In support of this hypothesis, Brenchley et al. (26) found that the level of glutamate was largely responsible for controlling the synthesis of GDH in *S. typhimurium*. When glutamate was limiting for growth the level of GDH was high; an excess of glutamate, on the other hand, severely depressed the GDH concentration. Their studies were not able to demonstrate whether glutamate itself or a derivative of it was responsible for controlling the level of GDH in the cell. 2-Oxoglutarate also appeared to be important in controlling the level of GDH in chemostat cultures of *E. coli* (11). As the specific growth rate of an ammonia-limited culture was increased there was a parallel increase in GDH and the 2-oxoglutarate pool, supporting Cole's contention that 2-oxoglutarate stimulates GDH systems in *E. coli*. The level of GDH in *K. aerogenes* under nitrogen limitation, however, was undetectable.

Control of Ammonia Metabolism in Vivo   It must be borne in mind that the model proposed above has been made largely from studies of reconstructed systems in vitro and may not be truly representative of the system in vivo. Several studies have been made to determine the levels and state of the enzymes and metabolites in vivo, with limited success. One of the shortcomings is that measurements of the variations in the level of 2-oxoglutarate in vivo are unreliable following a pulse of ammonia, although ATP, glutamate, and glutamine can be measured with some degree of accuracy. A further problem arises when one tries to extend the interpretation of the in vitro results to the situation in vivo.

Holzer and colleagues (27) have pointed out that a false picture is painted if one measures the cellular concentrations of various metabolites and tries to relate these to the K_m values of the enzymes for which the metabolites are substrates. The concentration of ammonia in the cell was measured as 3.8 mM when proline served as the nitrogen source for growth and the measured K_m for ammonia by glutamine synthetase was 1.8 mM[2] (28). It was concluded, assuming Michaelis-Menten kinetics, that GS would only be expected at best to increase in activity by one-third if given an extra supply of ammonia because the enzyme would only be functioning at three-fourths V_{max} at an ammonia concentration of 3.8 mM. When 10 mM ammonia was added to the culture there was a rapid expansion in the glutamine pool [from 0.5 mM to 8mM in 20 sec (29)] and a depletion of the glutamate pool. This should not have been expected from the K_m and NH_3 measurements, which predicted that the glutamine pool would increase from 0.5mM to only 0.65 mM.

One obvious conclusion from this is that the measurement of the free NH_3 level is a gross overestimate (or the K_m is an underestimate). Wohl-hueter et al. (27) favored the former explanation, particularly since they counted over 40 enzymes in the cell that could bind ammonia, all of which had different K_m values and were present in varying concentrations. The measurement of cellular concentrations of metabolites would therefore include metabolites bound to enzymes and would not be a true estimate of the free metabolite level in the cell. With these reservations in mind one can still make serious attempts at unravelling the complex series of reactions that occur in vivo when cells are confronted with a change in the environmental level of ammonia. The model proposed above has been based on the results of Holzer and Stadtman and their colleagues, who have studied the reactions in *E. coli* both in vivo and in vitro.

The in vivo experiments have generally relied upon pulsing a nitrogen-limited batch culture with an excess of ammonia and following changes in metabolites and enzyme levels in the cell. Recently Senior (11), working in Stadtman's laboratory, has looked at the effects on nitrogen metabolism of changes in the nutritional status of both *E. coli* and *K. aerogenes* using the continuous culture technique. He has confirmed the in vitro results from that laboratory (21, 30), which indicated that the ratio of 2-oxogluta-rate: glutamine determined the average state of adenylylation of GS in *E. coli*. A low value (below 1.5:1) favored adenylylation, whereas above a ratio of 1.5:1 the GS was mostly de-adenylylated. A very small change in the value caused a marked change in the adenylylation state; between ratios of 1.5

---

[2] In general, lower values than this have been determined, e.g., 0.2 mM (6).

and 1.75 there was a change in the adenylylation state of GS from 10 to 2. One significant finding of this study was that the state of adenylylation of GS in *E. coli* was dependent upon the specific growth rate of the culture in cultures in which ammonia was in excess; at low specific growth rates GS was adenylylated but at high specific growth rates GS was largely unadenylylated. The implication of this is that if one pulses an excess of ammonia to an ammonia-limited batch culture of *E. coli* (in which GS is unadenylylated) one would *not* see an increase in adenylylation if the culture was growing at a high specific growth rate prior to the pulse. This, of course, is contrary to the findings of Holzer and his colleagues, who did find an increase in adenylylation after excess ammonia was added to cultures that were in mid-exponential growth phase and therefore were growing close to μ_{max}.

This sort of paradox often occurs when attempts are made to match results from continuous and batch culture studies. Kavanagh and Cole have drawn attention to the anomalies that occur when one analyzes the results for the regulation of nitrogen-metabolizing enzymes in both batch and continuous cultures of *E. coli* such that the proposed methods of regulation that apply in batch culture do not apply in continuous culture and vice versa (31). The correlation between in vitro and in vivo results may become a little clearer if one bears in mind the profound changes that can occur within the cell during the time taken to separate the cells from the growth medium. Indeed, the actual method (centrifugation or filtration) can have a considerable influence over metabolite levels in the cell (32). Many researchers have noted that GS is largely unadenylylated after harvesting from an ammonia-rich medium, although the in vitro results would predict the opposite. Presumably the state of adenylylation has changed during harvesting.

There is now a measure of agreement among most researchers that glutamine synthetase is regulated both in vivo and in vitro by the levels of 2-oxoglutarate and glutamine and there is some evidence (and much supposition) that these levels are controlled by the ammonia concentration. This rather elegant cascade system, in which these metabolites control not only the ratio of P_{IIA} to P_{IID} but also the type of ATase activity specified by each protein and the UTase activity, appears to have definite advantages over an allosteric system. A small change in the activity of an early member of the series can have a profound effect upon the activity of the final enzyme (in this case GS).

This cascade system for control of GS is essentially similar in both *E. coli* and *K. aerogenes*, whereas the control over the low affinity GDH system is not. GS appears to be the repressor for GDH synthesis in *K. aerogenes* as well as a regulator for its own synthesis. The situation in *E. coli*, however, is less clear because GDH will function in its biosynthetic form during ammonia-limited growth and as a catabolic enzyme during glutamate- and proline-limited growth. Under these circumstances GS is also synthesized in high amounts. Another complicating factor with *E. coli* is the observation that the state of adenylylation of GS is independent of the concentration of the nitrogen source but is dependent upon specific growth rate, indicating that it is not solely the availability of the nitrogen source that is responsible for the regulation of GS, as had previously been supposed.

Transport of Ammonia into Cells   Early investigations into transport of inorganic nitrogen compounds into microbial cells led to the conclusion that, in the case of fungal hyphae, ammonia entered by free diffusion (33). Very few studies had in fact been undertaken to investigate how inorganic nitrogen compounds entered microbial cells, largely because there was no convenient isotopic method available. It was theoretically possible to use ^{15}N-labeled substrates, but these were expensive and required a mass spectrometer for analysis. Furthermore, the use of any simple assay for ammonia was complicated by the fact that ammonia was rapidly incorporated into organic matter in vivo and thus would not accumulate to any measurable extent.

In an attempt to see if there was a specific transport system for the ammonium ion in fungi, Segal and co-workers (34) investigated the possibility of using a labeled analog of ammonium, the [^{14}C]methylammonium ion, to monitor inorganic nitrogen uptake in *Penicillium*. They found that there was a derepression of an energy-dependent uptake system for the methylammonium ion that would concentrate the ion over 800 times faster under nitrogen-limited conditions than when nitrogen was in excess. Furthermore, they concluded that a permease was responsible for the transport because inhibitors of protein synthesis also inhibited the development of a transport system. Their most significant finding was that the ammonium ion was an extremely potent inhibitor of methylammonium transport. The K_i was about 2×10^{-7} M, whereas the K_m for methylammonium transport was about 2×10^{-5} M with a V_{max} of about 10 nmol/min/mg dry weight of cells. The kinetic determinations clearly revealed that the affinity of the transport systems was much greater for NH_4^+ than for $CH_3NH_3^+$ and that a shared permease was responsible for the transport of both ions into the cell. Asparagine and glutamine acted as very effective inhibitors of methylammonium transport and because their effect was so rapid the authors suggested that they acted as feedback inhibitors of the transport system.

Since these observations were made a number of other researchers have established that similar systems operate in two other eukaryotes, *Aspergillus* (in mycelia and germinating conidia) (35, 36, 37) and *Saccharomyces* (38). Pateman and co-workers found that there were four genetic loci that effected transport of the methylammonium, and hence ammonium, ion in *Aspergillus*. Mutations at these loci produced strains that were unable to transport nitrogen compounds into the cell. Three of the classes of mutants are thought to possess abnormal structural proteins and the fourth is probably a defect in the regulation of the transport system (36). These authors also found that a fifth mutation resulted in cells that lacked normal glutamate dehydrogenase activity (gdhA) and as such these cells did not show the inhibition by ammonia of many of the systems normally affected by this compound. From their studies of the gdhA and other mutants, Pateman and co-workers postulated that GDH played a dual role in ammonia regulation. First, it could complex with extracellular ammonia at one site on the enzyme and thus it regulates those systems affected by the external ammonia concentration (i.e., thiourea, L-glutamate, and urea uptake, nitrate reductase, and xanthine dehydrogenase activities). Second, GDH could complex with intracellular ammonium ions at a second site on the enzyme to form a different regulatory complex that was responsible for controlling the level of ammonium uptake.

Measurements of the intracellular level of ammonium ions revealed that as this parameter increased the transport of methylammonium ions decreased in the wild type and in the mutants (in the case of the gdh mutants the same trend was observed but the organisms would still transport methylammonium ions even when the intracellular concentration of ammonium ions was four times that required to inhibit completely transport in the wild type). The system in Aspergillus therefore differs from that in Penicillium, in which the intracellular ammonium concentration played a minor role in controlling methylammonium uptake.

The results of experiments in germinating conidia of Aspergillus (37) and in cells of Saccharomyces (38) also point to the fact that ammonium and methylammonium ions are transported by a single permease whose affinity for ammonia is greater than that for methylammonium. When the effects of glutamine and asparagine were tested, they were both found to inhibit methylammonium transport in Aspergillus conidia (as in Penicillium), but only asparagine completely inhibited uptake in Saccharomyces, while glutamine inhibited uptake by only 60%.

Recently $CH_3NH_3^+$ has been used in studying NH_4^+ uptake in a prokaryotic system, E. coli (39). Although NH_4^+ effectively inhibited the energy-dependent concentrative uptake of $CH_3NH_3^+$, conclusions regarding $CH_3NH_3^+$ as an alternative substrate to NH_4^+ in the transport process were difficult to make. This was because there appeared to be two systems for $CH_3NH_3^+$ uptake, one with a pH optimum at 7 that was inhibited by cyanide or m-chlorophenyl carbonylcyanide hydrazone (CCCP) and another with a pH optimum at 9 that was resistant to cyanide or CCCP. The response of these two systems to different concentrations of NH_4^+ was that generally they were inhibited by it, but the inhibition pattern observed was not reproducible in either system.

METABOLISM OF DINITROGEN

Of the many inorganic nitrogen compounds studied in recent years, dinitrogen has aroused the greatest interest, largely because of the importance of dinitrogen fixation by microbes in determining the biological productivity of many soils. Clearly an understanding of how microbes can convert (at ordinary temperatures and pressures) the seemingly inert dinitrogen molecule into ammonia could be important in determining the future trends in research at both the fundamental and applied levels [see Postgate (40)].

Nitrogenase

Although the existence of nitrogen-fixing bacteria was known in 1862 (41), it is only in the last 15 years or so that any progress has been made toward an understanding of the mechanism of nitrogen fixation at the enzyme level. The name "nitrogenase" has been given to the enzyme system that converts N_2 to NH_3. In all bacteria from which the enzyme has been isolated, it is comprised of two proteins, a MoFe protein that is a high molecular weight species (around 220,000) containing two atoms of molybdenum and about 24 atoms of non-heme iron per molecule, and an Fe protein that is a low molecular weight species (between 55,000 and 65,000 depending on the species) containing 4 atoms of non-heme iron per molecule. Both proteins are extremely sensitive to oxygen and so purification of the components has necessitated the use of rigorous anaerobic techniques throughout. The one exception to this is the enzyme from the aerobe *Azotobacter*, which can be extracted in an oxygen-insensitive particulate form but assumes oxygen sensitivity when resolved into its components (42,43).

The characterization of the nitrogenase components has been studied in only six of the 17 or so organisms from which active nitrogenase preparations have been made. Those characterized are from *Azotobacter chroococcum, A. vinelandii, Klebsiella pneumoniae, C. pasteurianum, Rhizobium lupini* bacteroids, and *R. japonicum* bacteroids (Table 1). The casual observer of the progress of characterization of these proteins may be slightly perplexed by the variance in the values for metal content and specific activity over the years. This has been due largely to the presence of an inactive species of the MoFe protein. Removal of this contaminant from apparently homogeneous preparations of *C. pasteurianum* MoFe protein containing 1 Mo and 14 Fe atoms/molecule yielded an inactive contaminant protein containing 0.5 Mo and 8 Fe atoms/molecule and a species of higher specific activity containing 2 Mo and 24 Fe atoms/molecule. The inactive species was characterized by an EPR spectrum of the $g = 1.94$ type, which is absent from the EPR spectrum of the highly active species (44, 45). The preparations from *A. vinelandi* (41, 47), *A. chroococcum* (48), and *K. pneumoniae* (49) have also been extensively purified and do not now contain the $g = 1.94$ species, although preparations from other organisms that have a lower specific activity and molybdenum content still have this type of EPR spectrum. Many of the MoFe proteins from other organisms still have a $g = 1.94$ EPR spectrum, suggesting that they are contaminated by this inactive species.

In an attempt to determine the mechanism of N_2 fixation in biological

Table 1. Properties of MoFe protein from different sources[a]

Property	Clostridium pasteurianum	Azotobacter vinelandii	Azotobacter chroococcum	Klebsiella pneumoniae	Rhizobium japonicum bacteroids	Rhizobium lupini bacteroids
Molecular weight	220,000	280,000	220,000	218,000	200,000	194,000
Subunits	50,000 and 59,500	70,000	60,000 (approx.)	51,300 59,600	50,000	57,000
Molybdenum content	2.0	2	2	2	1.3	1.0
Iron content	24	34–38	20–24	27–33	28.8	18–20
S = content	24	26–28	28–22	no data	26.2	
Sp. activity	2,250	1,400	2,000	2,150	1,000	704
References	(76)	(46, 77)	(48, 74)	(74)	(78)	(75)

[a] Most of the data in the table have been taken from the references indicated, although there is some conflict in the literature on the exact numbers. Published data of protein with highest activity have been used. For further details the reader is referred to the review by Eady and Smith (79).

systems, a great deal of attention has been focused in recent years on the use of EPR spectroscopy (50). It is now firmly established that a transition metal is involved in the coordination of the dinitrogen at the active site of the enzyme (chemical model systems also demonstrate this) and, because it is difficult to study bound transition metals in proteins by most techniques, EPR has proved invaluable in this respect. Consequently most of our current understanding on the mechanistic aspects has required a good working knowledge of this technique and so I have briefly included below some of the most salient EPR features of the proteins.

The MoFe Protein   Pure MoFe preparations have EPR spectra with *g* values around 4.3, 3.7, and 2.01 at temperatures below 30 °K. The observed spectrum appears to be due to the iron atoms in the protein and not the molybdenum species (51). The *g* values and line widths of the EPR spectrum are pH dependent and when the protein was incubated with acetylene (an alternative substrate to N_2 for the enzyme) under argon there was an increase in the high pH form of the protein, confirming an earlier but inconclusive report that this protein would specifically bind cyanide, another reducible substrate (52).

The nature and the role of the molybdenum in the MoFe protein is not known. Its presence in the protein is necessary for activity (see below) because derepressed protein extracted from cells grown in a molybdenum-deficient medium with added tungsten was inactive, even though active Fe protein was synthesized (53). Activity could be restored by adding molybdate to the medium (54). This restoration of activity by molybdate would only occur in vivo and not in vitro. It was possible, however, to restore some activity to this protein in vitro by adding back a low molecular weight molybdenum-containing peptide prepared from an acid-treated active MoFe preparation. Furthermore, this Mo-containing peptide would also restore activity to a *Neurospora crassa* mutant (*nit*-1) that lacked a functional assimilatory nitrate reductase. Indeed, many experiments of this kind have now been done with a number of acid-treated Mo enzymes (55) and it has been confirmed that molybdenum is the functional species that is incorporated into the *Neurospora* protein (56). The molybdenum co-factor has now been purified from *A. vinelandii* and its activity measured by its ability to activate MoFe protein produced by a nitrogenaseless mutant of *Azotobacter* (UW45) (57). A similar peptide has also been prepared from *A. vinelandii* by dissociation of the MoFe protein at pH 8.9 for 100 min (58). This peptide has a molecular weight of 1,300 and contains 1 atom of molybdenum that has EPR characteristics of Mo(V).

Clearly the isolation of such a co-factor in quantity may well provide valuable information on the role of molybdenum in the MoFe protein.

The Fe Protein   The Fe protein has been obtained in a high degree of purity from those organisms in which the MoFe protein has been characterized. The Fe protein is more sensitive to oxygen than the MoFe protein—the half-life of the MoFe protein from *K. pneumoniae* is 10 min in air, whereas the half-life of the Fe protein is 45 s (49). In *A. vinelandii* (59) and *C. pasteurianium* (60) the Fe protein is cold labile, but the protein from *Klebsiella* is stable for at least 15 days at 1 °C under anaerobic conditions (49). Both the MoFe and the Fe proteins are usually stored in bead form in liquid nitrogen, conditions in which the proteins are perfectly stable.

The Fe proteins appear to be monomeric dimers (see Table 2) with a similar number of Fe and $S^=$ contents. At temperatures below 30°K they exhibit a rhombic EPR spectrum in the reduced form with a g_1 2.05, g_2 1.94, and g_3 1.86. Addition of ATP and Mg^{2+} to the Fe proteins from *C. pasteurianium, K. pneumoniae,* and *A. vinelandii* changed the rhombic EPR signal to one that had nearly axial symmetry with g_1 2.04 and $g_\perp$ 1.93. Similar changes were also observed when 5 M urea was added to the protein (61), suggesting that ATP caused a conformational change in the protein.

The Nitrogenase Reaction   Neither the MoFe nor the Fe protein has any catalytic activity alone in the normal assay. When combined anaerobically however, in the presence of the assay components, (MgATP, a reductant such as sodium dithionite, and substrate N_2) a number of reactions occur: the N_2 is reduced to ammonia, dihydrogen is evolved, and ATP is hydrolyzed.

The ATP-dependent evolution of dihydrogen in the nitrogenase reaction is due to the reduction of protons in the aqueous environment; however, the amount is greatly reduced in the presence of another reducible substrate (Table 3), such as N_2, and reflects competition between substrates for reduction by the enzyme. This reduction of protons does not appear to be catalyzed at the same site as N_2 reduction, because hydrogen evolution is not inhibited by CO, a strong inhibitor of N_2 reduction. The inhibition of reduction of various substrates by CO is rather complex, however, and tends to confuse the overall picture. Based on their findings of the kinetics of substrate reductions, Hwang et al. (62) have proposed that there are 5 substrate or inhibitor binding sites on the enzyme. These are: 1) an N_2 or H_2 site; 2) a C_2H_2 site; 3) an N_3^-, CN^-, and CH_3 NC site; 4) a CO site; and 5) an H^+ site. Furthermore, the ATPase activity of the enzyme requires reductant for its activity and has been called "reductant-dependent ATPase activity" to distinguish it from most other ATPase reactions that do not require reductant for activity. This mandatory requirement for ATP is probably a little surprising in view of the fact that thermodynamically the reduction of N_2 to NH_3 is an exergonic reaction. Consequently the role of ATP in the nitrogenase reaction has been the subject of intense interest in recent years and its role in the nitrogenase reaction is considered below. By studying the two proteins separately and in combination it is now possible to propose a scheme for enzymic N_2 fixation that is based largely on fact and less on hypothesis.

The Fe protein is readily reduced by a number of electron donors that include both physiological (ferredoxin, flavodixin, or NADPH) and artificial (dithionite or reduced methyl viologen) types. In its reduced form the

Table 2. Properties of the Fe protein

Property	Clostridium pasteurianum	Klebsiella pneumoniae	Azotobacter vinelandii	Azotobacter chroococcum	Rhizobium lupini bacteroids
Molecular weight[a]	56,000	62,000	64,000	63,000	65,000
Subunits	27,500	34,600 ± 2,000	33,000	30,800	32,000
Iron content	4	4	3.45	4	32
S = content	4	3.85	2.85	3.9	
Sp. activity	2,708	980	1100	2000	434
References	(71, 72)	(49)	(73)	(48, 74)	(75)

[a]The values for the molecular weight were obtained by gel filtration because ultracentrifugation can lead to dissociation inactivation.

Table 3. Products from reduction of various substrates by nitrogenase

Substrate	Products	Number of electrons required for reduction	Relative rate
N_2	$2NH_3$	6	1
N_3^-	NH_3, N_2	2	3
N_2O	H_2O, N_2	2	3
C_2H_2	C_2H_4	2	3-4
HCN	CH_4, NH_3	6	0.6
	CH_3NH_2	4	
CH_3CN	C_2H_6, NH_3	6	0.2
CH_3NC	CH_3NH_2, CH_4	6	0.8
	C_2H_4, C_2H_6	8,10	
	C_3H_6, C_3H_8	12,14	
CH_2CHCN	C_3H_6, C_3H_8, NH_3	6,8	0.2
CH_2CCH_2	$CH_2CH CH_3$	2	0.05
$2H_3O^+$	$H_2 + 2H_2O$	2	3

Fe protein complexes with MgATP, which affects its conformation (61,63, 64), its accessibility to iron chelating agents (65), and its redox potential (66). In this last respect it was observed that the midpoint potential of the dithionite-reduced protein was lowered from -294 mV to -402 mV. By studying the interaction between the reduced Fe-MgATP protein and the MoFe protein, it appears that the Fe protein then passes its electrons to the MoFe protein. The MoFe protein appears to bind the reducible substrate N_2, C_2H_2, or protons (see above), producing a small change in the EPR spectrum. When combined with the reduced MgATP-Fe protein, the EPR signals attributed to the MoFe protein are diminished to produce a super-reduced form that is essentially EPR signal-free (51). Furthermore, the EPR signals resulting from the Fe protein were altered from its reduced form to the oxidized signal-free form, indicating that the Fe protein was giving up its electron to the MoFe-N_2 complex. Because the supply of reductant (sodium dithionite) was allowed to run out as the enzyme reaction proceeded, both proteins became progressively more oxidized.

These results have been summarized in Figure 4. As it stands this scheme only accounts for the transfer of one electron in the process. Because 6 electrons are required to reduce N_2 to $2NH_3$ and no intermediates, such as diimide ($2e$ reduction) or hydrazine ($4e$ reduction), have been detected during turnover, the exact mechanism whereby the N_2 is reduced is still unclear. Furthermore, the scheme does not ascribe any role for molybdenum. In view of the fact that dinitrogen, when bound to molybdenum, can be reduced to ammonia (67,68), it seems probable that molybdenum may be responsible for binding the reducible substrate in the enzyme. Although the EPR changes observed upon addition of substrate to the MoFe proteins have been ascribed to Fe this does not rule out the possibility that coordi-

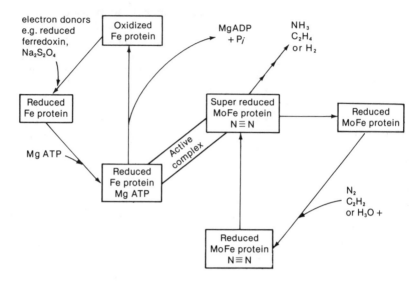

Figure 4. Possible scheme for sequence of events during reduction of substrate by nitrogenase. The substrate, N_2, is only reduced to NH_3 when an active complex between the Fe protein MgATP and the MoFe protein N_2 is formed. In this scheme the arrows leading to reduced product represent three 2-electron transfer steps before ammonia is released. If either C_2H_2 or $2H_3O^+$ is the substrate, then only 2 electrons are required to effect its reduction to C_2H_4 or H_2. (Adapted from ref. 79.)

nation of dinitrogen could involve both Mo and Fe, with the Mo being in an oxidation state that is not detected by EPR.

As stated earlier, studies on the substrate versatility of the enzyme complex have revealed that a variety of doubly and triply bonded substrates are reduced to products that involve only 2-electron steps (Table 3). If molybdenum in the resting MoFe protein is in oxidation VI [one possible interpretation of X-ray absorption data as discussed recently by Smith (69)] then a 2-electron reduction of the molybdenum species to Mo(IV), followed by reduction of the substrate, would not be expected to produce any molybdenum signals that were EPR active.

During turnover of the enzyme complex ATP is hydrolyzed, but the number of ATP molecules required for reduction of N_2 in vitro depends on many factors (70). The role of ATP in the reaction appeared to facilitate transfer of electrons from the Fe protein to the MoFe protein (see above) but recent data suggest that ATP may have more than one role to play. By making use of the ability of the MoFe protein from one organism to form a partially functional complex with the Fe protein from a different organism, Smith et al. have managed to uncouple ATP hydrolysis from electron transfer (70). Normally ATP:2e rates of about 4 are found, but with the MoFe protein from *Klebsiella* (Kp$_1$) and the Fe protein from *Clostridium* (Cp$_2$) values around 50 were found. Clearly ATP was being hydrolyzed in a reaction that involved limited electron transfer, but it is not yet clear what significance this reaction has.

Physiological Reductants for Nitrogenase Because the nature of the physiological reductant for nitrogenase in bacteria has proved elusive in some cases, most in vitro studies with nitrogenase have relied upon the use of sodium dithionite as an electron donor for the conversion of N_2 to NH_3. In most instances in which a physiological carrier has been identified, it appears that either a ferredoxin or flavodoxin or both is involved. The existence of ferredoxin has been known for about 15 years and its requirement for N_2 fixation by *C. pasteurianum* for almost as long (80). It is only recently, however, that it was shown to link directly between hydrogenase and nitrogenase in this organism (81). Although the immediate source of reductant for nitrogenase is reduced ferredoxin, the primary source of electrons could come either from the metabolism of pyruvate in the phosphoroclastic system (Figure 5) or from H_2 if an exogenous supply of ATP was available. Ferredoxin, a non-heme iron sulfur protein, has also been found in the aerobic nitrogen-fixing organisms *Azotobacter* spp. and *Mycobacterium flavum* and is capable of transferring electrons to nitrogenase (82,83). Flavodoxins,

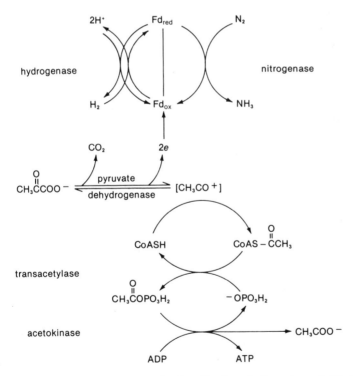

Figure 5. The metabolism of pyruvate in the phosphoroclastic system of *Clostridium*. Electrons from the decarboxylation of pyruvate reduce ferredoxin, which couples to hydrogenase and nitrogenase.

FMN-containing proteins that do not possess iron atoms, have also been isolated from aerobic nitrogen-fixing organisms and implicated in electron transfer to nitrogenase (84,85).

In the case of the system from *A. vinelandii,* Benemann et al. (86) have proposed a scheme that involves both ferredoxin and flavodoxin in electron transfer. Here the primary source of electrons is NADPH, which could be generated from a number of $NADP^+$-linked dehydrogenases, and which reduces ferredoxin in the following sequences:

$$NADPH \rightarrow ferredoxin \rightarrow x \rightarrow flavodoxin \rightarrow nitrogenase$$

A similar scheme has also been proposed to operate in soybean bacteroids (87). Component x has not been identified but is soluble and stable to heating at 55°C for 10 min. Although this scheme has had some support (88), it has been pointed out that there are still a number of problems that have to be resolved (89). In particular, explanations are required for the low activities with physiological substrates when compared with rates using sodium dithionite and the observation that both ferredoxin and flavodoxin will mediate electron transfer to nitrogenase independently (89). Furthermore, isolation of an NADPH-ferredoxin reductase from *Azotobacter* or bacteroids has not yet been possible. An alternative explanation has been put forward by Veeger and co-workers (90,91), who suggest that a pyruvate phosphoroclastic system in *A. vinelandii* furnishes the enzyme with both ATP and electrons in a similar fashion to that found in *Clostridium.* Such an explanation, however, cannot be true for all *Azotobacter* species because acetokinase is not present in other *Azotobacter* species (92).

The presence of an intact chemiosmotic gradient has also been suggested as a requirement for nitrogenase activity (91) and may explain the low in vitro rates for nitrogenase activity with physiological substrates.

Control of Nitrogenase   Nitrogenase, like most enzymes, is subject to a wide range of controlling factors that affect both its activity and its synthesis. In many instances one factor will affect both functions, but for convenience factors affecting activity and synthesis have been separated.

Regulation of Activity by Oxygen   There is now a considerable amount of evidence to suggest that nitrogen fixation is an anaerobic and reductive process both in vivo and in vitro. At the molecular level little is known about how oxygen inactivates the nitrogenase proteins. The Fe protein is certainly the more sensitive of the two nitrogenase proteins and its association with ATP renders the protein even more susceptible to O_2 damage (93,94). The possibility that the superoxide anion or hydrogen peroxide is the species that inactivates proteins has been reviewed elsewhere (95,96, 97), but it is not known whether these or the OH radical are responsible for nitrogenase inactivation. It is known, however, that the activity of both superoxide dismutase (98) and catalase (99) are much greater at high than at low oxygen tensions in *A. chroococcum* and may be important in protecting nitrogenase from oxygen damage.

Clearly anaerobic nitrogen-fixing organisms are not faced with the same problems that confront their aerobic counterparts because the metabolism of the former is geared to an existence in the absence of oxygen. Aerobic organisms, however, have a mandatory requirement for oxygen and so to survive and still fix dinitrogen they have had to develop a variety of processes for protecting the nitrogenase proteins from dioxygen.

In addition to the increased synthesis of protective enzymes (SOD, catalase, etc.) the aerobic Azotobacters, given sufficient carbon substrate, can significantly augment their respiration in response to increased oxygen supply and thereby maintain a low and nonlethal internal level of oxygen (100,101). This respiratory protection of nitrogenase has been developed further to reveal a branched respiratory chain (102) in *Azotobacter* that allows for a considerable variation in electron flow and phosphorylating efficiency in response to different levels of oxygen (Figure 6.). Under conditions of high aeration, respiration occurs via uncoupled NADH dehydrogenase and a pathway of low phosphorylating efficiency, terminating in cytochrome a_2. This has two beneficial effects on the cell. First, it ensures that large quantities of oxygen are used as a terminal electron acceptor in a pathway of high electron flux, thereby ensuring that concentrations of oxygen that are lethal for nitrogenase do not accumulate. Second, because the pathway is of low phosphorylating efficiency, the increased electron flux does not allow the ATP:ADP ratio to become too high and inhibit nitrogenase activity (103).

At low oxygen tensions the opposite occurs. A pathway of low electron

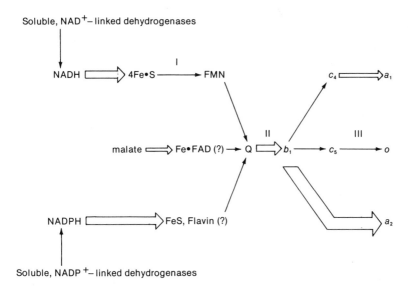

Figure 6. Cytochrome-linked electron transport pathways in *Azotobacter*. Three phosphorylation sites are indicated by I, II, and III. Thickness of the arrows represents the effective electron flux through a particular pathway. (Adapted from ref. 102).

flux and high phosphorylating efficiency predominates, producing sufficient ATP for nitrogenase function. A similar branched chain electron transport system has also been observed in the legume bacteroid system, in which two terminal oxidases have been observed (104).

Previous studies (105) had established that oxygen consumption and nitrogenase activity in bacteroids would proceed maximally at one twenty-fifth the concentration of oxygen if the assays were done in the presence of leghemoglobin. It was suggested that the role of the oxygen carrier leghemoglobin was to maintain a low but stable concentration of oxygen in the bacteroid that permitted normal respiration, and therefore ATP supply for nitrogenase, but that prevented an excess of oxygen from damaging nitrogenase. These researchers subsequently showed (104) that this low concentration of oxygen was optimal for ATP production and hence nitrogenase activity in the bacteroids. Under these conditions of low oxygen concentration, a terminal oxidase with a high affinity for oxygen that produced ATP efficiently was operative. The role of leghemoglobin in this system was effectively to extend the range of oxygen concentrations down to very low levels and still maintain active respiration and hence nitrogenase activity. In this respect leghemoglobin acts as an oxygen carrier by facilitating diffusion of oxygen into the bacteroids, which would otherwise be starved of oxygen.

When the concentration of free oxygen was high, respiration in the bacteroid was mediated by a terminal oxidase that had a low affinity for oxygen and produced ATP rather inefficiently. Therefore, inhibition of nitrogenase activity at high concentrations of oxygen was more probably due to lack of ATP than to a direct toxic effect of oxygen on nitrogenase.

A second, and apparently much cruder, form of control over nitrogenase activity was proposed by Dalton and Postgate (100) in which oxygen caused a conformational change in the nitrogenase complex. When cultures of *Azotobacter* were exposed to a sudden increase in the oxygen supply the enzyme adopted a conformation [possibly brought about by association with some cellular component(s)] in which it was oxygen insensitive and inactive. This conformationally protected form of nitrogenase can be readily extracted as a particulate fraction from cells but assumes oxygen sensitivity when resolved into its soluble components (42). When the oxygen solution rate was reduced to its original low value the enzyme activity is "switched on," a situation that apparently does not require the *de novo* synthesis of nitrogenase (106). There are now a number of classic "switch-off, switch-on" responses by aerobic nitrogen fixers (89) and although there is much speculation concerning the mechanism of such a phenomenon, there is a dearth of experimental data on which to base any sound conclusions.

Other Regulators of Nitrogenase Activity   Although there are a number of physiological inhibitors of nitrogenase activity in vitro, it has been difficult to extend these studies to in vivo systems. ADP produced during turnover of nitrogenase in vitro is an inhibitor of activity (107) and will bind to the

Fe protein (52). Recent measurements of the K_m for ATP and the K_i for ADP (108,109) show that the former is approximately 20 times higher than the latter, so that under physiological conditions the ATP:ADP ratio would have to be at least 20 to ensure that nitrogenase was active. Accurate measurements of the ATP:ADP ratio in bacteria are difficult and this could be a reflection of the conflicting reports on the effect of this ratio on nitrogenase activity (cf. refs. 110 and 111).

Ammonium ions do not inhibit purified nitrogenase but L'Vov et al. (112) found that freshly disrupted particulate preparations from *A. vinelandii* were inhibited by NH_4^+ ions. Drozd et al. (113) found that nitrogenase activity in vivo was reduced by 30% at concentrations of ammonium ions that did not repress synthesis of the enzyme. Such an effect, however, can only represent very fine control on activity and is clearly not as marked as the effect of oxygen. In *C. pasteurianum,* carbamyl phosphate caused a maximum inhibition of 50% of the in vitro nitrogenase activity and, at 1 mM, caused a 30% inhibition of the in vivo activity as well as a repression of its synthesis (114). The carbamyl phosphate bound only to the MoFe protein but required the presence of the Fe protein and MgATP. An interesting observation was that the endogenous level of carbamyl phosphate in N_2-fixing cells was equal to the binding constant and strongly suggested that nitrogenase activity in vivo is partially repressed. As such, control of nitrogenase by carbamyl phosphate would have to be very sensitive.

Control of Nitrogenase Synthesis   All nitrogen-fixing bacteria, when grown in the presence of excess ammonium ions, fail to synthesize the nitrogenase proteins. This repression by NH_4^+ has been studied in sulfate-limited chemostat cultures of *K. pneumoniae* (115). Addition of an excess of ammonium ions to a nitrogen-fixing culture resulted in a progressive decay of nitrogenase activity after an initial lag phase of about 75 min. This decay in activity proceeded at a faster rate than the specific growth rate of the culture. Similarly there was a delay in the appearance of activity when fully repressed cultures were derepressed by transferring cells to an ammonium-free medium. Experiments on repression and derepression of nitrogenase conducted in the presence of rifampicin or chloramphenicol indicated that the effects of NH_4^+ on nitrogenase synthesis acted at the level of mRNA synthesis (transcriptional) and not at the level of protein synthesis (translational).

When nitrogen-fixing bacteria are grown in media containing an excess of ammonium ions, the major route of nitrogen assimilation is via glutamate dehydrogenase (GDH), but when grown under nitrogen-fixing conditions, i.e., ammonia limitation, the glutamine synthetase/glutamate synthase (GS/GOGAT) pathway is functional (116). Studies on mutants of *K. pneumoniae* have led Tubb (117) and Streicher et al. (118) to suggest that the nonadenylylated form of GS is a positive controller of nitrogenase synthesis. These findings have been confirmed using both wild-type organisms (119) and the inhibitor of adenylylation, methionine sulfoximine

(120). Clearly ammonia appears to regulate nitrogenase synthesis through its effect upon GS, but whether GS in either its adenylylated or de-adenlylated form interacts at the transcriptional or translational level is a mystery. Indeed, there is some evidence to suggest that GS may not be involved in this regulation at all, since Shanmugam et al. (121) have obtained mutants of K. pneumoniae that were deficient in GS but were capable of fixing dinitrogen and Drozd et al. (113) observed no difference in GDH or GOGAT in A. chroococcum whether fixing N_2 or assimilating NH_3, conditions that produced significant changes in the levels of these enzymes in other nitrogen fixers (116). Unfortunately the Sussex workers did not measure GS and so it is not possible to conclude whether or not this enzyme had any role to play in controlling nitrogenase in that organism.

Oxygen also appears to regulate nitrogenase synthesis and, like the situation with ammonia, there appear to be several clearly distinguishable effects. Ammonia-grown K. pneumoniae, which only fixes N_2 anaerobically, was transferred to a nitrogen-free medium and aerated. If oxygen acted as an inactivator of nitrogenase activity then nitrogenase would be synthesized but would be inactive and detectable by its antigenic reactions. If it inhibited synthesis then no nitrogenase would be formed. St. John, Shah, and Brill (122) did not find any antigenically detectable nitrogenase components and, because complementary experiments on O_2 inactivation of nitrogenase components in vivo revealed that inactivation was slower than synthesis during derepression, they concluded that oxygen inhibited synthesis of the enzyme. Unfortunately the Wisconsin group could not determine whether O_2 per se or an effector produced by exposure to oxygen caused the inhibition of nitrogenase synthesis. It is of interest to note that in careful study of the effect of oxygen on K. pneumoniae in chemostat culture Hill (123) found evidence that this organism was capable of a limited amount of respiratory protection as well as some form of conformational protection sensu Dalton and Postgate (100).

Oxygen also appears to exert its effect on nitrogenase synthesis through GS in cultures of free-living Rhizobium. In chemostat culture, increasing the concentration of oxygen in the steady state resulted in an equivalent decline in nitrogenase activity and a progressive increase in the level of adenylylation of GS (124). It was postulated that when the oxygen supply was restricted to such cultures there would be low levels of ATP production and therefore little ATP would be available for adenylylation of GS. As a result GS would be largely unadenylylated and active, thereby allowing normal nitrogenase synthesis. Conversely, high oxygen produced adenylylated GS that would inhibit nitrogenase synthesis.

It is quite clear that the control of nitrogenase activity and synthesis is an extremely complex affair involving a wide variety of regulators. Such a situation should be expected if one bears in mind that the vast amount of energy in terms of ATP and reducing power required to fix N_2 would be largely wasted if the cell had an intracellular excess of ammonium ions.

In many instances up to 10% of the protein content of a cell is comprised of the nitrogenase proteins and, clearly, for an organism to compete successfully with non-nitrogen fixers, it should be able to regulate carefully the activity of the enzymes in order to adapt to even minor fluctuations in demand for fixed nitrogen.

NITRATE METABOLISM

The reduction of nitrate can be accomplished by two distinct mechanisms, but only one of these leads to a net incorporation of nitrogen in the cell. The dissimilatory metabolism of nitrate is performed by anaerobic or facultatively anaerobic bacteria in which nitrate serves as a terminal electron acceptor. The nitrate is reduced to nitrite or even N_2, but rarely to ammonia, and is not incorporated into the cell. This respiratory nitrate reductase is generally a particle-bound enzyme system and may be associated with cytochromes (125). In keeping with its role in anaerobic respiration is its sensitivity to repression by oxygen, an observation that led Pichinoty to distinguish between two types of nitrate reductase (126). He observed that *Paracoccus denitrificans* when grown anaerobically produced the particulate system that would reduce chlorate as well as nitrate (type A). When these cells were grown aerobically using oxygen as the terminal electron acceptor, a second type of nitrate reductase was produced with no evidence for the type A enzyme. This enzyme was soluble and would reduce nitrate but would not reduce chlorate (type B). The type B enzyme was responsible for the assimilation of nitrate nitrogen into the cell and is the major topic of discussion in the remaining section of this chapter.

Assimilatory Nitrate Reductase

Distribution   To the author's knowledge there has not been a systematic survey of the ability of microorganisms to assimilate nitrate. Consequently our knowledge of the distribution of this enzyme system has been rather sketchy. One review (127) listed 23 genera of chemosynthetic bacteria, 15 genera of filamentous fungi, 16 genera of yeasts, and a number of photosynthetic bacteria and algae in which there had been reports of the presence of assimilatory nitrate reduction. Because it has been possible to distinguish between the two types of nitrate reductase, this list has been extended slightly and, as more groups of organisms are researched, we expect this list to increase.

The Enzyme System   Organisms that can assimilate nitrate nitrogen have first to reduce the substrate to the level of ammonia before the nitrogen is incorporated into the cell by one or more of the mechanisms outlined above. In those organisms in which the enzyme systems have been characterized, it appears that the nitrate is first reduced to nitrite by nitrate reductase and the product is then reduced to ammonia by nitrite reductase.

Nitrate Reductase   Only the enzyme systems from the alga *Chlorella vulgaris* (128) and the yeast *Rhodotorula glutinis* (129) have been purified to homogeneity with a high specific activity. The enzyme systems from the filamentous fungi *Neurospora crassa* (130) and *Aspergillus nidulans* (131, 132, 133) have been obtained reasonably pure, whereas the system from *A. chroococcum* has been only partially purified (134, 135), although a number of similarities between the bacterial and fungal forms have been noted.

The enzyme from *N. crassa* is soluble, contains molybdenum, a cytochrome b_{557} component and ready dissociable FAD (130). Very few detailed investigations of this enzyme have been made but, based on the evidence to date, it appears that within the protein complex the sequence of transfer of electrons to nitrate can be envisaged as:

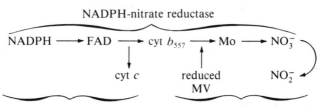

NADPH-nitrate reductase

NADPH ⟶ FAD ⟶ cyt b_{557} ⟶ Mo ⟶ NO_3^-

cyt c reduced NO_2^-
            MV

NADP cyt c reductase   reduced MV nitrate reductase

As isolated, the enzyme shows a number of other reactions, all of which are inducible by nitrate and purify along with the NADPH–nitrate reductase. These are NADPH–cytochrome c reductase, reduced FAD–nitrate reductase and reduced methyl viologen–nitrate reductase activities. Studies with mutants have revealed that the enzyme is a complex of at least two proteins, one of which is flavin-dependent, has NADPH-cytochrome c reductase activity, and is presumed to be involved in the early part of the electron transport pathway; this nitrate-inducible protein is found in mutants unable to reduce nitrate (*nit*-1). The other was presumed to be absent from the *nit*-1 mutant but constitutively produced in all other mutants, including uninduced wild type. It was inferred to be responsible for the latter part of the nitrate reductase pathway and includes the molybdenum species. In vitro recombination of *nit*-1 extracts with extracts from other mutants or the uninduced wild type resulted in restoration of full nitrate reductase activity. Furthermore, it was possible to supply the second component from a number of other molybdenum-containing enzymes, either microbial or mammalian in origin, that had previously been broken down into subunits by acidification (55, 136).

Several hypotheses were put forward by Nason's group to account for the in vitro assembly of the nitrate reductase complex. They favored the view that the protein species containing molybdenum existed as a small polypeptide that could be extracted from other molybdenum-containing enzymes, possibly in association with a higher molecular weight species. When added to the *nit*-1 extracts the molybdenum co-factor was released from the "protein carrier" with which it was associated to form a fully functional nitrate reductase complex. The observation that no acidification of the *Neurospora* mutant extracts was necessary to complement the *nit*-1 extract suggested that the co-factor was more readily accessible in these circumstances.

The functional role of molybdenum in *Neurospora* has also been investigated by growing cells in the absence of molybdenum or in the presence of tungstate (137). The protein produced under such circumstances appeared to have a different tertiary structure (presumably because of the absence of molybdenum or presence of tungsten) to normal molybdenum-grown cells and only possessed NADPH-cytochrome *c* reductase activity. As in the case with the mutant work in Nason's laboratory, there was no reconstitution of activity in vitro when molybdenum was added to the protein, suggesting that molybdenum is probably bound to a protein moiety before it can recombine with the inactive protein. It appears also that in addition to serving a functional role molybdenum could be important in maintaining the structural complex of nitrate reductase. Unfortunately the enzyme extracted from cells grown on tungsten plus molybdenum was not analyzed for the presence of either metal and so it is not possible to say whether tungsten prevented molybdenum incorporation into the protein or the protein contained an inactive molybdenum species. It would be of interest to investigate whether or not acid-treated molybdenum enzymes would recombine with the protein extracted from cells grown in the absence of that metal. Such a system would clearly provide some interesting data on the role of molybdenum in this and other molybdoproteins.

The nitrate reductase complex from another filamentous mould, *Asp. nidulans*, is also a molybdoflavoprotein with a molecular weight of around 200,000 (131) but, unlike the *Neurospora* enzyme, there is no evidence for the involvement of any heme compounds in the functional system. This organism has been the subject of extensive genetic analysis by both Cove and co-workers at Cambridge and Pateman and co-workers at Glasgow. The genes that are known to be involved in nitrate reduction are: *niaD*, the structural gene for nitrate reductase; *niiA*, probably the structural gene for nitrite and hydroxylamine reductase; five or six *cnx* (co-factor, nitrate reductase, xanthine dehydrogenase) genes that are involved in coding for a co-factor common to both nitrate reductase and xanthine dehydrogenase; and *nirA*, which is presumed to be a regulator gene concerned with the induction of nitrate and nitrite reductases.

Based on the observation that any mutation affecting the *cnx* region results in the simultaneous loss of ability to utilize nitrate and hypoxanthine, Pateman et al. (138) proposed that the *cnx* genes determined a common component necessary for the activity of nitrate reductase and xanthine dehydrogenase. The *cnx* mutants, however, have not lost cytochrome *c* reductase activity, suggesting that they are impaired in some component of nitrate reductase required for later stages in the transport of electrons to nitrate. It has been possible to partially restore nitrate reductase and xanthine dehydrogenase activities by adding high concentrations of molybdate to growing cultures of *cnxE* mutants (139), and it was argued that the *cnxE* gene product may be concerned with insertion of molybdenum into a co-factor specified by the other *cnx* genes. In a study of mutants of the *cnx* region, McDonald and Cove (140) have tentatively concluded that the *cnxE* and *-F* genes were involved in insertion of molybdenum into the polypeptide coded for by the *cnxH* gene. The role of the other genes in the *cnx* region is still unknown.

Recent experiments with mutants of *N. crassa* and *Asp. nidulans* have confirmed that the *nit*-1 mutants in the former are the same as the *cnx* mutants in the latter (Cove, personal communication) and that the co-factors are indeed similar. Such experiments clearly pave the way for a more detailed investigation into the mechanism of insertion of metals in proteins, and, with the recent identification of molybdenum co-factors for the MoFe nitrogenase, protein may well prove to be a universal mechanism for molybdenum at least.

Nitrate reductase has been purified to homogeneity from the alga *Chl. vulgaris* (128). The molecular weight, determined by a variety of techniques, is around 356,000 and contains a minimum of 2 mol each of FAD, heme, and molybdenum distributed on at least 3 subunits. The enzyme, like other nitrate reductases, also possesses NADH cytochrome *c* reductase and reduced methyl viologen–nitrate reductase. The enzyme system from the yeast *Rho. glutinis* has now been obtained in a purified form (129). The enzyme had a molecular weight of 230,000 and was composed of two subunits. It contained a b_{557}-type cytochrome and very loosely bound FAD. Although the enzyme was not assayed for molybdenum, it possessed all the characteristics of other nitrate reductases, and because of its high specific activity and recovery it could provide an extremely useful system for further biochemical studies.

The only prokaryotic nitrate assimilating system studied so far has been that from *Azotobacter*, in which some attempt has been made to purify the enzyme. The molecular weight was estimated to be around 100,000 (134) and contained molybdenum (135). A recent report (141) gave evidence that the soluble enzyme with the characteristics of Pichinoty's type B nitrate reductase seemed to be supplemented by a nitrate reductase of the type A class after the cells had been repeatedly subcultured on a nitrate medium. The physiological significance of this change remains a mystery.

Nitrite Reductase   One of the distinguishing features between organisms that assimilate nitrate, as opposed to those that use nitrate as an alternative acceptor to oxygen, is the presence of nitrite reductase in the former. Besides reducing nitrite to ammonia, the enzyme also appeared to catalyze the reduction of hydroxylamine to ammonia (142, 143). In a study of 123 mutants, Pateman, Rever, and Cove (142) found that only two structural genes were involved in the reduction of nitrate to nitrite.

The first was the gene for nitrate reductase and the second for the reduction of nitrite to ammonia. Furthermore, the *niiA* gene (structural gene for nitrite reductase) in *Asp. nidulans* is the only gene known in which a mutation affects both the nitrite and hydroxylamine reductase activities. Although an extensive search for other enzymes with altered activities has been made, none has been found. Using crude extracts of *N. crassa*, Garrett (143) found that the relative specific activities of the NADPH nitrite reductase, the reduced benzyl viologen–nitrite reductase and the NADPH hydroxylamine reductase were identical under a variety of conditions. In the absence of any purified preparation of nitrite reductase from fungi, this biochemical and genetic evidence strongly suggests that both nitrite and hydroxylamine reduction are catalyzed by the same enzyme. Recently the enzyme has been purified 90-fold from *N. crassa* (144) and found to catalyze five reactions, i.e., NAD(P)H-nitrite reduction, NAD(P)H-hydroxylamine reduction, and reduced benzyl viologen–nitrite reduction. The enzyme was still contaminated with NADPH nitrate reductase and was associated with at least two contaminating proteins, as judged by gel electrophoresis. Nevertheless the five activities listed above co-purified together and were not associated with the nitrite reductase. The molecular weight of the nitrite reductase was estimated to be around 290,000. This value is much higher than the enzyme purified from the bacterium *Achromobacter fischeri*, which was comprised of 2 subunits with a weight of about 40,000 each in size (145), or from the alga *Chlorella* (146), which was a single polypeptide with a weight of 63,000. The involvement of sulfhydryl groups in catalytic activity was inferred from pCMB inhibition experiments in which only the reduced pyridine nucleotide–dependent activities were affected. The reactions also required the addition of FAD for full activity in all but crude extracts, implying that the flavin, although present in the enzyme, was loosely bound and readily lost on purification.

All activities were sensitive to inhibition by cyanide and it was suggested that a metal (not identified) may have been important for activity. Based on these observations it was proposed that the scheme for electron transport was as follows:

$$\text{NAD(P)H} \rightarrow (-\text{SH, FAD}) \rightarrow (\text{metal?}) \rightarrow \text{NO}_2, \text{NH}_2\text{OH}$$

$$\uparrow \qquad\qquad\qquad \downarrow$$

$$\text{reduced} \qquad\qquad \text{NH}_3$$
$$\text{benzyl}$$
$$\text{viologen}$$

The K_m for nitrite was two orders of magnitude lower than that for hydroxylamine, indicating that the enzyme was primarily a nitrite-reducing system, with the remote possibility that hydroxylamine was a free intermediate. Nitrite reductases from bacteria appear to be NADH-specific (147, 148). The enzyme from *A. chroococcum* is FAD-dependent and, like the enzyme from *N. crassa*, was sensitive to pCMB and cyanide. It had a molecular weight of about 67,000 but was not purified.

The pure nitrite reductase from *Chlorella* appeared to exist in two forms that may have been isoenzymes (146). The enzyme did not have any bound flavin nor was it required for activity, but 2 mol of iron were present. This enzyme, like that from plants and other algae, did not utilize NAD(P)H as a direct reductant (see ref. 149) but did use reduced ferredoxin, benzyl viologen, or methyl viologen as electron donors.

Control of Nitrate Reduction

Genetic Studies   The regulation of nitrate reduction has been one of the most intensely studied aspects of nitrogen metabolism in *Aspergillus* in recent years. The synthesis of both nitrate and nitrite reductase (and the attendant hydroxylamine reductase) is induced in wild-type mycelia by nitrate ions and repressed by ammonium ions (150), although it is not known whether these ions act at the level of transcription or the level of translation or both. The effects of ammonium and nitrate ions are independent, because it is possible to overcome ammonium inhibition by increasing the ratio of nitrate to ammonium ions (151). Based on genetic and biochemical evidence gathered over many years from both Pateman's and Cove's laboratories, it has been possible to propose a model for the regulation of synthesis of the enzymes involved in nitrate assimilation in *Aspergillus*. This scheme is illustrated in Figure 7. At present there is no biochemical evidence concerning the regulatory function of the *nirA* gene product, or even its site of action, but genetic evidence does suggest that it is a region close to the *niaD* and *niiA* structural genes. Similarly, there is genetic evidence that indicates that the gene products of the *tamA* locus (thiourea, aspartic hydroxamate, and methylammonium resistance) and the *areA* locus (ammonium regulation enzymes) act in the same region as the *nirA* gene product.

In the presence of ammonium ions the toxic effects of thiourea, aspartic hydroxamate, and methylammonia are relieved and so it is presumed that the gene products from the *tamA* locus can no longer interact with the *nii* region, thereby resulting in repression of the expression of the *niiA* and *niaD* gene products. The product of the regulatory gene *nirA* is presumed to play a role in both permitting and preventing expression of the structural genes for nitrate and nitrite reductase (152). In the presence of nitrate the nitrate reductase is complexed with nitrate and therefore will not complex with the *nirA* gene product. The *nirA* gene product can now interact with the *nii* region and allow expression of nitrate reductase. In the absence of nitrate a complex is formed between endogeneous nitrate reductase and the *nirA* gene product, which now acts as a repressor of nitrate reductase synthesis.

Regulation of Activity in Aspergillus   The demand for NADPH by the nitrate-utilizing enzyme system is extremely high, so it is of importance

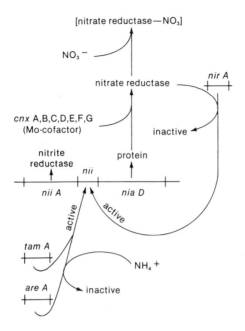

Figure 7. Proposed model for the action of the regulator genes *nirA*, *areA*, and *tamA* in *Aspergillus nidulans*. The *nii* control region, adjacent to the structural genes for nitrite and nitrate reductases, is presumed to be the site of interaction for the *nirA* gene product and the *tamA* and *areA* gene products. In the presence of ammonia indirect inactivation of the *tamA* and *areA* gene products occurs, leading to ammonium repression. Nitrate reductase per se, in the presence of nitrate, complexes with nitrate to effect its reduction. In the absence of nitrate the enzyme interacts with the *nirA* gene product and inactivates it. [After Pateman (153)].

that some mechanism exists to provide this necessary reducing power without seriously upsetting the delicate metabolic balance within the cell.

In *Aspergillus* it has been observed that the level of the enzymes in the pentose phosphate pathway (PPP) are present in greater concentration under nitrate-assimilating conditions than they are when grown with other nitrogen sources (154). This pathway includes two NADPH-generating enzymes that are involved in the catabolism of glucose. Furthermore, in vitro and in vivo studies of *Asp. nidulans* have revealed that nitrate reductase activity can be correlated with the NADPH concentration (155, 156). When assimilating urea, the NADPH:NADP$^+$ ratio was low and the level of pyridine nucleotide-linked enzymes of the PPP was also low. Nitrate-induced cells, on the other hand, had a much higher NADPH: NADP$^+$ ratio and pyridine nucleotide–linked enzyme activity was 50% higher (Table 4). These results suggest, therefore, that nitrate reductase activity is carefully regulated by the relative concentration of reduced-to-oxidized pyridine nucleotide, thus providing a fast-acting and reversible control system. In this

Table 4.   Effect of growth conditions on level of pentose phosphate pathway enzymes, nitrate reductase, and NADPH:NADP$^+$ ratio in *Aspergillus nidulans*

Growth conditions	NADPH:NADP$^+$ ratio	G.6.P. deH^a	6.P.G. deH^b	NO$_3$R^c
20 mм NO$_3$	2.39	1925	1160	89
20 mм urea	1.41	1209	790	9

^{a}G.6.P. deH = Glucose 6-phosphate dehydrogenase.
^{b}6.P.G. deH = phosphogluconate dehydrogenase.
^{c}NO$_3$R = nitrate reductase.

context it is also interesting to note that extracts from a mutant with altered levels of PPP enzyme had reduced levels of NADPH lose their nitrate reductase activity more rapidly than wild-type extracts and that this loss in activity can be reversed only by adding NADPH. It also appears that the *nirA* gene product may be involved in regulation either by complexing with nitrate reductase per se (see above) or by controlling production of a nitrate reductase inactivating enzyme (157, 158).

Photosynthetic Organisms   As in other systems, ammonium ions repress the synthesis of the enzymes of nitrate assimilation in photosynthetic microbes (159). The lack of vast numbers of genetic markers has prevented the sort of study undertaken for *Aspergillus* and attention has been firmly focused on control of activity.

In the early 1970s work on *Chlorella* in both Vennesland's (160, 161, 162) and Losada's (163, 164, 165) laboratories revealed that the nitrate reductase in freshly prepared cell extracts was inactive but could be reactivated by nitrate and ferricyanide. Furthermore, the ferricyanide-activated form of the enzyme could be inactivated by either NADH or NADPH. Losada's group (165) also observed that addition of ammonium ions to *Chlorella* cells growing in the light on nitrate led to a rapid inactivation of nitrate reductase that was alleviated upon removal of ammonia. They proposed (163, 164) that uncoupling of photophosphorylation by ammonia, which would lead to an increased level of reduced pyridine nucleotides in the cell, caused the inactivation of nitrate reductase.

This proposal was later challenged when workers in Vennesland's group observed that the active form of pure nitrate reductase from *Chlorella* was reversibly inactivated by a combination of cyanide and NADH (166). Furthermore, the partially purified form contained bound cyanide that was released upon conversion to the active form (167). They believed that cyanide, which was present in cells in amounts that were sufficient to account for the presence of inactive nitrate reductase (166), was the regulator of activity. They later observed that purified preparations of the enzyme were found to bind 2 mol of cyanide per mol of enzyme at independent but equivalent sites, suggesting that the enzyme contained 2 separate catalytic centers (168) and that it was the binding of cyanide that was responsible for the conversion of the active to the inactive form. Unfortunately there is no evidence to implicate ammonium ions in this process or indeed any indication as to how cyanide is synthesized.

Recent investigations (169) in Losada's laboratory using the blue-green alga *Nostoc* have revealed that both methylamine and arsenate, known uncouplers of photophosphorylation, behave like ammonia in inhibiting photoreduction of nitrate. Furthermore, inactivation of the enzyme depends on photosynthetically generated reducing power that can be prevented by incubation with nitrate (cf. refs. 170 and 171). The enzyme from *Nostoc* was inactivated not only by physiologically generated reducing power but also by sodium dithionite, an effect that could be prevented by cyanate, which binds to the same site as nitrate and protects against inactivation. These results, taken with those from *Chlorella*, substantiate the view that ammonia controls the activity of nitrate reductase through its effect on uncoupling photophosphorylation. This leads to an increase in reduced pyridine nucleotides, which causes a reduction of the enzyme protein and hence inactivation. This effect is prevented by nitrate, which preserves the oxidized, active form of the enzyme by binding to the protein. Clearly this effect appears to be directly opposite to that observed in *Aspergillus* and *Neurospora*, in which reduced pyridine nucleotides maintain the enzyme in a reduced but active form.

A recent observation (172) of the effect of ammonia on nitrate reductase activity in another blue-green alga, *Anabaena cylindrica*, suggests that neither Losada's nor Vennesland's explanations for regulation of the enzyme in this organism are tenable. Ammonia appeared to inhibit only the uptake of nitrate. No inhibition of activity of the enzyme was observed either in vitro or after 30 min exposure in vivo. The uptake of nitrate was competitively inhibited by ammonia at concentrations greater than 1×10^{-5} M, with ammonia uptake being favored in relation to nitrate uptake. In view of this uptake, competition between the two species, and the fact that active transport may be involved in the process, the authors postulated that ATP availability would be the controlling influence of nitrate metabolism. They argued that ammonia was probably assimilated by the GS system, which would require ATP and therefore diminish the supply of energy for nitrate uptake. Armed with this hypothesis as a basis for future research, one would expect to see a resolution of several of the problems concerning regulation of nitrate metabolism in algae in the near future.

ACKNOWLEDGMENTS

I should like to thank Professor John A. Pateman, Dr. David J. Cove, Dr. M. Geoffrey Yates, Dr. Robert R. Eady, and Dr. Barry E. Smith for allowing me to see copies of their papers prior to publication.

REFERENCES

1. Tempest, D. W., Meers, J. L., and Brown, C. M. (1970). Biochem. J. 117:405.
2. Meers, J. L., Tempest, D. W., and Brown, C. M. (1970). J. Gen. Microbiol. 64:187.
3. Meister, A. (1962). *In* P. D. Boyer, H. Lardy, and K. Myrback (eds.), The Enzymes, Vol. 6, p. 193. Academic Press, New York.
4. Umbarger, H. E. (1969). Annu. Rev. Biochem. 38:323.
5. Shapiro, B. M., and Stadtman, E. R. (1970). Annu. Rev. Microbiol. 24:501.
6. Denton, M. D., and Ginsberg, A. (1970). Biochemistry 9:617.
7. Meers, J. L., and Pedersen, L. K., (1972). J. Gen. Microbiol. 70:277.
8. Le John, H. B., and McCrae, B. E. (1968). J. Bacteriol. 95:87.
9. Kramer, J. (1970). Arch. Mikrobiol. 71:226.
10. Brown, C. M., Macdonald-Brown, D. S., and Stanley, S. O. (1973). Antonie Van Leeuwenhoek 39:89.
11. Senior, P. J. (1975). J. Bacteriol. 123:407.
12. Brown, C. M., and Stanley, S. O. (1972). J. Appl. Chem. Biotechnol. 22:363.
13. Brown, C. M., Burn, V. J., and Johnson, B. (1973). Nature New Biol. 246:115.
14. Johnson, B., and Brown, C. M. (1974). J. Gen. Microbiol. 85:169.
15. Roon, R. R., Even, H. L., and Larimore, F. (1974). J. Bacteriol. 118:89.
16. Brown, C. M. (1976). *In* A. C. R. Dean, D. C. Ellwood, C. G. T. Evans, and J. Melling (eds.)., Continuous Culture 6: Applications and New Fields, p. 170. Ellis Horwood, Chichester, England.
17. Brown, C. M., Macdonald-Brown, D. S., and Meers, J. L. (1974). Adv. Microb. Physiol. 11:1.
18. Wulff, K., Mecke, D., and Holzer, H. (1967). Biochem. Biophys. Res. Commun. 28:740.
19. Kingdon, H. S., Shaphiro, B. M., and Stadtman, E. R. (1967). Proc. Natl. Acad. Sci. USA 58:1703.
20. Brown, M. S., Segal, A., and Stadtman, E. R. (1971). Proc. Natl. Acad. Sci. USA 68:2949.
21. Magnum, J. H., Magni, G., and Stadtman, E. R. (1973). Arch. Biochem. Biophys. 158:514.
22. Brenchley, J. E., Prival, M. J., and Magasanik, B. (1973). J. Biol. Chem. 248:6122.
23. Dainty, R. H. (1972). Biochem. J. 126:1055.
24. Miller, R. E., and Stadtman, E. R. (1972). J. Biol. Chem. 247:7407.
25. Cole, J. A., Coleman, K. J., Compton, B. E., Kavanagh, B. M., and Keevil, C. W. (1974). J. Gen. Microbiol. 85:11.
26. Brenchley, J. E., Baker, C. A., and Patil, L. G. (1975) J. Bacteriol. 124:182.
27. Wohlhueter, R. M., Schutt, H., and Holzer, H. (1973). *In* S. Pruisner and E. R. Stadtman (eds.), The Enzymes of Glutamine Metabolism, p. 45. Academic Press, New York and London.
28. Woolfolk, C. A., Shapiro, B. M., and Stadtman, E. R. (1966). Arch. Biochem. Biophys. 116:177.
29. Schutt, H., and Holzer, H. (1972). Eur. J. Biochem. 26:68.
30. Segal, A., Brown, M. S., and Stadtman, E. R. (1974). Arch. Biochem. Biophys. 161:319.
31. Kavanagh, M. B., and Cole, J. A. (1976). *In* A. C. R. Dean, D. C. Ellwood, C. G. T. Evans, and J. Melling (eds.), Continuous Culture 6: Applications and New Fields, p. 184. Ellis Horwood, Chichester, England.
32. Lowry, O. H., Carter, J., Ward, J. B., and Glaser, L. (1971). J. Biol. Chem. 246:6511.

33. MacMillan, A. (1956). J. Exp. Bot. 7:113.
34. Hackette, S. L., Skye, G. E., Burton, C., and Segal, I. H. (1970). J. Biol. Chem. 245:4241.
35. Pateman, J. A., Kinghorn, J. R., Dunn, E., and Forbes, E. (1973). J. Bacteriol. 114:943.
36. Pateman, J. A., Dunn, E., Kinghorn, J. R., and Forbes, E. (1974). Molec. Gen. Genet. 133:225.
37. Cook, R. J., and Anthony, C. (1973). J. Gen. Microbiol. 77:vii.
38. Roon, R. J., Even, H. L., Dunlop, P., and Larimore, F. L. (1975). J. Bacteriol. 122:502.
39. Stevenson, R., and Silver, S. (1977). Biochem. Biophys. Res. Commun. 75:1133.
40. Postgate, J. R., (1974). J. Appl. Bacteriol. 37:185.
41. Jodin, C. R. (1862). C. R. Acad. Sci. (Paris) 55:612.
42. Bulen, W. A., and Le Comte, J. R. (1966). Proc. Natl. Acad. Sci. USA 56:979.
43. Kelly, M. (1969). Biochim. Biophys. Acta 191:527.
44. Zumft, W. G., Cretney, W. C., Huang, T. C., Mortenson, L. E., and Palmer, G. (1972). Biochem. Biophys. Res. Commun. 48:1525.
45. Zumft, W. G., and Mortenson, L. E. (1973). Eur. J. Biochem. 35:401.
46. Burns, R. C., Holsten, R. D., and Hardy, R. W. F. (1970). Biochem. Biophys. Res. Commun. 39:90.
47. Shah, V. K., and Brill, W. J. (1973). Biochem. Biophys. Acta 305:445.
48. Yates, M. G., and Planque, K. (1975). Eur. J. Biochem. 60:467.
49. Eady, R. R., Smith, B. E., Cook, K. A., and Postgate, J. R., (1972). Biochem. J. 128:655.
50. Bray, R. C. (1969). FEBS Lett. 5:1.
51. Smith, B. E., Lowe, D. J., and Bray, R. C. (1973). Biochem. J. 135:331.
52. Bui, P. T., and Mortenson, L. E. (1968). Proc. Natl. Acad. Sci. USA 61:1021.
53. Nagatani, H. H., and Brill, W. J. (1974). Biochem. Biophys. Acta 362:160.
54. Nagatani, H. H., Shah, V. K., and Brill, W. J. (1974). J. Bacteriol. 120:697.
55. Nason, A., Lee, K. Y., Pan, S. S., Ketchum, P. A., Lamberti, A., and DeVries, J. (1971). Proc. Natl. Acad. Sci. USA 68:3242.
56. Nason, A., Lee, K. Y., Pan S. S., and Erickson, R. H. (1973). In P.C.H. Mitchell (ed.), Chemistry and Uses of Molybdenum, p. 233. Climax Molybdenum Co. Ltd., London.
57. Shah, V. K. (1976). In Abstracts of the 2nd International Symposium on Nitrogen Fixation, Salamanca, Spain. (in press).
58. Ganelin, V. L., L'Vov, N. P., Veinova, N. K., and Sergeev, N. S. (1973). In Mechanism of Biological Nitrogen Fixation, p. 31. 4th N. A. Basha Symposium. Dokl. NAAS. Moscow, U.S.S.R.
59. Moustafa, E. (1970). Biochim. Biophs. Acta 206:178.
60. Moustafa, E., and Mortenson, L. E. (1969). Biochim. Biophys. Acta 172:106.
61. Zumft, W. G., Palmer, G., and Mortenson, L. E. (1973). Biochim. Biophys. Acta 292:413.
62. Hwang, J. C., Chen, C. H., and Burris, R. H. (1973). Biochim. Biophys. Acta 292:256.
63. Tso, M. -Y., and Burris, R. H. (1973). Biochim. Biophys. Acta 309:263.
64. Thorneley, R. N. F., and Eady, R. R. (1973). Biochem. J. 133:405.
65. Walker, G. A., and Mortenson, L. E. (1974). Biochemistry 13:2382.
66. Zumft, W. G., Mortenson, L. E. and Palmer, G. (1974). Eur. J. Biochem. 46:525.
67. Brulet, C. R., and Van Tamelen, E. E. (1975). J. Amer. Chem. Soc. 97:911.
68. Chatt, J., Pearman, A. J., and Richards, R. L. (1975). Nature 253:39.
69. Smith, B. E. (1977). J. Less. Comm. Met. 54:465.

264   Dalton

70. Smith, B. E., Thorneley, R. N. F., Eady, R. R., and Mortenson, L. E. (1976). Biochem. J. 157:439.
71. Tso, M. -Y. (1974). Arch. Microbiol. 99:71.
72. Nakos, G., and Mortenson, L. E. (1971). Biochemistry 10:455.
73. Kleiner, D., and Chen, C. H. (1974). Arch. Microbiol. 98:100.
74. Smith, B. E., Thorneley, R. N. F., Yates, M. G., Eady, R. R., and Postgate, J. R. *In* International Symposium on Nitrogen Fixation, Pullman, Washington. (in press).
75. Whiting, M. J., and Dilworth, M. J. (1974). Biochim. Biophys. Acta 371:337.
76. Huang, T. C., Zumft, W. G., and Mortenson, L. E. (1973). J. Bacteriol. 113:884.
77. Stasny, J. T., Burns, R. C., Korant, B. D., and Hardy, R. W. F. (1974). J. Cell. Biol. 60:311.
78. Israel, D. W., Howard, R. L., Evans, H. J., and Russell, S. A. (1974). J. Biol. Chem. 249:500.
79. Eady, R. R., and Smith, B. E. (1977). *In* R. W. F. Hardy and R. C. Burns (eds.), Treatise on Dinitrogen Fixation, Wiley, New York. (in press).
80. Mortenson, L. E. (1964). Proc. Natl. Acad. Sci. USA 52:272.
81. Walker, M. N., and Mortenson, L. E. (1974). J. Biol. Chem. 249:6356.
82. Yoch, D. C., and Arnon, D. I. (1972). J. Biol. Chem. 247:4514.
83. Bothe, H., and Yates, M. G. (1976). Arch. Microbiol. 107:25.
84. Benemann, J. R., and Valentine, R. C. (1972). Adv. Microb. Physiol. 8:59.
85. Yoch, D. C. (1974). J. Gen. Microbiol. 83:153.
86. Benemann, J. R., Yoch, D. C., Valentine, R. C., and Arnon, D. I. (1971). Biochim. Biophys. Acta 226:205.
87. Evans, H. J., and Phillips, D. A. (1975). *In* W. D. P. Stewart (ed.), Nitrogen Fixation by Free-Living Micro-Organisms, p. 389. Cambridge University Press, Cambridge, England.
88. Yates, M. G. (1972). FEBS Lett. 27:63.
89. Yates, M. G. (1976). *In* Abstracts of the Second International Symposium on Nitrogen Fixation, Salamanca, Spain. (in press).
90. Bresters, T. W., Krul, J., Scheepens, P. C., and Veeger, C. (1972). FEBS Lett. 22:305.
91. Haaker, H., Bresters, T. W., and Veeger, C. (1972). FEBS Lett. 23:100.
92. Campbell, F. C., and Yates, M. G. (1973). FEBS Lett. 37:203.
93. Biggins, D. R., and Postgate, J. R. (1971). Eur. J. Biochem. 19:408.
94. Yates, M. G. (1972). Eur. J. Biochem. 29:386.
95. Morris, J. G. (1975). Adv. Microb. Physiol. 12:169.
96. Cole, J. A. (1976). Adv. Microb. Physiol. 14:1.
97. Harrison, D. E. F. (1976). The Oxygen Metabolism of Microorganisms. Meadowfield Press Ltd., Shildon, Co. Durham, England.
98. Lees, H. Personal communication.
99. Dalton, H. (1968). Ph.D. thesis, University of Sussex, England.
100. Dalton, H., and Postgate, J. R. (1969). J. Gen. Microbiol. 54:463.
101. Dalton, H., and Postgate, J. R. (1969) J. Gen. Microbiol. 56:307.
102. Jones, C. W., and Redfearn, E. (1967). Biochim. Biophys. Acta 143:340.
103. Yates, M. G. (1970). J. Gen. Microbiol. 60:393.
104. Bergersen, F. J., and Turner, G. L. (1975). J. Gen. Microbiol. 91:345.
105. Bergersen, F. J., and Turner, G. L. (1975). J. Gen. Microbiol. 89:31.
106. Drozd, J. A., and Postgate, J. R. (1970). J. Gen. Microbiol. 63:63.
107. Bulen, W. A., Burns, R. C., and Le Comte, J. R. (1965). Proc. Natl. Acad. Sci. USA 53:532.
108. Thorneley, R. N. F., and Willison, K. R. (1974). Biochem J. 139:211.

109. Thorneley, R. N. F., and Cornish-Bowden, A. (1977). Biochem. J. 165:255.
110. Haaker, H., DeKok, A., and Veeger, C. (1974). Biochim. Biophys. Acta 357:344.
111. Haaker, H., and Veeger, C. (1976). Eur. J. Biochem. 63:499.
112. L'Vov, N. P., Sergeer, N. S., Veinova, M. K., Shaposhnikov, G. L., and Kretovich, W. L. (1971). C. R. Acad. Sci. (USSR) 206:1493.
113. Drozd, J. W., Tubb, R. S., and Postgate, J. R. (1972). J. Gen. Microbiol. 73:221.
114. Seto, B. L., and Mortenson, L. E. (1974). J. Bacteriol. 117:805.
115. Tubb, R. S., and Postgate, J. R. (1973). J. Gen. Microbiol. 79:103.
116. Nagatani, H., Shimazu, M., and Valentine, R. C. (1971). Arch. Microbiol. 79:164.
117. Tubb, R. S. (1974). Nature 251:481.
118. Streicher, S. L., Shanmugam, K. T., Ausubel, F., Morandi, C., and Goldberg, R. B. (1974). J. Bacteriol. 120:815.
119. Kleiner, D. (1976.) Arch Microbiol. 111:85.
120. Bishop, P. E., McParland, R. H., and Evans, H. J. (1975). Biochem. Biophys. Res. Commun. 67:774.
121. Shanmugam, K. T., Chan, I., and Morandi, C. (1975). Biochim. Biophys. Acta 408:101.
122. St. John, R. T., Shah, V. K., and Brill, W. J. (1974). J. Bacteriol. 119:266.
123. Hill, S. (1976). J. Gen. Microbiol. 93:335.
124. Bergersen, F. J., and Turner, G. L. (1976). Biochem. Biophys. Res. Commun. 73:524.
125. Van't Reit, J., Knook, D. L., and Planta, R. L. (1972). FEBS Lett. 23:44.
126. Pichinoty, F. (1964). Biochim. Biophys. Acta 89:378.
127. Payne, W. J. (1973). Bacteriol. Rev. 37:409.
128. Solomonson, L. P., Lorimer, G. H., Hall, R. L., Borchers, R., and Bailey, J. L. (1975). J. Biol. Chem. 250:4120.
129. Guerro, M. G., and Gutierrez, M. (1977). Biochim. Biophys. Acta 482:272.
130. Garrett, R. H., and Nason, A. (1969). J. Biol. Chem. 244:2870.
131. Downey, R. J. (1971). J. Bacteriol. 105:759.
132. Downey, R. J. (1973). Biochem. Biophys. Res. Commun. 50:920.
133. McDonald, D. W., and Coddington, A. (1974). Eur. J. Biochem. 46:169.
134. Guerrero, M. G., Vega, J. M., Leadbetter, E., and Losada, M. (1973). Arch. Microbiol. 91:287.
135. Guerrero, M. G., and Vega, J. M. (1975). Arch Microbiol. 102:91.
136. Ketchum, P. A., Cambier, H. Y., Frazier, W. A., III, Madansky, C. H. and Nason, A. (1970). Proc. Natl. Acad. Sci. USA 66:1016.
137. Subramanan, K. N., and Sorger, G. J. (1972). Biochim. Biophys. Acta 256:533.
138. Pateman, J. A., Cove, D. J., Rever, B. M., and Roberts, D. B. (1964). Nature 201:58.
139. Arst, H. N., MacDonald, D. W., and Cove, D. J. (1970). Mol. Gen. Genet. 108:129.
140. McDonald, D. W., and Cove, D. J. (1974). Eur. J. Biochem. 47:107.
141. Vila, R., Barcena, J. A., Llobell, A., and Paneque, A. (1977). Biochem. Biophys. Res. Commun. 75:682.
142. Pateman, J. A., Cove, D. J., Rever, B. M., and Roberts, D. B. (1964). Nature 201:58.
143. Garrett, R. H. (1972). Biochim. Biophys. Acta 264:481.
144. Lafferty, M. A., and Garrett, R. H. (1974). J. Biol. Chem. 249:7555.
145. Husain, M., and Sudana, J. (1974). Eur. J. Biochem. 42:283.

146. Zumft, W. G. (1972). Biochim. Biophys. Acta 276:363.
147. Kemp, J. D., and Atkinson, D. E. (1966). J. Bacteriol. 92:628.
148. Vega, J. M., Guerrero, M. G., Leadbetter, E., and Losada, M. (1973). Biochem. J. 133:701.
149. Hewitt, E. J. (1975). Annu. Rev. Plant. Physiol. 26:73.
150. Cove, D. J., and Pateman, J. A. (1963). Nature 198:262.
151. Cove, D. J., and Pateman, J. A. (1969). J. Bacteriol. 97:1374.
152. Cove, D. J. (1969). Nature 224:272.
153. Pateman, J. R., and Kinghorn, J. R. (1977). *In* J. E. Smith and J. A. Pateman (eds.), The Genetics and Physiology of *Aspergillus,* p. 203. Academic Press, London.
154. Hankinson, O., and Love, D. J. (1974). J. Biol. Chem. 249:2344.
155. Dunn-Coleman, N. S., and Pateman, J. A. (1975). Biochem. Soc. Trans. 3:531.
156. Dunn-Coleman, N. S., and Pateman, J. A. (1977). Mol. Gen. Genet. 152:285.
157. Wallace, W. (1975). Biochim. Biophys. Acta 377:234.
158. Kadam, S. S., Sawhney, S. K., and Naik, M. S. (1975). Ind. J. Biochem. Biophys. 12:81.
159. Syrett, P. J., and Morris, I. (1963). Biochim. Biophys. Acta 67:566.
160. Vennesland, B., and Jetschmann, K. (1971). Biochim. Biophys. Acta 227:554.
161. Jetschmann, C., Solomonson, L. P., and Vennesland, B. (1972). Biochim. Biophys. Acta 275:276.
162. Solomonson, L. P., Jetschmann, K., and Vennesland, B. (1973). Biochim. Biophys. Acta 309:32.
163. Moreno, C. G., Aparicio, P. J., Palacian, E., and Losada, M. (1972). FEBS Lett. 26:11.
164. Maldonado, J. M., Herrera, J., Paneque, A., and Losada, M. (1973). Biochem. Biophys. Res. Commun. 51:27.
165. Losada, M., Paneque, A., Aparicio, P. J., Vega, J. M., Cardenas, S., and Herrera, J. (1970). Biochem. Biophys. Res. Commun. 38:1009.
166. Solomonson, L. P. (1974). Biochim. Biophys. Acta 334:297.
167. Lorimer, G. H., Gervitz, H. S., Volker, W., Solomonson, L. P., and Vennessland, B. (1974). J. Biol. Chem. 249:6074.
168. Solomonson, L. P., Lorimer, G. H., Hall, R. L., Borchers, R., and Bailey, J. L. (1975). J. Biol. Chem. 250:4120.
169. Ortega, T., Castillo, F., Cardenas, J., and Losada, M. (1977). Biochem. Biophys. Res. Commun. 75:823.
170. Losada, M. (1976). J. Molec. Catal. 1:245.
171. Pistorius, E. K., Gervitz, H., Voss, H., and Vennesland, B. (1976). Planta 128:73.
172. Ohmari, M., Ohmari, K., and Strotmann, H. (1977). Arch. Microbiol. 114:225.

International Review of Biochemistry
Microbial Biochemistry, Volume 21
Edited by J. R. Quayle
Copyright 1979   University Park Press   Baltimore

7
Microbial Biochemistry of Methane— A Study in Contrasts

R. S. WOLFE
Dept. of Microbiology
University of Illinois, Urbana, U.S.A.

I. J. HIGGINS
Biological Laboratory
University of Kent at Canterbury, England

EDITORIAL INTRODUCTION

Over the years methane has usually been regarded as something of an oddity; it tends to turn up in odd places with odd results, e.g., in coal mines, with the ever-present threat of its explosion, in marshes (it is said to be the spirit behind the romantic and evanescent Will o' the Wisp), and in the bellies of ruminants, to the accompaniment sometimes of spectacular flatulence. However, serious recognition of the extent of the role of methane in the cycling of carbon in the environment has only emerged over the last 10 years as a result of measurements by "atmospheric chemists" using miniaturized automatic instruments capable of measuring gas changes in environments ranging from the submarine to the stratospheric. The results of these studies are dramatic. At the Symposium on Microbial Production and Utilization of Gases, held at Göttingen in 1974, Ehhalt estimated that 500–800 million tons of biologically generated methane are released into the atmosphere per year. This corresponds to as much as 0.5% of the total annual production of dry organic matter by photosynthesis. The methane so released is photolyzed in the upper atmosphere and is returned to earth as CH_2O, CO, and CO_2, thus completing a terrestrial methane-carbon cycle of unexpectedly large dimensions.

The methane is generated mainly from acetate, CO_2, and H_2 by a specialized group of anaerobic bacteria—the methanogens, functioning at the end of an anaerobic food chain. Their true biological activity as estimated above is masked by the presence of methane-utilizing bacteria—the methanotrophs, which oxidize methane to CO_2. Hence much of the methane generated at the bottom of a lake may not emerge in the surface bubbles because it will have been extracted from the water as a source of carbon and energy by the methanotrophs, thus constituting a second methane-based carbon cycle, in this case operating below, rather than above, the earth's surface.

Thus the microbial biochemistry of methane may be divided into two parts: 1) methanogenesis; and 2) methanotrophy. In the past, study of methanogenesis has been the province of only those microbial biochemists brave enough to try to handle these nutritionally exacting, strictly anaerobic organisms. Meanwhile a burgeoning literature on methanotrophy has emerged from laboratories not normally dealing with highly anaerobic systems. The studies in these two fields often get separated in both writing and discussion, and yet the paper biochemistry of these two processes can be made to look almost like mirror images:

Methanogenesis

$$CO_2 \xrightarrow{2H} OHC \cdot X \xrightarrow{2H} HOH_2C \cdot X \xrightarrow{2H} H_3C \cdot X \xrightarrow{2H} CH_4$$

$$\underbrace{\qquad\qquad\qquad\qquad}_{?} \downarrow$$

Cell constituents

$$X = \text{Coenzyme M?}$$

Methanotrophy

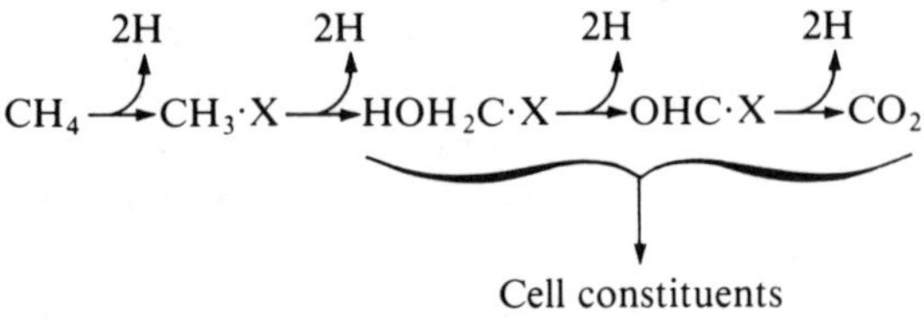

$$CH_4 \xrightarrow{\;2H\;} CH_3 \cdot X \xrightarrow{\;2H\;} HOH_2C \cdot X \xrightarrow{\;2H\;} OHC \cdot X \xrightarrow{\;2H\;} CO_2$$

Cell constituents

X = ·OH, and/or a
pteridine derivative,
and/or a folate,
and/or glutathione

Biochemical experience nevertheless tells us that to go from B to A does not usually mean retracing the steps that are used to go from A to B; to what extent does this paradigm apply when A is CH_4 and B is CO_2? This is the theme that is explored in this chapter, where the present state of knowledge of methane biochemistry is presented in two separate parts— methanogenesis by R. S. Wolfe and methanotrophy by I. J. Higgins. Much still remains to be discovered, but as of now, the overall chapter is indeed best described as a study in contrasts.

PART I.   METHANOGENESIS   *R. S. Wolfe**

Since methanogenic bacteria have been studied in only a few laboratories, an overview of the habitats and the ecological niche of the methanogens is first considered. For a more detailed view of these organisms, reviews by Barker (1), Stadtman (2), Wolfe (3), Zeikus (4), and Mah et al. (5) should be consulted.

Four rather well-defined groups of anaerobes may be operative in methanogenic habitats. Group I includes those anaerobes that excrete polymer-hydrolyzing extracellular enzymes and that ferment products of polymer hydrolysis to fatty acids, CO_2, and H_2. Group II organisms are poorly represented in pure culture at the present time, but their important role in anaerobic biodegradation is becoming increasingly apparent; they convert fatty acids or alcohols to acetate and H_2. (In addition to acetate and H_2, CO_2 is produced from fatty acids with odd-numbered carbon chains.) These organisms may be included in the group proposed by

---

*Original results reported here were obtained under grants from the National Science Foundation PCM-02652 and the U.S. Public Health Service AI 12277.

Thauer et al. (6) under the term "obligate proton reducers." That is, their major role is the oxidation of fatty acids or alcohols, with the reduction of protons to molecular hydrogen. Group III organisms are represented by *Acetobacterium* and *Clostridium aceticum,* organisms that oxidize H_2 anaerobically with the reduction of CO_2 to acetate. Group IV organisms are the methanogens. They have a very narrow ecological niche, i.e., the formation of methane from H_2 and CO_2 or from acetate. At the present time the conversion of acetate to methane is the least-understood step in anaerobic biodegradation. Pure cultures of methanogens are reluctant to convert acetate to methane in the laboratory, and there is a good possibility that in nature a microbial interaction between methanogenic bacteria and unknown organisms may be required for the efficient conversion of acetate to methane.

The habitats in which methanogenic bacteria occur may be grouped into three types. Type A includes sediments, bogs, tundra, sewage sludge digesters, and decaying heartwood of certain trees (Figure 1). Biodegradation in these habitats may involve all four groups of organisms, because polymers are converted eventually to CH_4 and CO_2. This is the system that nature has evolved for biodegradable compounds; it avoids the build-up of carboxyl groups and eventual cessation of biodegradation by anaerobic acid–preservation. In the sludge digester, for example, fatty acids and other intermediate products cannot "escape" readily. In Type A habitats acetate plays a major role in the flow of carbon from polymers to methane and carbon dioxide; 60 to 70% of the methane produced in the sewage sludge digester is estimated to arise from acetate (7, 8, 9). Carbon moves through higher fatty acids to acetate. Recently, in the

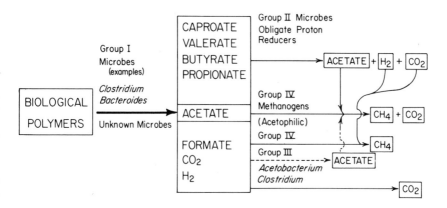

METHANOGENIC HABITAT TYPE – A
(Aquatic Sediments, Swamps, Bogs, Tundra,
Decaying Heartwood, Sludge Digester)

Figure 1. Flow diagram of the routes of polymer degradation (cellulose) in habitats where CO_2 and CH_4 are the final products.

laboratory of M. P. Bryant, an organism that converts butyrate to acetate and hydrogen has been isolated. This is the first example of a Group II organism besides the "S" organism (10, 11, 12) to be isolated (M. P. Bryant, personal communication.) In these habitats the primary polymers may differ somewhat from the rumen; the sludge digester may receive more fats and protein, but anaerobic biodegradation is believed to follow the same general route. Our knowledge of the bacterial species in Type A habitats is rather primitive compared to knowledge of the rumen.

Type B habitats (Figure 2) are confined to the alimentary tracts of animals, especially herbivores that have evolved specialized structures, the rumen and cecum, where food passage is delayed. The rumen is the most thoroughly studied methanogenic habitat (13, 14, 15, 16). Polymers that serve as major microbial energy sources include cellulose, pentosan, pectin, and starch. *Butyrivibrio, Ruminococcus,* and *Bacteroides* are examples of important genera in Group I that are involved in the hydrolysis of cellulose and production of fatty acids, CO_2, and H_2. Acetate, succinate, formate, lactate, propionate, butyrate, valerate, and caproate are examples of fatty acids that are produced in the rumen, the major end products being acetate, propionate, and butyrate. These fatty acids enter the bloodstream and serve as the major energy source for the ruminant. Thus in these habitats biodegradation is catalyzed by organisms that can be placed in Groups I and IV, there being no need to degrade fatty acids to acetate, CO_2, and H_2. Acetate-fermenting methanogens appear to play no role in this habitat. For technical reasons the cecum cannot be as readily studied as the rumen, where fistulated animals have been of great value; however, the overall fermentation appears to be similar.

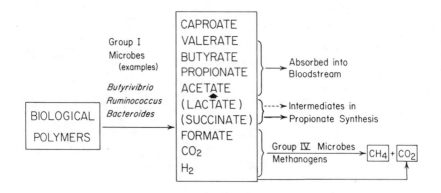

METHANOGENIC HABITAT TYPE - B
(Rumen, Caecum, Intestine)

Figure 2. Flow diagram of the routes of polymer degradation (cellulose) in the rumen or cecum. The short heavy arrow above lactate indicates that this intermediate is mainly metabolized to the fatty acids above it.

Even less is known about biodegradation in the intestine, but it may represent a miniature version of the rumen and cecum.

Type C habitats (Figure 3) include geothermal springs and certain unusual lakes, such as Lake Kivu, that may be fed by such springs. Geothermal springs are known to contain molecular hydrogen, juvenile carbon in the form of carbon dioxide, sulfide, and mineral salts (17). Strains of thermophilic methanogens have been isolated recently from such habitats in Yellowstone Park by Zeikus (4). Results of studies on the unusual anaerobic lake, Lake Kivu, have indicated that there is not sufficient synthesis of biomass from photosynthesis to account for the formation of the enormous quantities of methane dissolved in its waters. The bottom of the lake is a few degrees warmer than the surface, and the lake is believed to be fed by geothermal springs rich in hydrogen, carbon dioxide, and sulfide, with hydrogen serving as the substrate for growth of mesophilic methanogens (18). The role of hydrogen-oxidizing anaerobes, such as *Acetobacterium* (19), has not been established for Lake Kivu.

MICROBIOLOGY

Organisms

As pointed out by Bryant (20) the taxonomy of the methanogenic bacteria is badly in need of revision. Since such a study is in progress we shall not consider the various problems in this area. It should be noted, however, that some of the "type species" have never been obtained in pure cultures. It is possible, but not established scientifically, that some of the proposed species (for example, *Methanobacterium suboxydans*) (1) may not exist and may actually have been mixed cultures composed of group II and IV organisms. The organisms listed in Table 1 represent species, strains,

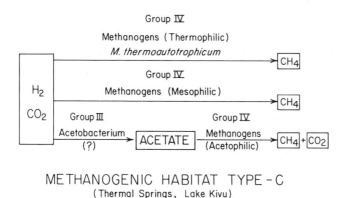

METHANOGENIC HABITAT TYPE-C
(Thermal Springs, Lake Kivu)

Figure 3. Routes of methane formation in thermal springs. Lake Kivu is believed to be fed by thermal springs; however, the role of *Acetobacterium* in this lake has not been established.

Table 1. Methanogens

Short rods and coccobacillus-like cells, gram (+)
  Methanobacterium arbophilicum
  Methanobacterium strain AZ
  Methanobacterium ruminantium strain M1
  Methanobacterium ruminantium strain PS
Long rods, gram (+)
  Methanobacterium thermoautotrophicum
  Methanobacterium formicicum
  Methanobacterium M.o.H. (Bryant)
  Methanobacterium M.o.H. (Göttingen)
Gram (−) cells
  Methanospirillum hungatii
  Methanobacterium mobile
  Methanococcus vannielii
  Black Sea Isolate JR 1
  Cariaco Isolate JR 1
Gram (+) Sarcina
  Methanosarcina barkeri
  Methanosarcina (gas vacuoles) Zhilina
  Methanosarcina (gas vacuoles) Mah

---

and new isolates that are now available in pure culture. It is almost certain that there are many undiscovered methanogens in nature.

Physiology

Substrates and Nutrition   For methanogens now in pure culture the universal substrate of choice is molecular hydrogen, the final electron acceptor being carbon dioxide. Formate is readily converted to hydrogen and carbon dioxide by many methanogens. *Methanosarcina* has the ability to convert methanol to methane (1). Because the formation of methanol in nature has not been well documented for anaerobic habitats, it appears unlikely that the conversion of methanol to methane is of ecological significance. The conversion of acetate to methane has been mentioned previously as a major source of methane in Type A habitats, but the formation of methane from acetate remains an unresolved problem of methanogenesis. Although Zeikus et al. (21) proposed that methanogens like *Methanobacterium thermoautotrophicum* could convert acetate to methane at rates that could be of ecological importance, the acetate pool pro- duced by *Clostridium thermocellum* from degradation of cellulose was not con- verted to methane by *M. thermoautotrophicum* when the two organisms were grown in mixed culture (22). The conversion of carbon monoxide to methane was studied by Kluyver and Schnellen (23). Recently Daniels et al. (24) have shown that *M. thermoautotrophicum* is able to grow, although poorly, on carbon monoxide. Because the growth rate on the substrate was only about 1% of that on H_2/CO_2, and because carbon monoxide-producing reactions have not been documented in methanogenic habitats, this reaction, too, appears not to be of ecological significance.

The methanogens are essentially chemolithotrophs in their metabolism, and their nutrition basically reflects this physiology. Growth of most of these organisms in a mineral salts medium is stimulated by addition of acetate, yeast extract, or tryptic digests of casein. It is of interest to note that certain strains have lost some synthetic abilities. For example, *Methanobacterium ruminantium* strain M1 from the rumen requires 2-methylbutyrate (25) and coenzyme M (2-mercaptoethanesulfonic acid) (26), as well as amino acids, as growth factors. *Methanobacterium mobile* requires an unknown growth factor that is present in rumen fluid (27). This growth factor has been recently shown not to be coenzyme M (W. E. Balch, personal communication). A general characteristic of most methanogens is that they require or are greatly stimulated by a source of reduced sulfur; sulfide or cysteine may satisfy this requirement.

pH and Redox Requirements   Methanogenic bacteria have the most stringent anaerobic requirements among anaerobes. Not only must oxygen be excluded, but these organisms can carry out methanogenesis only where the redox potential is lower than -330 mV (28). Although methane is produced readily in acid peat bogs, methanogenic bacteria in pure culture are physiologically most active in the pH range of 6.7 to 8.0. There is one reported exception and that is *Methanobacillus kuzneceovii* (29, 30), which has been reported to grow at pH 4.0. So far this organism has not been available for other workers to study. The buffer in nature is carbonate-bicarbonate-carbon dioxide. In natural habitats oxygen is removed by aerobes and facultative anaerobes, and a reducing potential is formed by the production of sulfide and molecular hydrogen by the anaerobic metabolism of other microbes.

Interspecies Hydrogen Transfer   The separation of "*Methanobacillus omelianskii*" into its component organisms about 10 years ago opened a new era in our understanding of the methanogens and of the importance of interspecies hydrogen transfer (31). By use of chemostats, Wolin and co-workers extended this concept to other systems (32, 33) and pointed out the dependence of the free-energy change in certain hydrogen-producing reactions on the partial pressure of hydrogen. Production of hydrogen by Group I organisms does not prevent their growth, but more reduced products are formed as hydrogen accumulates (34, 35). Chung (36) has studied this problem in *Clostridium cellobioparum*. Hungate (37) pioneered studies into the dynamic role of hydrogen as an intermediate in the rumen fermentation. Other researchers have obtained similar results on interspecies hydrogen transfer; Weimer and Zeikus (22) studied the coupling of cellulose degradation to methane at 60°C by mixed cultures of *C. thermocellum* and *M. thermoautotrophicum*. In this thermophilic co-culture the amount of ethanol formed was greatly reduced as compared to the monoculture, indicating that in the co-culture electron flow was displaced toward hydrogen formation by the *Clostridium* and toward hydrogen utilization by the methanogen; no hydrogen was detected in the co-culture, and acetate formation increased dramatically. When cellulose was used as the starting substrate for *C. thermocellum,* interspecies hydrogen transfer was excellent. In contrast, when cellobiose was employed, the soluble sugar was fermented quickly, and conditions became unfavorable for the methanogen; coupling could undoubtedly be effected in a chemostat where cellobiose could be made limiting.

Little is known about the anaerobic degradation of aromatic compounds to methane. Conversion rates are slow, and it takes time to establish the microbial food chain for this degradation. Tarvin and Buswell (38) studied the conversion of benzoate to methane by sewage sludge; Fina and Fiskin (39) later showed that when ^{14}C ring-labeled benzoate was used as a substrate by sewage sludge, $^{14}CH_4$ was produced. Nottingham and Hungate (40) also studied anaerobic benzoate degradation. Such degradations occur in the absence of sulfate or nitrate. Recently Ferry and Wolfe (41) studied the conversion of benzoate to methane by a stabilized consortium of microbes that was derived from sewage sludge and that was maintained at a redox potential near the hydrogen electrode in a benzoate–mineral salts medium. Intermediates in this fermentation were shown to be acetate, formate, H_2 and CO_2. When uniformly ring-labeled [^{14}C]benzoate was used, the radioactivity was evenly divided between the $^{14}CH_4$ and $^{14}CO_2$. As shown in Figure 4, CH_4 was produced as benzoate disappeared; the concentration of acetate as an intermediate increased and then disappeared. If [*methyl*-^{14}C]acetate was added, it was converted stoichiometrically to $^{14}CH_4$. To gain information on the direct conversion of benzoate to methane the inhibitor *o*-chlorobenzoate was added. As shown in Figure 5, the rate of benzoate degradation decreased dramatically after the addition of the inhibitor; however, acetate degradation and methane formation were not inhibited. Two methanogenic organisms were isolated from this stabilized culture; neither isolate could use benzoate as a substrate for methanogenesis. It was not possible to isolate an acetate utilizing methanogen, although the large filamentous rod, which has been observed in acetate-degrading methanogenic enrichments by a number of investigators (42), was one of the major constituents of the mixed culture. So it appears that by removing the intermediates, acetate, formate, and hydrogen the methanogenic bacteria displace the equilibrium of the overall reaction in favor of the complete degradation of benzoate to methane. The calculations presented in Figure 6 indicate that for reaction 1 (below) equilibrium would be reached when benzoate was only half degraded. The following equations describe the events that occurred in the consortium:

Degradation of benzoate to acetate, formate, and hydrogen:

$$4 \text{ C}_7\text{H}_6\text{O}_2 + 24 \text{ H}_2\text{O} \rightarrow 12 \text{ C}_2\text{H}_4\text{O}_2 + 4 \text{ CH}_2\text{O}_2 + 8 \text{ H}_2 \qquad (1)$$

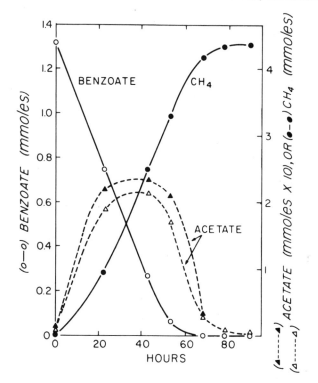

Figure 4. Appearance and disappearance of acetate during the degradation of benzoate and production of methane in a 100-ml culture. Acetate was assayed by acetokinase (▲) and by gas chromatography (△). (Reproduced from Archives of Microbiology with permission.)

Conversion of acetate to methane and carbon dioxide:

$$12\ C_2H_4O_2 \rightarrow 12\ CH_4 + 12\ CO_2 \tag{2}$$

Conversion of formate to CO_2 and H_2:

$$4\ CH_2O_2 \rightarrow 4\ H_2 + 4\ CO_2 \tag{3}$$

Reduction of carbon dioxide by hydrogen:

$$12\ H_2 + 3\ CO_2 \rightarrow 3\ CH_4 + 6\ H_2O \tag{4}$$

Sum:

$$4\ C_7H_6O_2 + 18\ H_2O \rightarrow 15\ CH_4 + 13\ CO_2 \tag{5}$$

To summarize the physiology of the methanogenic bacteria is to say that they occupy a very narrow ecological niche, i.e., the conversion of simple substrates (H_2 and CO_2, formate, and acetate) to methane in a highly reducing environment. Evidence increasingly supports the conten-

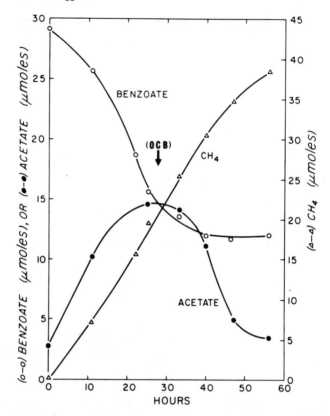

Figure 5. The effect of *o*-chlorobenzoic acid (OCB) on the degradation of benzoate and acetate. OCB was added (arrow) 28 h after benzoate was added. (Reproduced from Archives of Microbiology with permission.)

tion that biodegradation of organic compounds to methane in anaerobic habitats involves a microbial metabolic food chain, the complexity of which depends on the habitat. In the absence of nitrate, sulfate, or elemental sulfur, carbon dioxide becomes a major electron sink for anaerobic respiration. The ability of obligate proton-reducing bacteria to carry out the oxidation of fatty acids and alcohols in such environments is absolutely dependent on the metabolism of the methanogenic bacteria; by rapidly oxidizing and removing hydrogen from the anaerobic habitat, conditions thermodynamically favorable for the more complete anaerobic oxidation of carbon skeletons are produced.

Isolation and Cultivation

The Pressurized Culture Vessel Since the 1940s the Hungate technique (43, 44, 45) has been a standard procedure for isolation and cultivation of fastidious non-sporeforming anaerobes. This technique has many

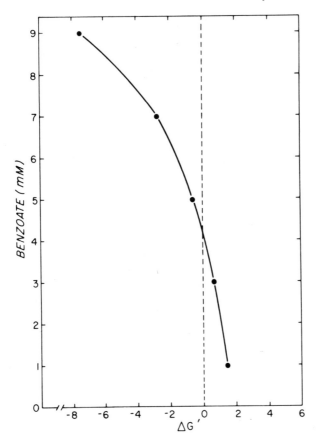

Figure 6. Relationship of $\Delta G'$ and benzoate concentration for the degradation of benzoate if the products are allowed to accumulate. Calculations were made for a 10-ml culture with a 15-ml gas phase. Initial conditions: 10 mM benzoate, 0.0 atmospheric H_2 pressure, 0.0 mM acetate and formate. (Reproduced from Archives of Microbiology with permission.)

modifications and is finding widespread use in clinical applications. For microbes that produce a positive pressure within the culture tube or that alter the internal pressure in a negligible way, the technique provides a reliable procedure once the operator has become competent. However, for the culture of organisms like the methanogens or other hydrogen-oxidizing anaerobes that produce a strong negative pressure within the culture tube, the technique has serious drawbacks. Substrate depletion may limit growth. When the stopper is removed, atmospheric pressure quickly reaches equilibrium with the pressure within the culture tube, bringing oxygen and possibly contaminating bacteria from the environment into the tube. Even when syringes are used the competence of a seasoned expert in the Hungate technique is severely taxed by the methanogenic bacteria, and only a few workers have reached the level of competence where they are able to maintain a culture collection of methano-gens. These limitations of the Hungate technique have placed severe constraints on the use of methanogens as experimental organisms. Our laboratory for many years was completely dependent upon the ability of M. P. Bryant to supply us with viable cultures.

During our initial attempts to use *M. ruminantium* strain M1 in a growth assay for coenzyme M, it became painfully evident that a more reliable method of cultivating the methanogens was required. In the hands of W. E. Balch the techniques that employ pressurized atmospheres have evolved into reliable and convenient procedures that can be acquired readily by persons with little background in microbiology (46). Use of these techniques has made it possible to handle methanogens in experiments with the ease that one would encounter in handling ordinary anaerobes. Contamination problems as well as variabilities due to oxygen have vir-tually disappeared, and stock cultures are maintained with little difficulty.

The culture tube is diagrammed in Figure 7. It features a standard serum-seal tube and an aluminum seal (47). The special feature is a solid tight-fitting black rubber stopper, durometer hardness 45, that is designed for repeated passage of syringe needles without loss of anaerobiosis within

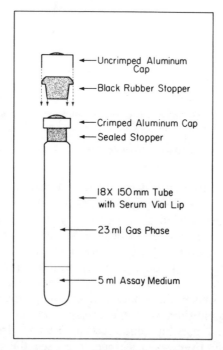

Figure 7. Growth tube developed for growing methanogens under a pressurized atmosphere of H_2/CO_2 (80:20) (46). (Reproduced from Applied and Environmental Microbiology with permission.)

the tube. To use this technique, the medium is degassed by boiling it in a round-bottomed flask under a stream of oxygen-free N_2/CO_2 (80:20), a standard procedure in the conventional Hungate technique (44). Cysteine sulfide is added as the redox buffer (44). The flask is sealed and transferred through an airlock into an anaerobic chamber where the medium is dispensed into serum tubes or vials. A black rubber stopper is inserted into the opening of each tube or vial, and the stopper is crimped into place. The atmosphere in each tube is identical at the time each tube is sealed. The culture vessels are then transferred out of the chamber. Prior to sterilization the atmosphere in each tube may be replaced with H_2/CO_2 (80:20) by means of a gassing manifold, as diagrammed in Figure 8. By means of this apparatus gassing channels may be used so that the atmosphere in each of 8 tubes is replaced simultaneously by operation of the three-way valve. Tubes are routinely pressurized to 2 atm. After sterilization all additions to or withdrawals from each tube are made by use of a sterile syringe that has been flushed free of oxygen prior to use. To repressurize the gas atmosphere during growth of a culture the entering gas is filtered through sterile cotton, as illustrated in Figure 9. Although this illustration shows a three-way valve for gassing a single tube, we routinely use the apparatus shown in Figure 8, where a sterile cotton-filled

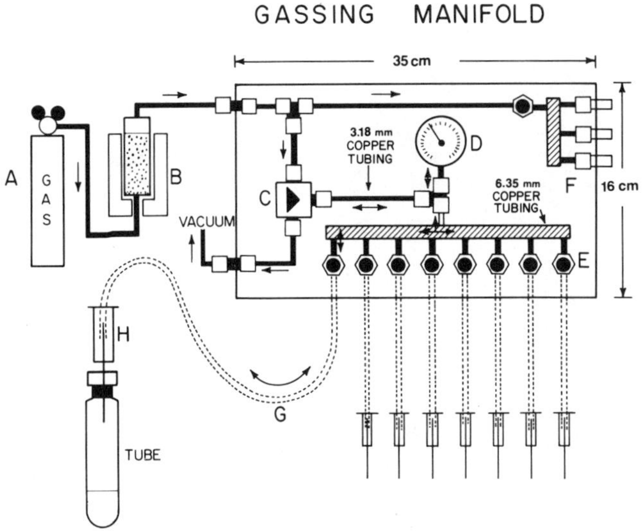

GASSING   MANIFOLD

Figure 8. Gassing manifold for evacuating and pressurizing growth vessels with oxygen-free gas. (A) Gas mixture tank; (B) reduced copper column (oxygen scrubber) with heater; (C) three-way valve with Swagelok fittings; (D) vacuum-pressure gauge; (E) Nupro Fine Metering Valves; (F) alternate tubing connectors for gassing probes that may be used in the normal Hungate procedure; (G) thin-bore polyethylene tubing; (H) Vacutainer-Holder Needle. Squares (□) indicate Swagelok brass fittings. Sections of copper tubing 6.35 and 3.18 mm thick are indicated by the shaded and solid regions, respectively. (Reproduced from Applied and Environmental Microbiology with permission.)

syringe is attached to each gassing channel. Results of a typical growth experiment with *M. ruminantium* in an 18 × 150 mm culture tube are shown in Figure 10. The atmosphere within the tube was repressurized twice during the experiment.

To scale up the culture volume a 1-liter bottle that contains 200 ml of medium is fitted with a rubber stopper in which is placed the upper section of a standard serum tube (W. E. Balch, personal communication). This culture vessel makes the open 200-ml flask (48, 49) obsolete and provides all the convenience of the aluminum seal tube for a 200-ml culture. Loss of sulfide from the medium is eliminated by this procedure. Some of our more sensitive organisms could not be cultivated in the open apparatus where a gas atmosphere was passed over the culture. Cultures in these bottles may be shaken as desired at a given temperature and may be repressurized quickly during growth of the organism.

Use of Petri Dishes   Edwards and McBride first reported the successful cultivation of methanogenic bacteria in Petri dishes (50). Plates that were poured and streaked in an anaerobic chamber were then placed in a

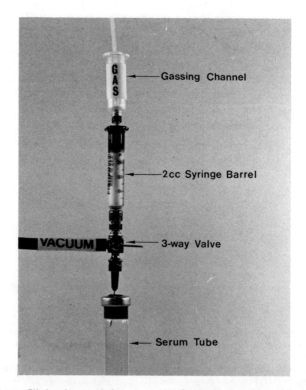

Figure 9. Cotton-filled syringe and three-way valve for flushing or pressurizing a single tube aseptically and anaerobically. (Reproduced from Applied and Environmental Microbiology with permission.)

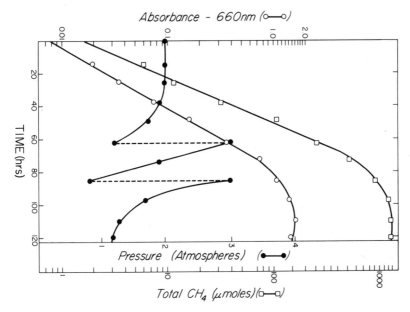

Figure 10. Relationship between growth (o), CH₄ formation (□), and pressure (●) in pressurized cultures of *M. ruminantium* strain M1. (Reproduced from Applied and Environmental Microbiology with permission.)

separate container within the anaerobic chamber. A gas atmosphere of hydrogen and carbon dioxide was then added to the isolated container. This container could be opened easily and colonies on the plates could be examined in the outer chamber. For our purposes this technique has two disadvantages: working space in an anaerobic chamber quickly becomes limiting, and some of our cultures require incubation at 60 °C. A modification of this method was developed that removed these constraints (46). Small domestic pressure cookers were modified, the size of the cooker being limited by the size of the airlock to the anaerobic chamber. Plates that have been streaked in the anaerobic chamber are placed in the pressure cooker, which is then sealed and brought through the air lock into the laboratory atmosphere. By use of the gassing manifold (Figure 8) the desired gas mixture is added to the cooker to a pressure of 2 atm; the vessel may then be incubated at the desired temperature. To remove oxygen from glass Petri dishes they should be transferred into the anaerobic chamber several hours before use; sterile plastic Petri dishes require several days of equilibration within the anaerobic chamber prior to use.

The above details have been presented for the simple reason that by use of these techniques the methanogens are now available as experimental organisms without the trauma formerly associated with their cultivation.

Titanium Citrate   Titanium chloride has been used as a chemical reducing agent for many decades; it has not been used in biologic systems, because at physiological pH titanium hydroxide is insoluble. Recently Zehnder and Wuhrmann have employed titanium citrate successfully as a reducing agent in the cultivation of a methanogen, *Methanobacterium* strain AZ (51). It is too early to predict whether or not this strong reducing agent will find widespread use in anaerobic culture media. Because it is inhibitory to certain organisms, it might be used as a selective agent as well as a source of low potential electrons.

METABOLISM

In the process of reducing the most oxidized of carbon compounds, CO_2, to the most reduced of carbon compounds, CH_4, 8 electrons are derived from 4 mol of H_2, and these electrons are used to reduce 1 mol of CO_2; the mechanism of this process has not yet been fully elucidated.

$$4 H_2 + HCO_3^- + H^+ \rightarrow CH_4 + 3 H_2O; \Delta G_0' \; -32.4 \, kcal/reaction$$

In our studies of CO_2 reduction to methane the area of methyl-group transfer and reduction has yielded most information.

Coenzyme M

Coenzyme M was discovered by McBride and Wolfe (52), and its structure was determined by Taylor and Wolfe (53). The coenzyme is distinguished by its small size, high sulfur content, strong acidity, and stability in the presence of heat or acid. This compound, 2-mercaptoethanesulfonic acid (HS-CoM), is readily oxidized in air to form the disulfide 2,2'-dithio-diethanesulfonic acid (S-CoM)$_2$. The role of the coenzyme as the methyl carrier in methanogenesis is well established (54), 2-(methylthio)ethane-sulfonic acid (CH$_3$-S-CoM) being the methylated form of the coenzyme. The abbreviations, chemical names, and structures of the three forms of the coenzyme are presented in Figure 11.

Prior to our studies on extracts of hydrogen-grown methanogens it appeared as though a methylcobamide could be the active methylated intermediate in methanogenesis, since the methyl group of methylcobala-min was readily reduced to methane by extracts of methanogenic bacteria (55). However, the more carefully we examined extracts of methanogens the less evidence was obtained that methylcobalamin or cobamide enzymes were involved in methanogenesis. At present there is no evidence to support the role of a cobamide in the activation and reduction of CO_2 to CH_4 by hydrogen-grown cells. In elucidating the structure of coenzyme M, Taylor and Wolfe employed a methyltransferase from *Methanobacterium* that activated the methyl group of methylcobalamin and transferred it to HS-CoM. This enzyme was purified 100-fold and proved to be stable in

FORMS OF COENZYME M		
HS-CoM	(S-CoM)$_2$	CH$_3$-S-CoM
HSCH$_2$CH$_2$SO$_3^-$	$^-$O$_3$ SCH$_2$CH$_2$ S - SCH$_2$ CH$_2$ SO$_3^-$	CH$_3$SCH$_2$CH$_2$SO$_3^-$
2-mercaptoeth-anesulfonic acid	2, 2'-dithiodiethanesulfonic acid	2-(methylthio)-ethanesulfonic acid

Figure 11. Forms of coenzyme M that can be isolated from living cells.

air at this stage of purity (56). Whether or not methylcobalamin is the natural methyl donor for this enzyme has not been determined, and we have obtained no evidence that suggests that this enzyme is involved in methanogenesis. Methylcobalamin is a strong alkylating agent and may simply play the role of a nonspecific methylating agent. With methyl-cobalamin as substrate and HS-CoM as methyl acceptor, this enzyme provided the tool by which the structure and role of CH$_3$-S-CoM were determined. Commercial preparations of 2-mercaptoethanesulfonic acid are available, and the synthesis of CH$_3$-S-CoM or [*methyl*-^{14}C]CH$_3$-S-CoM is straightforward (53).

Role as Methyl Carrier CH$_3$-S-CoM was shown to be the first product of CO$_2$ reduction that accumulated in substrate amounts in cell extracts or whole cells (52); it is the only substrate for the methylreductase of methanogenic bacteria so far detected (54). Other methyl donors, for example, CH$_3$-B$_{12}$, require additional enzymes for the activation of the methyl group and its transfer to HS-CoM.

The methyl reductase of *M. thermoautotrophicum* has been resolved by R. P. Gunsalus (57). The reaction requires a hydrogen atmosphere, ATP, and Mg^{2+} as well as three components fractionated from the extract:

$$CH_3\text{-S-CoM} \xrightarrow[\substack{ATP \\ Mg^{2+} \\ H_2}]{\substack{Components \\ A, B, C}} CH_4 + HS\text{-CoM}$$

Component C is a heat-sensitive oxygen-stable protein. Component B is a heat-stable oxygen-sensitive co-factor with a molecular weight of about 1,000. It is not replaceable by any of the known coenzymes, and coenzyme activity is irreversibly lost upon exposure to oxygen. Component A has hydrogenase activity and is a large protein. The requirement for ATP appears to be catalytic; at the present stage of purity of the methylreduc-tase system the ratio of ATP added to CH$_4$ formed is 1:15 (57). Perhaps

ATP serves as an activator for protein-protein interaction or for a conformational change in one of the proteins. A more detailed study of the components of methylreductase is in progress.

Results of experiments carried out in our laboratory by Shapiro (54) implicate CH_3-S-CoM as an intermediate in the conversion of methanol to methane by *Methanosarcina barkeri*.

Specificity   A number of compounds that possess terminal sequences similar to that of CH_3-S-$CH_2CH_2SO_3^-$ were tested for their ability to serve as methyl donors in the methylreductase system (54, 57). For example, 3-(methylthio)propylamine (CH_3-S-$CH_2CH_2CH_2NH_2$), methionol (CH_3-S-$CH_2CH_2CHNH_2CH_2OH$), L- or D-methionine (CH_3-S-CH_2-CH_2-$CHNH_2$-COOH), 2-(methylthio)ethanol (CH_3-S-CH_2CH_2OH), or methyl-2-(methylthio)propionate (CH_3-S-$CH_2CH_2COOCH_3$) did not serve as methyl donors. Similarly, ethyl-2-(methylthio)acetate, or CH_3-S-CH_3, could not donate a methyl group for methane synthesis. When 3-(methylthio)propanesulfonic acid (CH_3-S-$CH_2CH_2CH_2SO_3^-$) and the butanesulfonic acid analog were tested, the extreme specificity of coenzyme M became apparent. If the C_2 moiety of CH_3-S-$CH_2CH_2SO_3^-$ is lengthened by a single CH_2 group, the specificity of the molecule as a substrate for methylreductase is completely lost; neither of these compounds serves as a substrate for methane formation. Taylor and Wolfe showed that 2-(dimethylsulfonium)ethanesulfonate was not active as a substrate (53). However, when an ethyl moiety was substituted for the methyl group CH_3-CH_2-S-CoM, ethane was formed at about 20% of the rate at which methane was formed from CH_3-S-CoM. The propyl analog was inactive (57).

Of a variety of compounds tested only three analogs of coenzyme M caused inhibition of methylreductase at significantly low concentrations. (S-CoM)$_2$ causes 50% inhibition of methylreductase at 10^{-3} M; the halogenated analogs $ClCH_2CH_2SO_3^-$ and $BrCH_2CH_2SO_3^-$ are potent inhibitors, causing 50% inhibition at 10^{-5} and 10^{-6} M, respectively (57).

Growth Factor Activity   McBride and Wolfe (52) first recognized the similarity of the properties of coenzyme M to the unidentified growth factor for *M. ruminantium*, which was studied by Bryant. When the coenzyme was synthesized by Taylor, it was tested in a growth assay by Bryant and was found to substitute at very low concentrations for the growth factor from rumen fluid (58). These results were repeated by Balch, using pressurized growth tubes (46). As shown in Figure 12, half maximal growth of *M. ruminantium* was achieved at a concentration of 25 nM HS-CoM. Cells of this organism also were able to use (S-CoM)$_2$ and CH_3-S-CoM to satisfy the growth requirement for coenzyme M. The results presented in Figure 13 clearly show that CH_3-S-CoM is used at about the same efficiency as HS-CoM; however, the disulfide form (S-CoM)$_2$, provided twice the amount of growth factor, because each nmol was reduced to 2 nmol of HS-CoM. Other derivatives that could be used to satisfy the growth factor requirements were CH_3CH_2-S-CoM, HOCH$_2$-S-

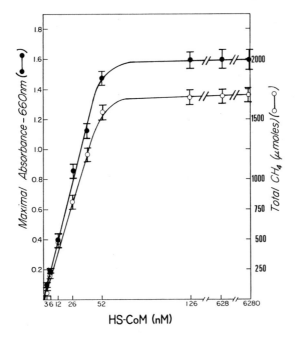

Figure 12. Relationship between total CH₄ accumulation (o) and maximal cell density (●) of *M. ruminantium* and increasing HS-CoM concentration. Each value presented is the mean of three tubes, with standard deviation indicated. (Reproduced from Applied and Environmental Microbiology with permission.)

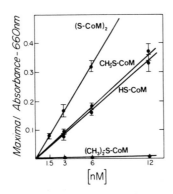

Figure 13. Growth response of *M. ruminantium* strain M1 to derivatives of coenzyme M. Each value presented is the mean of three tubes, with standard deviation indicated. (Reproduced from Applied and Environmental Microbiology with permission.)

CoM, and CH_3CO-S-CoM; other compounds discussed in the above section were inactive when tested in the growth assay. The halogenated ethane-sulfonic acids Cl-$CH_2CH_2SO_3^-$ and Br-$CH_2CH_2SO_3^-$ were strong inhibitors of growth (54). Of the methanogens in pure culture *M. ruminantium* strain

M1 is the only one that requires HS-CoM as a growth factor; thus this unique molecule exhibits classic vitamin coenzyme properties.

RPG Effect   One of the most significant observations to be made in recent years on the biochemistry of methanogenesis is the stimulation of CO_2 reduction to methane by CH_3-S-CoM. Gunsalus and Wolfe (59) observed that, when CH_3-S-CoM was added to cell extracts of *M. thermoautotrophicum* that were incubated under an atmosphere of H_2 and CO_2 (80:20), the rate of methane formation was stimulated thirtyfold, and for each mol of CH_3-S-CoM added 12 mol of CH_4 was produced. Neither HS-CoM nor (S-CoM)$_2$ replaced CH_3-S-CoM. The requirements for the RPG effect are presented in Table 2; ATP, Mg^{2+}, H_2, and CH_3-S-CoM are required. It should be noted that these components also are required for the methylreductase reaction. The RPG effect is presented dramatically in Figure 14, where under a H_2 atmosphere only CH_4 is formed in stoichiometric amounts after each addition of CH_3-S-CoM; however, the presence of CO_2 greatly stimulated the formation of CH_4. The effect could be reinitiated again and again by the addition of CH_3-S-CoM. Throughout the experiment no additional ATP was added. As shown in the insert (Figure 14) a straight line relationship is observed when the amount of methane produced is plotted against the amount of CH_3-S-CoM added. Other compounds that produce the RPG effect are formaldehyde, L-serine, thioproline and $HOCH_2$-S-CoM (J. Romesser, personal communication).

The RPG effect clearly demonstrates for the first time that the methylreductase reaction is coupled to the activation and reduction of CO_2. Thus, the terminal reaction appears to generate an intermediate that is involved

Table 2.   Requirements for CH_3-S-CoM-stimulated CO_2 reduction to CH_4

Reaction condition omissions[a]	CH_4 formed (nmol/ 30 min)
None	2,705
—Mg	24
—ATP	0
—CH_3-S-CoM	178
—cell extract	0
—H_2	3
—CO_2	306

[a]Each reaction vial received where indicated: 30 μmol TES; 5 μmol $MgCl_2$; 1 μmol ATP; 0.2 μmol CH_3-S-CoM; 50 μl cell extract; 2.7 mg protein in a reaction volume of 0.25 ml. The reaction time was 30 min at 60°C. The gas phase was H_2/CO_2 mixture (80:20 v/v) except when CO_2 was omitted (balance H_2) or H_2 omitted (balance N_2). (Reproduced from Biochemical and Biophysical Research Communications with permission.)

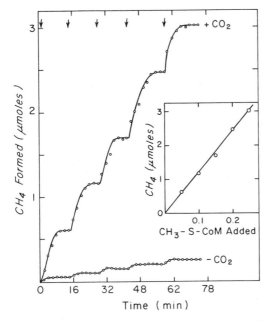

Figure 14. Effect of CH_3-S-CoM on the amount of methane formed. Reaction components and conditions were as described in Table 2 except that CH_3-S-CoM (50 nmol) was added at the times indicated by the arrows. Insert shows the total amount of methane produced after each addition versus the total amount of CH_3-S-CoM added. (Reproduced from Biochemical and Biophysical Research Communications with permission.)

in the primary step of CO_2 activation. We now must draw Barker's scheme as a definite cycle. Such a model is presented in Figure 15. The HS-CoM-enzyme complex is intended as an abbreviation for the active complex that

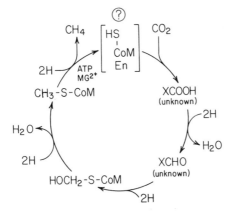

Figure 15. Modification of Barker's scheme for CO_2 reduction to CH_4 to emphasize a cycle where the unknown activated-intermediate produced by the methylreductase is involved in CO_2 activation.

we predict is formed in the methylreductase reaction. Hydroxymethyl-coenzyme M ($OHCH_2$-S-CoM) has been synthesized by James Romesser (54) and is reduced to methane by extracts, suggesting that it could be an intermediate. The concept that coenzyme M could be involved as the C_1 carrier at other sites than the methyl level must be considered seriously.

Coenzyme F_{420}

This coenzyme was first observed by Cheeseman et al. as a blue-green fluorescent compound in whole cells of *Methanobacterium* strain M.o.H.; the fluorescence disappeared when hydrogen was added to the cell suspension (60). The pure coenzyme is yellow in the oxidized state, as shown by the solid line in Figure 16. The large absorption peak at 420 nm disappears upon reduction of the coenzyme to the colorless form (60). The fluorescence of this coenzyme as well as that of other cell components serves as the basis of the method of Edwards and McBride for the detection of methanogens; cells in wet mount could be identified by their characteristic fluorescence when examined in the microscope under ultraviolet epifluorescence optics (61).

The coenzyme was shown by Tzeng et al. to serve as a low-potential electron carrier in *M. ruminantium* (62, 63). These F_{420}-dependent reactions are shown in Figure 17. The K_m for F_{420} in the F_{420}-dependent NADP-

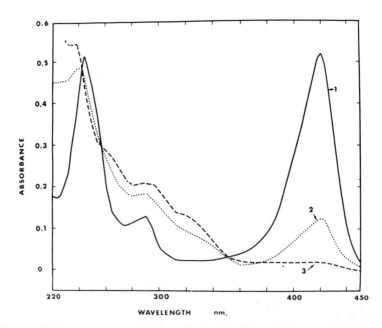

Figure 16. Absorption spectra of coenzyme F_{420}. (1) Oxidized compound 25.9 μg per ml in 0.05 M phosphate buffer (pH 7.3). (2) Spectrum of compound partially reduced by sodium borohydride. (3) Spectrum of completely reduced compound.

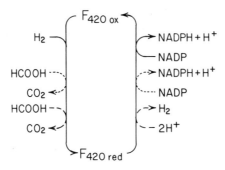

Figure 17. Reactions that require coenzyme F_{420}.

linked hydrogenase system of *M. ruminantium* was determined to be 5×10^{-6} M at pH 8. Similar reactions were studied in *Msp. hungatii* (64). A structure for this coenzyme (Figure 18) has been proposed recently by Eirich (65). The molecular weight of the tri-ammonium salt of F_{420} is 840. The coenzyme has an E_0' of -373 mV and has been shown to be a 2-electron carrier. The chromophore is a deazaflavin; to our knowledge this is the first report of this type of electron carrier. Another interesting feature of the molecule is the long side chain that ends in a diglutamyl moiety. Thus, the coenzyme has some similarity to flavins as well as folates: the chromophore is a modified flavin and the long side chain ends in glutamate moieties. This coenzyme is present in all methanogens and so far has not been found elsewhere.

ATP Synthesis

Because molecular hydrogen is a universal substrate for methanogens and because CO_2 serves both as a source of cell carbon and as the terminal electron acceptor, it is difficult to visualize how substrate-level phosphorylation could take place in these organisms. It appears that electron transport phosphorylation must be the source of ATP synthesis. Of various models available, probably the simplest to consider is one similar to that proposed by Thauer et al. (6) based on the proposals of Mitchell (66). Such a model is presented in Figure 19. A proton gradient could be established across a membrane by the action of hydrogenase and appropriate electron carriers. Reduced coenzyme F_{420} could be BH_2. These carriers have not yet been identified in methanogens, nor has a successful sub-cellular system been reported that carries out electron transport phosphorylation. Experiments have been carried out by Ferry and Peck and by Doddema and Vogels (personal communications).

The conversion of acetate to methane poses an even greater problem in explaining ATP generation. For over 20 years the results of the labeling experiments from Barker's laboratory (1) have served as a challenge to biochemists. If the hydrogen atoms of the methyl group of acetate remain

$$
\begin{array}{c}
\overset{\text{O}}{\underset{\parallel}{}}\overset{\text{CH}_3}{}\overset{\text{O}}{\underset{\parallel}{}}\overset{\text{COO}^\ominus}{}\overset{\text{O}}{\underset{\parallel}{}}\overset{\text{COO}^\ominus}{} \\
\end{array}
$$

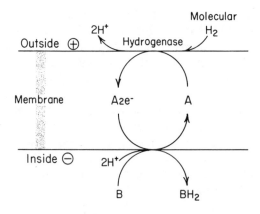

Figure 18. Structure of coenzyme F_{420}. (L. Dudley Eirich, Ph.D. thesis, 1977, University of Illinois, Urbana, Illinois.

associated with the methyl carbon atom and are transferred intact to methane, how does the organism make a living? These data preclude the oxidation of acetate to CO_2 and H_2. Results of experiments with crude systems have indicated that even in the sewage sludge digester most of the methane comes from the CH_3 group. Acetate is not oxidized to $2CO_2$. The labeling pattern was confirmed recently in the laboratory of Mah with *Methanosarcina* (5):

$$
\text{*CD}_3\text{COO}^- + \text{H}_2\text{O} \rightarrow \text{*CD}_3\text{H} + \text{HCO}_3^-; \Delta G_0' - 7.4 \text{ kcal/reaction} \qquad (6)
$$

The change in free energy for this reaction is not great enough to consider seriously substrate-level phosphorylation. Perhaps there is a way that electron transport phosphorylation could take place during conversion of acetate to methane and carbon dioxide. The scheme presented in Figure 20 is highly speculative and is presented with the hope that it may stimulate the design of experiments; certain aspects of the scheme are testable.

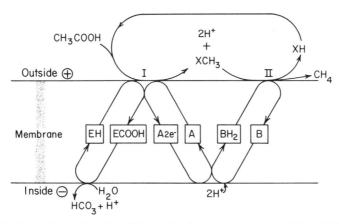

Figure 20. Proposed speculative model to explain how a methanogen could establish a proton gradient during acetate degradation and produce methane in a manner consistent with the labeling pattern obtained by Pine and Barker.

Even if the response of the reader is violent, this alone will justify its presentation, for this area has lain fallow far too long.

Cleavage of the carbon-carbon bond of acetate could generate charged moieties, such as $(CH^+)(^-COOH)$ or $(CH_3^-)(^+COOH)$, or neutral species, such as $(CH_3 \cdot)(\cdot COOH)$. During the enzymic cleavage appropriate co-factors could participate in the reaction by accepting the methyl and carboxyl moieties. Electron carriers also could participate, with component A carrying only electrons and component B carrying both electrons and protons. Component X could transfer the methyl moiety, and component E could be an enzyme, perhaps a transcarboxylase. The methyl and carboxyl moieties could remain charged or each charge could be transferred to a co-factor; for example, the XCH_3 moiety could be a neutral species. But for purposes of illustrating the possibilities here, it is simpler to use one possible system. We shall assume that the methyl moiety has the properties of a carbonium ion. At site I, XCH_3^+ is generated, component A accepts 2 electrons, which are transported through the membrane to a site where component B is reduced, and component E transports the carboxyl group. Two protons are picked up from the inside of the membrane, and reduced carrier BH_2 then transports $2e^- + 2H^+$ to the outside of the membrane. At site II, BH_2 reacts with XCH_3^+ to form CH_4 and XH. A proton gradient has been established across the membrane, and acetate has been converted to methane in a manner consistent with the labeling data of Pine and Barker (67).

CARBON ASSIMILATION

Very little is known about carbon assimilation by methanogens. Results of experiments on the incorporation of acetate into cell carbon by *M. rumi-*

nantium revealed that about 60% of the cell carbon was derived from acetate (25). Because *M. ruminantium* strain M1 is the most nutritionally fastidious of the methanogens, incorporation was expected, but the magnitude of incorporation exceeded expectations. Although some progress has been made on the route by which CO_2 is converted to methane, a breakthrough on the route of CO_2 reduction to cell carbon has not been reported. It is not known whether or not these two routes have a step in common.

   M. thermoautotrophicum has been used for studies of CO_2 reduction to cell carbon because it is among the fastest-growing methanogens and because it can be grown in a mineral medium with CO_2 as the only source of carbon. Ribulose 1,5-diphosphate carboxylase was not detected in this organism (4), indicating that the Calvin cycle probably is not operative here. In addition, Taylor et al. (68) were forced to conclude that for CO_2 fixation in this organism the route did not appear to involve serine or hexulose; nor was total synthesis of acetate from CO_2 deemed to be a feasible route. However, Daniels and Zeikus did report (69) that phosphoenolpyruvate carboxylase was active in cell extracts. This enzyme could play a role in the formation of carbon skeletons for amino acid synthesis; however, the major route of net CO_2 fixation into cell carbon would be expected to involve a cycle of some type, the precise nature of which remains to be elucidated.

NITROGEN METABOLISM

Essentially nothing is known about nitrogen metabolism in the methanogens. Evidence from a variety of nutritional studies suggests that NH_4^+ serves as the major nitrogen source for all methanogens (25). So far, nitrogenase has not been detected in methanogens. *M. ruminantium* has a growth requirement for amino acids, and tryptic digests of casein stimulate growth of certain organisms, such as *Methanobacterium* M.o.H. Addition of amino acids to growth media generally decreases generation time and increases the cell yield, *M. thermoautotrophicum* being an exception. Definitive studies have not been done on the transport of amino acids into cells or on the incorporation of amino acids into cell protein. Requirements of methanogens for other nitrogen-containing compounds, such as purines or pyrimidines, have not been demonstrated.

GENETICS

No genetic studies on the methanogens have been reported. At the present time not a single phage has been found for these organisms. It should be pointed out that the technology for developing genetic systems has only recently been evolved. Genetic studies were virtually impossible when the Hungate roll-tube was the only technique available, but with recent developments that allow the use of Petri dishes cultivation of these organisms is now rather conventional. It might be predicted that within the next few years dramatic chapters on the genetics and molecular biology of the methanogens will be written.

EVOLUTION

An approach to the study of evolution employed by Carl Woese has the advantage of probing very ancient events by analysis of the 16S rRNA molecule, which is generally regarded as one of the most conserved of cell structures. Yet it does change slowly, recording mutational events at a rate sufficiently low that comparative studies among microbial species are possible at a precision some orders of magnitude beyond the limits imposed by the relatively rapid mutational modifications of protein molecules, such as cytochrome c. The procedures developed in the laboratory of Carl Woese (70–75) may be summarized briefly as follows.

Cells of a growing culture are labeled with ^{32}P of high specific activity. The 16S rRNA is purified and digested with a specific nuclease. The resulting oligonucleotides (roughly 80 to 90 different oligonucleotides) are subjected to two-dimensional separation by electrophoresis. The position and base sequence of each oligonucleotide in each isopleth are determined. By comparison of the oligonucleotide fingerprint pattern ("signature") of each organism with that of other organisms a similarity coefficient (S_{AB}) is generated. By computer analyses of various "signatures" a dendogram may be constructed that reflects their relatedness. Such dendograms, which concern the methanogens and their relatedness to other microbes, have been published (76, 77), and one is presented in Figure 21. The startling picture that emerges from an examination of these data is that the methanogens very clearly constitute a well-defined group that appears to be unrelated to other microbial groups. *Escherichia coli,* for example, appears more closely related to the cyanobacteria than to the methanogens. The 16S rRNAs of the methanogens show a high degree of relatedness, indicating that methanogenesis is not coded by a plasmid and that methanogens are not a group of unrelated organisms that have received such a plasmid.

The divergence of the methanogens in the microbial line of descent is the most ancient event so far detected in the examination of over 70 species of bacteria by Woese and his associates. In such an old group one would expect to find evidence of divergence within the group. Such a deep cleft is apparent in the dendogram between methanogen groups I and II, and it is striking to note that this bifurcation is approximately as ancient as that which gave rise to the gram-negative/gram-positive split of conventional prokaryotes. These two groups of methanogens, which are generated by analyses of the 16S rRNA oligonucleotide signatures, correlate well with morphological groups. The rod-shaped methanogens fall into two

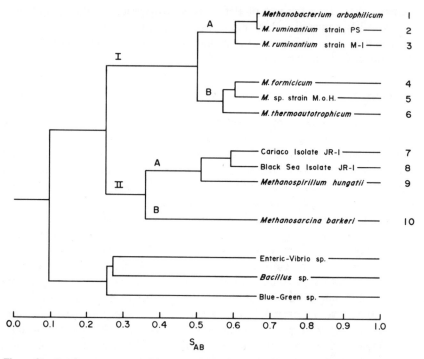

Figure 21. Dendogram generated by comparative analysis of the 16S rRNA of methanogens as well as other bacteria, and showing that methanogens are only distantly related to conventional prokaryotes. (Reproduced from Proceedings of the National Academy of Science USA with permission.)

groups. Group I is subdivided into two groups: group IA, the gram-positive coccobacillus-like organisms, and group IB, the long gram-positive rods. Group II contains morphologically diverse organisms, and is also subdivided into two groups. In group IIA, *Methanospirillum* clusters with Cariaco Isolate JR1 and Black Sea Isolate JRI. These marine organisms possess unusual cell morphology as well as cell fragility; a description of their properties as well as specific names will be published elsewhere.

Group IIB so far contains only *Methanosarcina*. Two isolates of *M. barkeri* have been examined. Although one strain was isolated by M. P. Bryant in Urbana, Illinois and the other strain by R. A. Mah from a sludge digester in Los Angeles, their 16S rRNA signatures are identical (C. R. Woese, personal communication). From examination of the dendogram it is apparent that the divergence of *Methanosarcina* occurred rather early; therefore, it could be expected that this group should show evidence of diversification. A gas-vacuolated isolate of *Methanosarcina* has been reported in Russia by Zhilina (78), and recently a similar organism has been isolated in the laboratory of R. A. Mah (R. A. Mah, personal communication). Perhaps the 16S rRNA oligonucleotide signature of this organism may prove to be different. In these studies a culture of *Methano-coccus* was not available for analysis. We would expect this organism to fall into group II.

Although we were aware of some unique properties among the methanogens, the results of the collaborative studies with Woese and his associates (76, 77) served to focus our attention on the unusual characteristics of the group. A number of additional facts support the concept that the methanogens are unrelated to conventional prokaryotes:

1. Absence of peptidoglycan: cell walls of 10 methanogenic organisms have been examined in the laboratory of Professor Otto Kandler, and no evidence of muramic acid has been obtained.

2. CoM: results of an examination of a wide variety of organisms and tissues by W. E. Balch for the presence of CoM have been negative. It appears that CoM is not present in tissues of conventional prokaryotes or eukaryotes; it is present in all cultures of methanogens available for examination. We predict that the possession of CoM will prove to be a property of all methanogenic organisms.

3. Unusual electron carriers: coenzyme F_{420} is found in relatively high concentrations in extracts of methanogenic bacteria. So far this compound has not been found in a variety of anaerobes and aerobes but is present in all methanogens that we have examined. In addition, new factors F_{430} and F_{342} await description.

4. Absence of cytochromes and quinones: no evidence for the presence of cytochromes or heme moieties has been obtained by the examination of whole cells or extracts of methanogens. Professor R. K. Thauer has exhaustively examined *M. thermoautotrophicum* for the presence of menaquinone and ubiquinone but has not found these compounds; this organism is the only species so far examined carefully for quinones.

5. Unusual ecological niche: all species of methanogens that have been studied in pure culture are able to oxidize hydrogen and to use carbon dioxide as the terminal electron acceptor. These organisms require a reducing potential near the hydrogen electrode, and frequently are found in nature in habitats high in sulfide. They are essentially chemolithotrophs.

There is one ecological habitat that probably has remained unchanged throughout geological time, i.e., the volcanic hot spring. Geothermal springs today may contain high amounts of sulfide, molecular hydrogen, juvenile carbon in the form of carbon dioxide, and mineral salts. *M. thermoautotrophicum* grows in this type of habitat at a temperature of 65°C. Perhaps methanogenic organisms evolved when the hydrosphere of the primitive earth was hot and highly reducing, e.g., high in sulfide, hydrogen, carbon dioxide, and ammonia. In the absence of oxygen (and its various oxides of nitrogen and sulfur) carbon dioxide would have been the electron acceptor of choice.

ACKNOWLEDGMENTS

I thank my students and colleagues for their contributions, and M. P. Bryant and J. R. Romesser for criticism of certain sections.

REFERENCES (Part I)

1. Barker, H. A. (1956). Bacterial Fermentations. John Wiley and Sons, Inc., New York.
2. Stadtman, T. C. (1967). Annu. Rev. Microbiol. 21:121.
3. Wolfe, R. S. (1971). Adv. Microb. Physiol. 6:107.
4. Zeikus, J. G. (1977). Bacteriol. Rev. 41:514.
5. Mah, R. A., Ward, D. M., Baresi, L., and Glass, T. L. (1977). Annu. Rev. Microbiol. 31:309.
6. Thauer, R. K., Jungermann, K., and Decker, K. (1977). Bacteriol. Rev. 41:100.
7. Jeris, J. S., and McCarty, P. L. (1965). J. Water Pollut. Control Fed. 37:178.
8. Smith, P. H., and Mah, R. H. (1966). Appl. Microbiol. 14:368.
9. Cappenberg, T., and Prins, H. (1974). Antonie van Leeuwenhoek 40:457.
10. Reddy, C. A., Bryant, M. P., and Wolin, M. J. (1972). J. Bacteriol. 109:539.
11. Reddy, C. A., Bryant, M. P., and Wolin, M. J. (1972). J. Bacteriol. 110:126.
12. Reddy, C. A., Bryant, M. P., and Wolin, M. J. (1972). J. Bacteriol. 110:133.
13. Hungate, R. E. (1966). The Rumen and Its Microbes. Academic Press, New York.
14. Hobson, P. N. (1971). Prog. Ind. Microbiol. 9:41.
15. Hobson, P. N., Bousfield, S., and Summers, R. (1974). Crit. Rev. Environ. Control 4:131.
16. Phillipson, A. T. (ed.) (1970). Physiology of Digestion in the Ruminant. Oriel Press, Newcastle-upon-Tyne, England.
17. Gunter, B. P., and Musgrave, B. C. (1966). Geochim. Cosmochim. Acta 30:1175.
18. Deuser, W. G., Degens, E. T., and Harvey, G. R. (1973). Science 181:51.
19. Balch, W. E., Schoberth, Z., Tanner, R., and Wolfe, R. S. (1977). Int. J. Syst. Bacteriol. 27:355.
20. Bryant, M. P. (1974). In R. E. Buchanan and N. E. Gibbons (eds.), Bergey's Manual of Determinative Bacteriology, p. 472. Williams and Wilkins Co., Baltimore.
21. Zeikus, J. G., Weimer, P. J., Nelson, D. R., and Daniels, L. (1975). Arch. Microbiol. 104:129.
22. Weimer, P. J., and Zeikus, J. G. (1977). Appl. Environ. Microbiol. 33:289.
23. Kluyver, A. J., and Schnellen, G. T. P. (1947). Arch. Biochem. 14:57.
24. Daniels, L., Fuchs, G., Thauer, R. K., and Zeikus, J. G. (1977). J. Bacteriol. 132:118.
25. Bryant, M. P., Tzeng, S. F., Robinson, I. M., and Joyner, A. E. (1971). In F. G. Pohland (ed.), Anaerobic biological treatment processes. Advances in Chemistry Series 105. American Chemical Society, Washington, D. C.
26. Taylor, C. D., McBride, B. C., Wolfe, R. S., and Bryant, M. P. (1974). J. Bacteriol. 120:974.
27. Paynter, M. J. B., and Hungate, R. E. (1968). J. Bacteriol. 95:1943.
28. Smith, P. H., and Hungate, R. E. (1958). J. Bacteriol. 75:713.
29. Pantskhava, E. S. (1969). Dokl. Biol. Sci. 188:699.
30. Pantskhava, E. S., and Pchelkina, V. V. (1968). Dokl. Biol. Sci. 182:552.
31. Bryant, M. P., Wolin, E. A., Wolin, M. J., and Wolfe, R. S. (1967). Arch. Mikrobiol. 59:20.

32. Iannotti, E. L., Kafkewitz, D., Wolin, M. J., and Bryant, M. P. (1973). J. Bacteriol. 114:1231.
33. Wolin, M. J. (1974). Am. J. Clin. Nutr. 27:1320.
34. Latham, M. J., and Wolin, M. J. (1977). Appl. Environ. Microbiol. 34:297.
35. Scheifinger, C. C., Linehan, B., and Wolin, M. J. (1975). Appl. Microbiol. 29:480.
36. Chung, K. T. (1976). Appl. Environ. Microbiol. 31:342.
37. Hungate, R. E. (1967). Arch. Mikrobiol. 59:158.
38. Tarvin, D., and Buswell, A. M. (1934). J. Amer. Chem. Soc. 56:1751.
39. Fina, L. R., and Fiskin, A. M. (1960). Arch. Biochem. Biophys. 91:163.
40. Nottingham, P. M., and Hungate, R. E. (1969). J. Bacteriol. 98:1170.
41. Ferry, J. G., and Wolfe, R. S. (1976). Arch. Microbiol. 107:33.
42. Pretorius, W. A. (1972). Water Res. 6:1213.
43. Hungate, R. E. (1950). Bacteriol. Rev. 14:1.
44. Hungate, R. E. (1969). In J. R. Norris and D. W. Ribbons (eds.), Methods in Microbiology 3B, p. 117. Academic Press, New York.
45. Bryant, M. P. (1972). Am. J. Clin. Nutr. 25:1324.
46. Balch, W. E., and Wolfe, R. S. (1976). Appl. Environ. Microbiol. 32:781.
47. Miller, T. L., and Wolin, M. J. (1973). J. Bacteriol. 116:836.
48. Bryant, M. P., McBride, B. C., and Wolfe, R. S. (1968). J. Bacteriol. 95:1118.
49. Daniels, L., and Zeikus, J. G. (1975). Appl. Microbiol. 29:710.
50. Edwards, T., and McBride, B. C. (1975). Appl. Microbiol. 29:540.
51. Zehnder, A. J. B., and Wuhrmann, K. (1977). Arch. Microbiol. 111:199.
52. McBride, B. C., and Wolfe, R. S. (1971). Biochemistry 10:2317.
53. Taylor, C. D., and Wolfe, R. S. (1974). J. Biol. Chem. 249:4879.
54. Gunsalus, R. P., Eirich, D., Romesser, J., Balch, W., Shapiro, S., and Wolfe, R. S. (1976). In H. Schlegel, G. Gottchalk, and N. Pfennig (eds.), Microbial Production and Utilization of Gases. Erich Goltze KG, Göttingen.
55. Blaylock, B. A., and Stadtman, T. C. (1963). Biochem. Biophys. Res. Commun. 11:34.
56. Taylor, C. D., and Wolfe, R. S. (1974). J. Biol. Chem. 249:4886.
57. Gunsalus, R. P. (1977). Ph.D. thesis, University of Illinois, Urbana, Illinois.
58. Taylor, C. D., McBride, B. C., Wolfe, R. S., and Bryant, M. P. (1974). J. Bacteriol. 120:974.
59. Gunsalus, R. P., and Wolfe, R. S. (1977). Biochem. Biophys. Res. Commun. 76:790.
60. Cheeseman, P., Toms-Wood, A., and Wolfe, R. S. (1972). J. Bacteriol. 112:527.
61. Mink, R., and Dugan, P. R. (1977). Appl. Environ. Microbiol. 33:713.
62. Tzeng, F. S., Bryant, M. P., and Wolfe, R. S. (1975). J. Bacteriol. 121:192.
63. Tzeng, F. S., Wolfe, R. S., and Bryant, M. P. (1975). J. Bacteriol. 121:184.
64. Ferry, J. G., and Wolfe, R. S. (1977). Appl. Environ. Microbiol. 34:371.
65. Eirich, D. (1977). Ph.D. thesis, University of Illinois, Urbana, Illinois.
66. Mitchell, P. (1961). Nature 191:144.
67. Pine, M. J., and Barker, H. A. (1956). J. Bacteriol. 71:644.
68. Taylor, G. T., Kelly, D. P., and Pirt, S. J. (1976). In H. Schlegel, G. Gottchalk, and N. Pfennig (eds.), Microbial Production and Utilization of Gases. Erich Gotlze KG, Göttingen.
69. Daniels, L., and Zeikus, J. G. (1974). Abstr. Annu. Meet. Am. Soc. Microbiol., p. 197.
70. Sogin, M. L., Pechman, K. J., Zablen, L., Lewis, B. J., and Woese, C. R. (1972). J. Bacteriol. 112:13.
71. Uchida, T., Bonen, L., Schaup, H. W., Lewis, B. J., Zablen, L., and Woese, C. R. (1974). J. Mol. Evol. 3:63.

72. Zablen, L., and Woese, C. R. (1975). J. Mol. Evol. 5:25.
73. Woese, C. R., Fox, G. E., Zablen, L., Uchida, T., Bonen, L., Pechman, K., Lewis, B. J., and Stahl, D. (1975). Nature 254:83.
74. Woese, C. R., Sogin, M., Stahl, D., Lewis, B. J., and Bonen, L. (1976). J. Mol. Evol. 7:197.
75. Fox, G. E., Pechman, K. R., and Woese, C. R. (1977). Int. J. Syst. Bacteriol. 27:44.
76. Balch, W. E., Magrum, L. J., Fox, G. E., Wolfe, R. S., and Woese, C. R. (1977). J. Mol. Evol. 9:305.
77. Fox, G. E., Magrum, L. J., Balch, W. E., Wolfe, R. S., and Woese, C. R. (1977). Proc. Natl. Acad. Sci. USA 74:4537.
78. Zhilina, T. N. (1971). Mikrobiologiia 40:674.

PART 2. METHANOTROPHY *I. J. Higgins*

The stability of the biosphere depends upon uninterrupted recycling of the elements and a considerable amount of carbon is recycled via methane. It follows, therefore, that methanogenesis should be balanced by processes reoxidizing methane to the major reservoir of carbon on earth, carbon dioxide. Much of the methane arising from methanogenesis in deeply submerged muds probably does not reach the atmosphere due to its oxidation to carbon dioxide by methane-utilizing bacteria (methanotrophs) in both aquatic and soil environments. This contention is supported by experiments using columns of sediments in which the indigenous methanotroph population was so effective in oxidizing methane generated in the lower anaerobic parts of the columns that none was released into the atmosphere above the sediment (1). Similarly, the methane content of bubbles released from the bottom of lakes decreases rapidly as they rise to the surface due to methanotrophic oxidation. Gas bubbles rising from the bottom of Lake Beloya, Russia, showed a decrease of methane content to one-quarter by the time they reached the surface, a distance of 10 m (2). The steady-state concentration of methane in the earth's atmosphere is about 1.4 ppm and its release into the atmosphere is balanced by its removal, in this case most likely by chemical reactions with hydroxyl radicals in the troposphere rather than as a result of biological activity (2, 3).

Although it has been known for over 70 years that methane oxidizing bacteria exist in the environment (4, 5), only four species had been well documented prior to 1970. The first, *Bacillus methanicus* was isolated and named by Söhngen in 1906 (5) and later re-isolated and renamed *Pseudomonas methanica* by Dworkin and Foster (6). The other species subsequently isolated were *Pseudomonas methanitrificans* (7), *Methano-*

monas methanooxidans (8, 9), and *Methylococcus capsulatus* (10). Whittenbury, Wilkinson, and their co-workers then isolated over 100 strains of methane-utilizing bacteria from various sources (11–13). All these bacteria proved to be obligate methylotrophs capable of growth only on methane, methanol, or dimethyl ether. At the time of this writing there is only one well-documented report of a methane-utilizing facultative methylotroph, *Methylobacterium organophilum* (14, 15). It is possible that some algae and fungi may have a limited capacity to utilize methane (16, 17), but the evidence is not conclusive. Although methane oxidation has been regarded by most researchers as an obligatorily aerobic phenomenon, recent evidence suggests that some microorganisms can oxidize it anaerobically (18).

It is the methanotrophs in the environment that are primarily responsible for the recycling of methane back to carbon dioxide, both directly by respiration and indirectly as a result of subsequent degradation of their biomass by aerobic and anaerobic heterotrophs (Figure 1). Methanotrophs are also involved in a much slower carbon-cycling process when oxidizing methane that is seeping from natural gas deposits after being trapped for considerable periods of time (19). Because oil deposits are usually accompanied by natural gas, it has been suggested that the relative number of methane utilizers in the soil might be used as an indicator of oil deposits in prospecting (20–23). Interestingly, such increases in methanotroph populations near natural gas leaks in Dutch towns have been implicated in associated death of trees resulting from oxygen deficiency and excess carbon dioxide in the soil, and perhaps ethylene formation (19).

Natural gas contains gaseous hydrocarbons other than methane and at least some of these will be partially oxidized by methanotrophic cooxidative metabolism. Cooxidation of higher hydrocarbons to the corresponding alcohols, aldehydes, ketones, and fatty acids by methane-utilizing bacteria is a well-known phenomenon (23–27). Other compounds that are not growth substrates but that can be oxidized include carbon monoxide (27–29), ammonia (12, 30, 33), methyl bromide (1, 31), methyl chloride, dichloromethane, ethyl bromide (1), acetamide, *N*-methylacetamide, acetone, methylacetate, butan-2-one, dimethoxymethane, dimethoxyethane, and tetramethyl-ammonium chloride (32).

In addition to their activities in the carbon cycle, it is becoming clear that the methanotrophs as a group play a significant role in the nitrogen cycle. Many of them oxidize ammonia to nitrite or nitrate (12, 30, 33). In this respect, the relative importance of methane utilizers as compared with the chemolithotrophs classically associated with this oxidative segment of the nitrogen cycle remains to be determined. However, recent studies of the distribution and numbers of methanotrophs in the environment (1, 34) suggest that they may make a major contribution to the process.

Since 1930 there have been a number of reports suggesting that methanotrophs fix atmospheric nitrogen (7, 12, 35–37). The first unam-

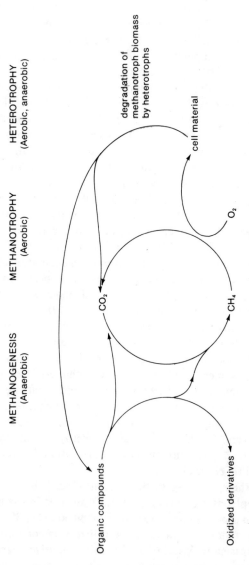

Figure 1. Carbon recycling in the biosphere via methane.

biguous demonstration of dinitrogen fixation by a pure culture of a methane-oxidizing bacterium using ^{15}N was described by de Bont and Mulder (24). It is now clear that the ability to fix dinitrogen is widely distributed among these bacteria (1).

During the last decade the considerable advances made in our knowledge of the methanotrophs have revealed a hitherto underestimated importance in their contribution to cycling of carbon and nitrogen in the biosphere. Also during this period they have attracted the attention of the fermentation industry as potential sources of single cell protein. Methane has some advantages over other novel substrates and industrial research and development is continuing (38). Other possibilities for commercial exploitation of methanotrophs include cooxidation of non–growth substrates to useful products and the use of immobilized organisms or immobilized enzymes derived from these products for low-temperature processes for conversion of methane to methanol, formaldehyde, or formate, detoxification of carbon monoxide, and a variety of other oxidative processes.

The following sections are devoted to an examination of the current state of biochemical knowledge of the methanotrophs, after a brief discussion of their microbiology. For details of the earlier literature and more general coverage of microbial carbon-1 metabolism the reader is referred to the excellent reviews of Quayle (39, 40), Ribbons, Harrison, and Wadzinski (41), and Anthony (42).

MICROBIOLOGY

As mentioned in the introduction to this section, only four species of methane-oxidizing bacteria were known before 1970. In the last few years there have been several reports of the isolation of obligate methanotrophs and this topic has been reviewed (43). Whittenbury, Phillips, and Wilkinson (12) described in detail their procedures for isolating such bacteria from mud, water, and soil samples collected worldwide and from ruminant buccal cavities. These authors possess the most extensive collection of methanotrophs, whose properties are summarized in Table 1. Although a satisfactory system for their classification has yet to be devised, the methanotrophs have been divided into two major groups (Type I and Type II), initially on a morphological basis involving their characteristic internal membrane structures (12) (Figure 2). This major subdivision was supported by the findings of Lawrence and Quayle (44) that methanotrophs of different membrane types employ different carbon incorporation pathways. More detailed examination of the properties of these isolates permitted further subdivision into five major groups, namely, *Methylomonas, Methylobacter,* and *Methylococcus* (Type I), and *Methylosinus* and *Methylocystis* (Type II) (12, 45). This classification is summarized in Table 2.

Other methanotrophs isolated recently (25, 46–50) generally seem to fall into one of these groups, although some isolates fail to show all

Table 1. Properties of obligate methanotrophs in Professor Whittenbury's collection[a]

---

1. Gram-negative; strictly aerobic; rods, vibrios, or cocci.
2. Catalase-positive, oxidase positive, and possess cytochromes of c, a, or b type.
3. To some degree sensitive to normal oxygen tension in air. Some will not grow in shaking flasks. If N_2 fixation is induced, they become extremely oxygen sensitive.

Carbon Sources
4. All use methane, methanol, dimethyl ether, methyl formate, dimethylcarbonate, and (those tested) formaldehyde as sole carbon and energy sources.
5. Carbon dioxide, acetate, and some amino acids are used as supplementing carbon sources. Carbon compounds cooxidized are also likely to provide supplementary carbon.

Nitrogen Sources
6. All use ammonia; most use nitrate and nitrite. Some use urea, amino acids, and yeast extract. All those with a serine pathway fix dinitrogen.
7. All reduce nitrate to nitrite, but will not grow anaerobically on CH_4 because O_2 is required.

Cooxidation
8. All oxidize ammonia, carbon monoxide, dimethyl ether, propane, ethane, ethanol, propanol, butanol, formaldehyde, and formate, among other compounds.

Morphology
9. All possess a complex internal arrangement of paired membranes.
10. All form a differentiated resting body (exospore or cyst).

---

[a]From Whittenbury et al. (1), by courtesy of Erich Goltze K. G.

characteristics required to classify them unambiguously on the basis of this scheme (25, 48). There does not seem to be any clear association between particular species and specific habitats; rather, a variety of species are found in all habitats (1). Thermophilic and thermotolerant isolates are well known (1, 12, 50).

Mtlb. organophilum, a facultative methanotroph, grows on a variety of sugars and organic acids in addition to methane and methanol (14, 15). As its name implies, it shows a preference for heterotrophic substrates, the pathways for methane oxidation and assimilation of carbon-1 units being repressed during heterotrophic growth. The bacterium is gram-negative, with a single polar flagellum, and contains membranes similar to those found in Type II methanotrophs only when growing on carbon-1 compounds. However, its $G + C$ content is rather different from either Type I or Type II obligate organisms and so its relationship to them remains uncertain (1, 15).

Davies and Yarborough (51) obtained preliminary evidence for anaerobic methane oxidation linked to sulfate reduction. Recently cultures of nonmotile gram-negative bacteria have been isolated from the surface of a lake sediment and shown to couple methane oxidation to carbon dioxide with sulfate reduction to hydrogen sulfide, while incorporating acetate into cell material (18). As yet there is no detailed information concerning this isolate.

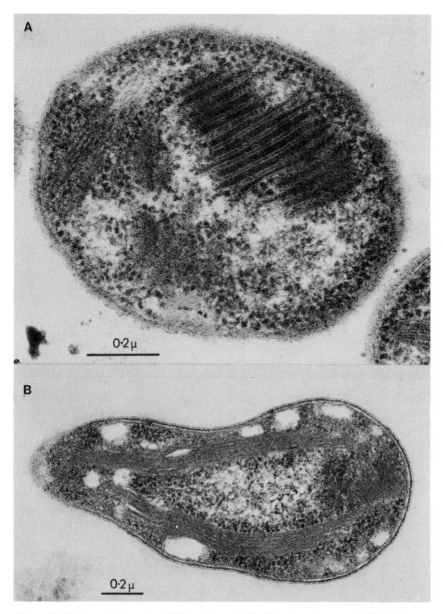

Figure 2. Electron micrographs of thin sections of (A) a Type I methanotroph (*Methylococcus capsulatus* × 125,000), (B) a Type II methanotroph (*Methylosinus trichosporium* × 75,000). (Courtesy of R. Newsam, E. M. Unit, University of Kent.)

Table 2.   Classification of obligate methanotrophs into groups[a]

	Type I (Methylococcus, Methylomonas, Methylobacter spp.)	Type II (Methylosinus, Methylocystis spp.)
Membrane arrangement	Bundles of vesicular discs	Paired membranes, layers around periphery
Resting stages	Cysts (Azotobacter-like)	Exospores or 'lipid' cyst—unique structures
Carbon assimilation pathway	All have ribulose monophosphate pathway (hexose phosphate synthetase$^+$)	A serine pathway (hydroxypyruvate reductase$^+$, hexose phosphate synthetase$^-$)
TCA cycle	Incomplete (α-ketoglutarate dehydrogenase negative—as in anaerobically grown Escherichia coli, many autotrophs and other ribulose monophosphate pathway C$_1$ utilizers)	Complete
Glucose and gluconate dehydrogenase	Present	Absent
Nitrogen fixation	Some fix	All fix

	Methylococcus capsulatus	Methylomonas spp.	
Cell shape	Coccus	Rod	Rod and vibrios
Isocitrate dehydrogenase	NAD specific	NAD and NADP	NADP specific
Malate dehydrogenase	Low activity	High activity	High activity
G + C base content	62.5	50–54	62.5

[a]From Whittenbury et al. (1), courtesy of Erich Goltze K. G.

CARBON METABOLISM

Methanotrophs are thought to oxidize methane to carbon dioxide via methanol, formaldehyde, and formate:

$$CH_4 \rightarrow CH_3OH \rightarrow HCHO \rightarrow HCOOH \rightarrow CO_2$$

In contrast to the process of methanogenesis, it is thought that the free intermediates are involved rather than their tetrahydrofolate derivatives. Much of the evidence for the operation of this pathway has been reviewed in detail elsewhere (40). It is from this oxidative sequence that the energy and reducing power for growth are generated, and the individual reactions involved are discussed under "Energy Metabolism," below.

The methanotrophs, and indeed all methylotrophs and methanogenic bacteria, when growing on carbon-1 compounds, share with the autotrophs a unique anabolic problem—the need to biosynthesize cell material entirely from 1-carbon units. These microorganisms, therefore, are characterized by possessing special biochemical pathways for synthesizing carbon-3 compounds from carbon-1 compounds. Thereafter their intermediary metabolism would seem to be essentially that of heterotrophs. The nature of the autotrophic solution to this anabolic problem in the form of the ribulose diphosphate cycle was revealed by Calvin and his co-workers in the early 1950s (52). The analogous methanotrophic solutions have been elucidated primarily by Quayle and his colleagues, who have demonstrated the existence of two incorporation pathways for reduced carbon-1 compounds, the ribulose monophosphate cycle (53) (used by Type 1 methanotrophs) and the serine pathway (40) (used by Type II methanotrophs). In the former case, formaldehyde alone is incorporated to form carbon-3 compounds, while in the latter instance 1 molecule of carbon dioxide is incorporated per 2 molecules of formaldehyde. The overall bioenergetics of these two pathways are quite different and this aspect, which has important cell yield implications, is discussed under "Energy Metabolism," below.

The Ribulose Monophosphate Cycle (Quayle Cycle)

Leadbetter and Foster (54), using $^{14}CO_2$, demonstrated that *P. methanica* does not employ the Calvin cycle for growth on methane. In addition, it was not possible to demonstrate ribulose diphosphate carboxylase activity in extracts of this bacterium (55). Nevertheless, subsequent pulse-labeling experiments using *P. methanica* and *M. capsulatus* (55–57) showed that radioactivity from [^{14}C]methane, [^{14}C]methanol, and [^{14}C]formaldehyde was incorporated mainly into sugar phosphates. [^{14}C]Formate and $^{14}CO_2$ were incorporated less rapidly into serine and malate, and into aspartate and malate, respectively. Detailed analysis of the data led to the conclusion that formaldehyde formed by oxidation of methane and methanol is incorporated into a hexose phosphate.

It was found that cell-free preparations of both *P. methanica* and *M. capsulatus* catalyzed the condensation of formaldehyde with ribose 5-phosphate to form a mixture of sugar phosphates, the major ones being fructose 6-phosphate and a compound that after exhaustive investigations was thought to be allulose 6-phosphate (the 3-epimer of fructose 6-phosphate) (57–59). This work led to the proposal of a cyclic scheme for formaldehyde fixation (40, 56, 58, 59) analogous to the Calvin cycle and in which 1 molecule of triose phosphate was synthesized from 3 molecules of formaldehyde:

$$3 \text{ HCHO} + \text{ATP} \rightarrow \text{triose phosphate} + \text{ADP}$$

Although originally called the ribose phosphate cycle, it has also appeared in the literature under various aliases, including the pentose phosphate, the hexose phosphate, and the allulose phosphate pathways.

During attempts to purify the enzyme responsible for the condensation reaction (hexose phosphate synthase) from *M. capsulatus*, two protein fractions were obtained, a soluble one that catalyzed the isomerization of ribose 5-phosphate into ribulose 5-phosphate and a particulate fraction that catalyzed the condensation of ribulose 5-phosphate with formaldehyde (60). In 1974, Kemp (61) reported that a more extensive investigation had shown that the condensation product is not, after all, allulose 6-phosphate but rather, D-*arabino*-3-hexulose 6-phosphate (D-*erythro*-L-glycero-3-hexulose 6-phosphate). The two key enzymes of the ribulose monophosphate cycle are, therefore, 3-hexulose phosphate synthase and phospho-3-hexulose isomerase, which together effect the synthesis of fructose 6-phosphate from ribulose 5-phosphate and formaldehyde:

ribulose 5-phosphate   D-*arabino*-3-hexulose 6-phosphate   fructose 6-phosphate

Both the synthase and the isomerase have been purified extensively from *M. capsulatus* (62) and have molecular weights of 310,000 and 67,000, respectively. The synthase dissociates into subunits of molecular weight 49,000 under conditions of low pH and low ionic strength and both enzymes are highly substrate-specific.

A formulation of the ribulose monophosphate cycle is depicted in Figure 3 that includes two possible routes for regeneration of the ribulose 5-phosphate acceptor. On the basis of the enzymological evidence, the transaldolase-transketolase route shown in the main part of the figure seems the more likely possibility. The alternative, involving sedoheptulose diphosphatase (but not transaldolase), has been regarded as less likely because the former enzyme activity is very low in *P. methanica* and *M. capsulatus* (53). However, the enzyme is difficult to assay. Two mechanisms for cleavage of fructose 6-phosphate are also possible (Figure 3), one involving glycolytic enzymes and the other involving those of the Entner-Doudoroff pathway. In *P. methanica* and *M. capsulatus* both sets of enzymes are present and again the relative importance of the two pathways in vivo is not known.

There is sound evidence, albeit less complete, that the ribulose monophosphate pathway operates in some other Type I methanotrophs (44), obligate methylotrophs (63, 64), a facultative methylotroph (65), and possibly (although not proven) in methanol-utilizing yeasts (66, 68). It is therefore not specific either to methanotrophs or to obligate methylotrophs.

Strøm, Ferenci, and Quayle (53) have pointed out that the presence of 6-phosphogluconate dehydrogenase in methanotrophs (53, 69) would, in theory, permit oxidation of formaldehyde to carbon dioxide by a route other than the one involving formate as an intermediate. This cyclic oxidation pathway, which would generate NADPH, is shown in Figure 4. Since none of the other reactions involved in oxidation of methane to carbon dioxide is NADP linked (see under "Energy Metabolism") this new pathway could represent a means for generation of NADPH required for biosynthesis. Recent isotopic labeling studies indicate that not only does such a cycle exist in vivo but that it is the predominant mechanism for formaldehyde oxidation in the facultative methylotroph *Pseudomonas* C (70); however, it remains to be demonstrated in a methanotroph. NADPH could also, of course, be generated by transhydrogenase activity, but to the author's knowledge this enzyme has not been reported in methanotrophs, so the mechanism of NADPH generation in those microorganisms remains to be clarified.

The Serine Pathway

Unlike studies of the ribulose monophosphate cycle, most of the detailed biochemical work on the serine pathway has been done with facultative methylotrophs, although enzymological data suggest that Type II obligate methanotrophs may all use this route for carbon incorporation (1, 44). The pathway was first proposed in 1961 by Large, Peel, and Quayle (71) on the basis of short-term isotope incubation studies using the facultative methylotrophs *Pseudomonas* AM1 and *Hyphomicrobium vulgare*, which had been grown on methanol. Similar studies have been done with two other facultative methylotrophs (72, 73) and one methanotroph, *Mtn. methanooxidans* (57).

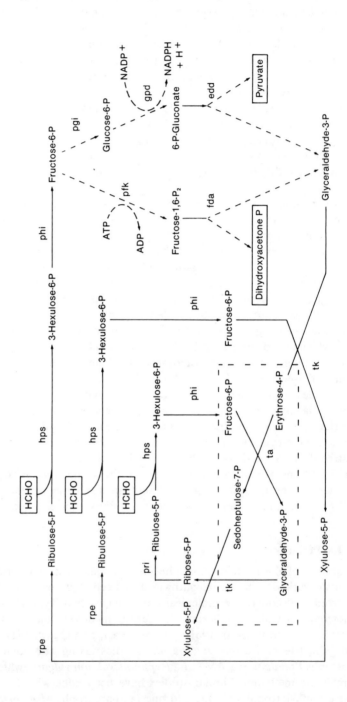

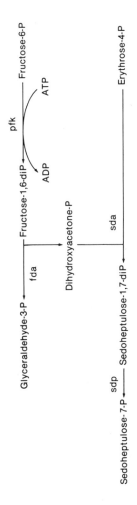

Figure 3. The ribulose monophosphate cycle: edd, 6-phosphogluconate dehydrase/phospho-2-keto-3-deoxygluconate aldolase; fda, fructose diphosphate aldolase; gpd, glucose-6-phosphate dehydrogenase; hps, 3-hexulose phosphate synthase; pkf, phosphofructokinase; pgi, phosphoglucoisomerase; phi, phospho-3-hexulose isomerase; pri, phosphoriboisomerase; rpe, ribulose-5-phosphate 3-epimerase; ta, transaldolase; tk, transketolase; sda, sedoheptulose diphosphate aldolase; sdp, sedoheptulose-1,7-diphosphatase. [From Strøm, Ferenci, and Quayle (53), by courtesy of Biochemical Journal.]

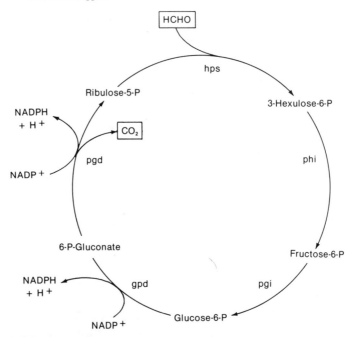

Figure 4. Scheme for cyclic oxidation of formaldehyde: pgd, 6-phosphogluconate dehydrogenase; other abbreviations as in legend to Figure 3. [From Strøm, Ferenci, and Quayle (53) by courtesy of Biochemical Journal.]

The two possible versions of the serine pathway originally conceived are shown in Figure 5 and the evidence for its overall operation has been reviewed in detail elsewhere (40, 42). One version involves direct synthesis of glycine, and the other involves cyclic regeneration of this carbon-1 acceptor. The key enzyme of the pathway, serine transhydroxymethylase, is an enzyme widely distributed in nature (74) that catalyzes the incorporation of carbon-1 units at the oxidation level of formaldehyde:

$$5,10\text{-methylenetetrahydrofolate} + glycine \rightleftharpoons serine + tetrahydrofolate$$

This enzyme has been detected in facultative methylotrophs (40), in the obligate methanotroph *Methylosinus trichosporium* (75) and in the facultative methanotroph *Mtlb. organophilum* (76).

A long-standing uncertainty concerning the serine pathway was the mechanism of glycine synthesis. It has taken approximately 15 years for it to become clear that a cyclic process does indeed operate and that in some, but not all, bacteria the mechanism of glycine regeneration has now been revealed in detail. The basic cycle thought to function in most serine pathway methylotrophs is shown in Figure 6. Evidence for cyclic regeneration of glyoxylate and glycine from a carbon-4 compound in vivo came initially from

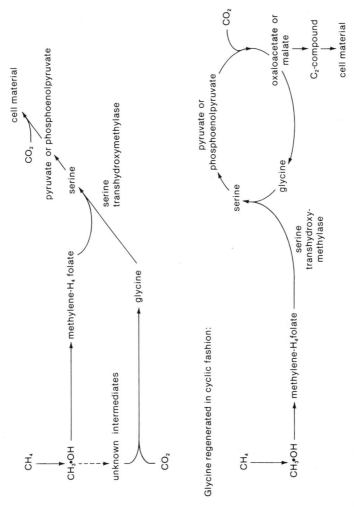

Figure 5. Serine pathway: possible modes of operation.

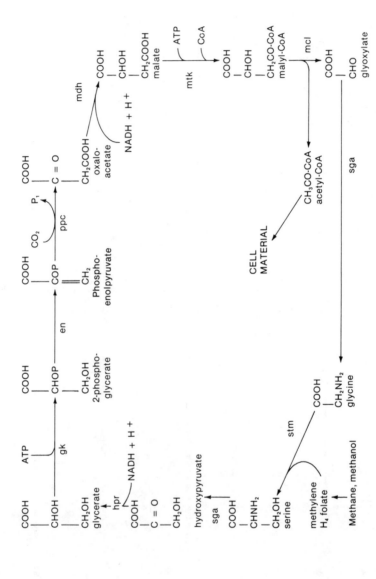

Figure 6. Reactions of the serine pathway: en, enolase; gk, glycerate kinase; hpr, hydroxypyruvate reductase; mcl, malyl-CoA lyase; mdh, malate dehydrogenase; mtk, malate thiokinase; ppc, phosphoenolpyruvate carboxylase; sga, serine-glyoxylate aminotransferase; stm, serine transhydroxymethylase.

the elegant labeling experiments of Salem, Large, and Quayle (77) using *Pseudomonas* AM1. The most important aspect of this work was the demonstration that label from 2,3-[^{14}C]succinate rapidly appeared in glycine and then in serine, the position of the label in serine being consistent with its having been formed by hydroxymethylation of glycine.

Enzymes of the Serine Pathway   The first demonstration of a carbon-4 cleavage enzyme in a methylotrophic organism was made by Hersh and Bellion (78) in a species of *Pseudomonas* grown on methylamine. They observed a reversible ATP- and CoA-dependent cleavage of malate into glyoxylate and acetyl-CoA. This led to the discovery of malyl-CoA lyase, which catalyzes the reaction:

$$\text{malyl-CoA} \rightarrow \text{glyoxylate} + \text{acetyl-CoA}$$

This enzyme is present in a variety of methylotrophs, including not only a Type II methanotroph (*Mts. trichosporium*) but also two Type I methanotrophs (*M. capsulatus* and *P. methanica*) (79). The significance of finding the enzyme in ribulose monophosphate cycle bacteria, albeit at somewhat lower specific activities, is uncertain, but it is probably required for glycine biosynthesis. The ATP- and CoA-dependent malate lyase activity found in other methylotrophs (78, 80) is due to a coupled reaction between thiokinase and malyl-CoA lyase (see ref. 42).

There are three other enzymes characteristic of the serine pathway, namely, serine-glyoxylate aminotransferase, hydroxypyruvate reductase, and glycerate kinase. Hydroxypyruvate reductase, however, is also involved in tartrate catabolism in pseudomonads (81, 82).

Serine-glyoxylate aminotransferase was first detected in *P.* AM1 and catalyzes the reaction (40):

$$\text{serine} + \text{glyoxylate} \rightleftharpoons \text{hydroxypyruvate} + \text{glycine}$$

There are few reports of this enzyme having been measured in methanotrophs, but interestingly it is present in some Type I organisms (50).

Hydroxypyruvate reductase catalyzes the reaction:

$$\text{hydroxypyruvate} + \text{NADH} + \text{H}^+ \rightarrow \text{glycerate} + \text{NAD}^+$$

It is present in high specific activities in extracts of *P.* AM1 (40, 83) and at similar activities in the Type II methanotrophs *Mts. sporium, Mts. trichosporium, Methylocystis parvus,* and *Mtn. methanooxidans* (44) as well as in the facultative methanotroph *Mtlb. organophilum* (14). The enzyme also appears to be present, albeit at low activities, in some type I organisms (44, 50) and in *Paracoccus denitrificans,* where its function (if any) is uncertain (84). While mere presence of low activities of this enzyme has been taken as evidence that a carbon-1–utilizing microorganism uses the serine pathway, it has recently been pointed out that such evidence alone is inadequate and further corroborating data is necessary before drawing this conclusion (84).

Glycerate kinase catalyzes the reaction:

$$glycerate + ATP \rightarrow phosphoglycerate + ADP$$

It has been found at high specific activity in extracts of methanol-grown P. AM1 (85) and also in *Hyphomicrobium* X, where detailed enzyme profiles also suggest operation of the serine pathway during growth on methanol (86).

Further evidence for operation of the serine pathway comes from the isolation of single- and double-metabolic mutants and revertants of P. AM1 (40, 42). Mutants lacking hydroxypyruvate reductase (40), serine-glyoxylate aminotransferase (40), glycerate kinase (87), or malyl-CoA lyase (42) do not grow methylotrophically. In this bacterium, these four enzymes are thought to be regulated coordinately, their synthesis being repressed by succinate or one of its metabolites (42, 87).

Details of the operation of the serine pathway in the facultative methylotroph P. MA (88) have recently been elucidated. This organism contains high activities of isocitrate lyase when grown methylotrophically and the involvement of this enzyme in the carbon incorporation mechanism has led to an understanding of the so-called icl-serine pathway, shown in Figure 7. There are three phases of the pathway (89), the first involving net synthesis of acetate (acetyl-CoA) from a reduced 1-carbon unit and carbon dioxide as shown in more detail in Figure 6. In phase 2, the acetate is oxidized to glyoxylate by a cyclic series of reactions catalyzed by citrate synthase, aconitase, and isocitrate lyase. In phase 3, the resulting glyoxylate is finally used for synthesis of phosphoglycerate via the serine pathway. Carbon-4 compounds may then be synthesized from phosphoglycerate via phosphoenolpyruvate or pyruvate by carboxylation. Such a detailed analysis of the pathway has not so far been undertaken for a Type II methanotroph and it is not known whether or not the icl-serine pathway operates in any of these bacteria.

A major problem still remains in those bacteria, exemplified by P. AM1, that use a pathway closely similar to the icl-serine pathway but that do not contain malate thiokinase or isocitrate lyase when grown on carbon-1 compounds (89). The mechanism by which these bacteria convert acetate to glyoxylate remains a mystery. It may be noted that a reexamination of the enzyme profile of several species of *Hyphomicrobium* (90) shows that these organisms also lack isocitrate lyase and, contrary to an earlier view (86), cannot therefore use the icl-serine pathway during growth on carbon-1 compounds.

Metabolism of Exogenous Compounds Containing Carbon-Carbon Bonds: The Nature of Obligate Methanotrophy

With the exception of *Mtlb. organophilum*, methanotrophs are unable to grow on compounds containing more than one carbon atom, dimethyl ether

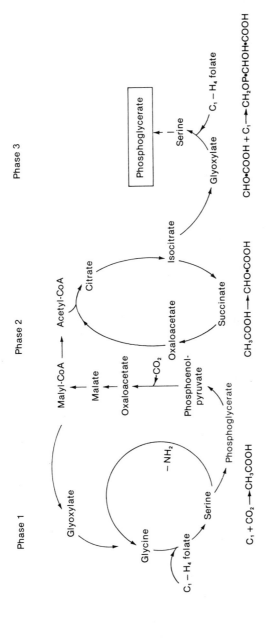

Figure 7. The icl-serine pathway. [From Quayle (89) by courtesy of the Society of Fermentation Technology, Japan.]

excepted. This is reminiscent of the properties of the so-called obligate autotrophs. One of the reasons proposed to explain why some microorganisms are obligate autotrophs has been their lack of NADH oxidase and 2-oxoglutarate dehydrogenase activities. They therefore have an incomplete tricarboxylic acid cycle and cannot use it for total oxidation of acetate (91). Similar studies with methanotrophs have revealed that all that have been examined possess NADH oxidase activity, but Type I organisms lack 2-oxoglutarate dehydrogenase while Type II organisms possess a complete tricarboxylic acid cycle (92–96). Clearly, therefore, an incomplete tricarboxylic acid cycle cannot be the sole reason for obligate methanotrophy. Nevertheless, it has been shown that lack of 2-oxoglutarate dehydrogenase alone can confer the characteristic of obligate methylotrophy. A mutant of *P.* AM1 lacking this enzyme could only grow on carbon-1 compounds. It seems unlikely that this was due to multiple lesions because two revertants were obtained, both of which had regained enzyme activity and grew on the same range of substrates as the wild type (97).

Other possible reasons for the characteristic of obligate methanotrophy include lack of suitable energy-transducing systems for heterotrophic substrates (41), accumulation of toxic metabolites from heterotrophic substrates, and lack of transcriptional control. The demonstration of NADH oxidase in these bacteria does not preclude the first possibility, because oxidative phosphorylation coupled to electron flow from NADH remains to be demonstrated unambiguously, although there is evidence for it from proton extrusion studies (see under "Energy Metabolism"). It is likely that the last two possibilities are responsible in some cases for failure of autotrophs to grow on heterotrophic substrates (98, 99). To test the toxic metabolite theory, attempts have been made to grow *Mtn. methanooxidans* on glucose in dialysis culture. These proved unsuccessful, which was not surprising in view of the finding that the bacterium lacks glucokinase activity (100). In conclusion, while lack of 2-oxoglutarate dehydrogenase in Type I methanotrophs may be adequate to account for their obligate nature, the reason for this property in Type II organisms is less clear.

In spite of inability to grow on heterotrophic substrates, obligate methanotrophs are capable of their utilization, incorporating some of these compounds into cell material during growth on carbon-1 substrates. There is also evidence that ATP is generated during the cooxidation of ethanol (1, 45) and this may occur during cooxidation of other compounds. When *Mtn. methanooxidans*, which possesses a complete tricarboxylic acid cycle, is supplied with acetate during growth on methane there is a marked increase in cell yield and acetate contributes up to 15% of the total carbon, indicating considerable assimilation of this compound (95). Not only is radioactivity from [^{14}C]acetate incorporated into the glutamate and aspartate families of amino acids, but it also appears in those derived from pyruvate, reflecting a functional tricarboxylic acid cycle. This is further emphasized by the oxidation of [^{14}C]acetate to $^{14}CO_2$ (95). In contrast, *M. capsulatus* shows restricted incorporation of [^{14}C] acetate into lipid and only four amino acids (glutamate, proline, arginine, and leucine) due to the incomplete tricarboxylic acid cycle in this species (94, 96). Similar results were obtained with [^{14}C] ethanol, suggesting initial cooxidation of the alcohol to acetate (94).

The interaction of *M. capsulatus* with exogenous organic compounds in general has been examined in some detail. Various alkanes and alcohols are cooxidized during growth on methane (10, 101). While some amino acids serve as nitrogen sources (101), others are growth inhibitory 102, 103), as are some common carboxylic acids and sugars (102). The effect of exogenous amino acids is thought to result from disturbance of systems involved in regulation of amino acid biosynthesis (103) and, in the case of threonine, the effect most likely results from interference with the biosynthesis of amino acids of the aspartate family by threonine-inhibiting aspartokinase (104). Some amino acids, such as histidine, inhibit methane oxidation and hence growth, probably by binding metal ions. (The involvement of metal ions in methane oxidation is discussed under "Energy Metabolism"). The reasons for growth inhibition by sugars are less clear. For example, radial growth of *M. capsulatus* colonies is completely inhibited by M/18 glucose, while logarithmic phase liquid cultures are less sensitive. However, when exposed to [^{14}C] glucose or [^{14}C] sucrose, only minute amounts of radioactivity are incorporated, suggesting either that these sugars are transported only poorly into this bacterium or that they cannot be phosphorylated or hydrolyzed (94). Lack of glucokinase in *Mtn. methanooxidans* is discussed above (100).

Although a variety of organic compounds can be cooxidized by methanotrophs there is little information concerning incorporation of cooxidation products into cell material, except in the cases of ethane and ethanol, where the product is acetate.

M. organophilum utilizes a range of organic substrates and grows most rapidly on glucose, succinate, or malate (105).

ENERGY METABOLISM

As mentioned in the previous section, the weight of evidence favors the following route for oxidation of methane by methanotrophs:

$$CH_4 \underset{(1)}{\rightarrow} CH_3OH \underset{(2)}{\rightarrow} H \cdot CHO \underset{(3)}{\rightarrow} H \cdot COOH \underset{(4)}{\rightarrow} CO_2$$

It is from this sequence of oxidative reactions that the energy for biosynthesis of cell components is derived. A brief summary of the overall energetics of the sequence and of the enzymes involved is recorded in Table 3. Each step is clearly sufficiently exergonic to permit, in theory at least, the generation of 1 or more ATP molecules. There is strong evidence for operation of this pathway, although it remains incomplete for any one species. Enzymes catalyzing each step have been detected in cell-free systems (Table 3) and,

Table 3.   Reactions involved in oxidation of methane by methanotrophs

Step in methane oxidation	Number in scheme in text	$\Delta G_0'$ pH 7 (Kcal mol^{-1}) for reaction in far left column (41)	Enzyme catalyzing reaction	Reaction catalyzed	Co-factor requirement	Microorganism in which the enzyme has been studied	References
CH$_4$ + ½ O$_2$ → CH$_3$OH	1	−26.12	methane monooxygenase (hydroxylase)	CH$_4$ + XH$_2$ + O$_2$ → CH$_3$OH + H$_2$O + X	NADH NADH, methanol, ascorbate (crude preparations) ascorbate (pure enzyme)	M. capsulatus P. methanica Mts. trichosporium	(107–113) (28, 29, 31, 114) (115–117)
CH$_3$OH + ½ O$_2$ → H · CHO + H$_2$O	2	−44.81	methanol dehydrogenase	CH$_3$OH + PMS → HCHO + PMSH$_2$	Phenazine methosulfate	P. methanica Mts. sporium M. capsulatus Mts. trichosporium	(29, 118) (40, 119) (101, 120–122) (117)
H · CHO + ½ O$_2$ → HCOO$^-$ + H$^+$	3	−57.15	formaldehyde dehydrogenase	HCHO + PMS → HCOOH + PMSH$_2$ HCHO + NAD$^+$ + H$_2$O → H · COOH + NADH + H$^+$ (GSH) HCHO + DCPIP + H$_2$O → HCOOH + DCPIPH$_2$	Phenazine methosulfate NAD$^+$-reduced glutathione Dichlorophenol indophenol	M. capsulatus Mts. sporium P. methanica P. methanica	(120, 122) (119) (29, 123, 118) (118)
HCOO$^-$ + H$^+$ + ½ O$_2$ → CO$_2$ + H$_2$O	4	−58.25	formate dehydrogenase	HCOOH + NAD$^+$ → CO$_2$ + NADH + H$^+$	NAD$^+$	P. methanica Mts. trichosporium M. capsulatus	(118) (215) (101, 111)

with the exception of formate dehydrogenase, have been extensively purified. These enzymes are discussed in detail later in this section. Other lines of evidence that have been reviewed previously (13, 40, 41) include:

1. The demonstration that cell suspensions of methanotrophs oxidize methane, methanol, formaldehyde, and formate (6, 8–10, 12, 54).
2. The detection of oxidative intermediates accumulating during the oxidation of methane or methanol by cell suspensions (8, 54, 124, 125).
3. Accumulation during methane oxidation of postulated intermediates in the presence of trapping agents or inhibitors (8, 124).

More recent evidence comes from studies of methane, methanol, and formaldehyde oxidation by particulate fractions derived from the intracytoplasmic membranes of *M. capsulatus*, in which the product was shown to be formate (110, 122).

Dimethyl Ether as a Possible Intermediate in Methane Oxidation

For some time, dimethyl ether has been considered as a possible intermediate between methane and methanol (13, 106). Many methanotrophs grow on the ether, it accumulates in the medium during growth on methane under oxygen-limiting conditions, and it is sometimes excreted by washed-cell suspensions oxidizing methane. Studies with whole organisms suggest that dimethyl ether is oxidized via methyl formate to methanol and formate. An alternative route for methane oxidation therefore seemed possible (106):

$$2CH_4 \rightarrow CH_3\text{-}O\text{-}CH_3 \rightarrow CH_3\text{-}O\text{-}CH_2OH \rightarrow CH_3\text{-}O\text{-}CHO$$
$$\nearrow \qquad\qquad\qquad \searrow$$
$$CH_3OH \rightarrow HCHO \rightarrow H\cdot COOH \rightarrow CO_2$$

However, studies of ethane oxidation suggest the following route for this analogue of methane:

$$C_2H_6 \rightarrow CH_3CH_2OH \rightarrow CH_3CHO \rightarrow CH_3COOH$$

Assuming involvement of the same enzyme system in the initial attack on both hydrocarbons, then by analogy with the above scheme for methane oxidation diethyl ether might be expected as an intermediate during ethane oxidation. Diethyl ether, however, does not give rise to intermediates known to be involved in ethane oxidation but is metabolized by the following route:

$$C_2H_5\text{-}O\text{-}C_2H_5 \rightarrow C_2H_5\text{-}O\text{-}CH_2CH_2OH \rightarrow C_2H_5\text{-}O\text{-}CH_2CHO \rightarrow C_2H_5\text{-}O\text{-}CH_2COOH$$

In addition, it has been shown that if carrier dimethyl ether or methyl formate are added to washed suspensions of methanotrophs oxidizing $^{14}CH_4$, negligible amounts of radioactivity appear in these compounds. However, carrier methanol does become labeled (106). These results make it highly unlikely that dimethyl ether is an intermediate in methane oxidation. It may, however, be produced in small amounts as a byproduct during enzymic oxidation of the hydrocarbon to methanol, and this is discussed further under "Methane Monooxygenase."

Enzymes of the Oxidative Pathway

Methane Monooxygenase   Leadbetter and Foster (126) proposed as long ago as 1959 that a monooxygenase mechanism was the most likely one for methane oxidation (X is a reducing agent):

$$CH_4 + O_2 + XH_2 \rightarrow CH_3OH + H_2O + X$$

It was shown that *P. methanica,* when incorporating methane, accumulated far more ^{18}O in cell material from ^{18}O-enriched atmospheres than when grown on methanol (126). Nevertheless, two arguments against monooxygenase involvement were made by Whittenbury in 1969 (11). The first involved the question of comparative molar growth yields, which at that time were thought to be somewhat higher with methane as substrate than with methanol (12). It was argued that a monooxygenase involving a reducing agent, probably in the form of NADH, would be energetically wasteful and hence inconsistent with the yield data. However, it was pointed out by Higgins and Quayle (124) that it is possible to explain these data if the exergonic oxygenation of methane were coupled to ATP synthesis; such a system would, of course, be novel. The second argument against involvement of such an enzyme came from the finding that iodoacetate inhibited methanol oxidation but not methane oxidation (8). It might be expected that inhibition of methanol oxidation would always be associated with inhibition of methane oxidation because the supply of reducing power for the monooxygenase would be derived from the further oxidation of methanol. However, several researchers have been unable to substantiate these findings with iodoacetate (106). Indeed, Higgins and Quayle (124) found, on the contrary, that iodoacetate inhibits methane oxidation more than methanol oxidation, results that are consistent with monooxygenase involvement.

It was suggested that perhaps an oxygenase was not involved after all (11), but very convincing evidence for such an enzyme came from measurements by Higgins and Quayle (124) of the incorporation of ^{18}O from ^{18}O-enriched atmospheres into methanol excreted by suspensions of *P. methanica* and *Mtn. methanooxidans* oxidizing methane (124). The oxygen in the alcohol was derived exclusively from dioxygen. These findings did not exclude the possibility that an unusual dioxygenase mechanism might be involved (124), which would be quite consistent with the cell yield data available at that time. Clearly the final resolution of these arguments awaited a cell-free methane oxidizing system. Such a system proved elusive in spite of the attentions of several groups of researchers until the report by Ribbons and Michalover (107) that described methane- and NADH-dependent oxygen consumption by cell-free particulate preparations containing fragmented intracytoplasmic membranes from *M. capsulatus* strain Texas. Oxygen and NADH were consumed in equimolar proportions, suggesting monooxygenase activity, although the product of the reaction was not identified. Later, more detailed studies with this system have shown that methane is oxidized to formate and the particles also oxidize methanol and formaldehyde to formate in the absence of added co-factors. Ribbons (110) found considerable variability in the methane monooxygenase activities of individual preparations and many showed very little activity. There was a specific requirement for NADH as electron donor and ethane was also oxidized by these preparations. The finding of these activities associated with particulate fractions suggests that the enzymes involved in oxidizing methane to formate are localized on the intracytoplasmic membranes in vivo. These membranes account for 40-60% of the total mass of *M. capsulatus*. Formate dehydrogenase activity, however was found in the soluble fraction (110).

In the light of this convincing evidence for NADH-dependent methane monooxygenase activity, it is appropriate to raise again the question of cell yields. While the data of Whittenbury and his colleagues (12) had suggested that molar growth yields were somewhat higher on methane than on methanol, earlier data of Vary and Johnson (127) had indicated that they are similar. Because methane is a gaseous substrate it is technically difficult to obtain highly accurate yield data and there are very considerable differences in reported yields even for the same bacterium, especially in the earlier literature. For example, yields on methane for *Mtn. methanooxidans* have been reported ranging between 0.44 and 1.0 g dry weight g^{-1} of methane (13, 128, 129) and the range for any methanotroph is 0.3-1.4 g dry weight g^{-1} of methane (128). Nevertheless, the most recent data indicate a yield in the region of 0.8-1.0 g dry weight g^{-1} of methane (38, 130-134). An NADH-linked methane monooxygenase in *M. capsulatus* suggests, therefore, that perhaps this organism cannot generate ATP from NADH or that the oxygenation of methane is indeed coupled to ATP synthesis (41, 124), possibilities considered by Ribbons (110). The implications of the nature of methane monooxygenase for growth yields is discussed further later in this section.

Methane monooxygenase enzyme systems have also been studied in another strain of *M. capsulatus* and in *P. methanica* and *Mts. trichosporium*. The enzyme system responsible in the last-named organism has now been highly purified and partially characterized. The nature of these enzymes are considered below.

Methane Monooxygenase of M. capsulatus (Bath) The methane monooxygenase system studied by Ribbons and his colleagues and discussed above was present in the Texas strain of *M. capsulatus*. Recently, Colby and Dalton (112) have examined *M. capsulatus* strain Bath for enzyme activity using procedures for cell breakage closely similar to those described by Ribbons (110). However, in *M. capsulatus* (Bath) there is no detectable methane monooxygenase activity in particulate fractions, but the soluble fraction of the cell-free extract catalyzes the NADH- or NADPH- and oxygen-dependent formation of methanol from methane. Like the system from *M. capsulatus* (Texas), preparations were quite stable for days at 0°C. Preliminary studies suggest that a multicomponent enzyme system is responsible for this activity. The enzyme system has recently been resolved into three components, one of which is an iron flavoprotein (241). These soluble preparations are able to oxygenate a quite extraordinary range of compounds in addition to methane (113), including methanol, a variety of methane derivatives, *n*-alkanes up to C_8, *n*-alkenes, ethers, and various alicyclic, aromatic, and heterocyclic compounds. Of particular interest is the finding that the enzyme is not a terminal alkane hydroxylase. Rather, it oxidizes *n*-alkanes to mixtures of the corresponding 1- and 2-alcohols and forms epoxides with both terminal and internal alkenes. There is evidence that the methane monooxygenase is responsible for all these oxidations, showing therefore a rare lack of substrate specificity. In some cases the rates of oxidation are of the same order as those for methane, although the K_m is probably much higher for most of these compounds.

Methane Monooxygenase of P. methanica   Ferenci (28) reported the isolation of cell-free particulate preparations from *P. methanica* capable of methane plus NADH-dependent oxygen consumption and carbon monoxide plus NADH-dependent oxygen consumption. The evidence suggests that the same monooxygenase enzyme is responsible for both activities. In most respects the system appears closely similar to the one described by Ribbons (110) in *M. capsulatus* (Texas), although the *Pseudomonas* system is rather less stable, most activity being lost on storage at 0°C for 24 h. The discovery that carbon monoxide is oxidized by the enzyme was particularly useful because it permitted determination of a complete stoichiometry, which was difficult in preparations catalyzing the further oxidation of methanol formed from methane. Careful examination of the stoichiometry strongly indicated a monooxygenase mechanism (29). These preparations also catalyzed oxidation of ethane and ammonium chloride. Activity was inhibited by dithiothreitol, reduced glutathione, and cyanide.

Colby, Dalton, and Whittenbury (31) have also studied this enzyme in crude cell-free systems. It was found to oxygenate the substrate analog bromomethane, and the disappearance of this halogenated derivative was used to assay enzyme activity. These researchers also found the activity unstable on storage unless it was frozen to −70°C when it was stable for several weeks.

Methane Monooxygenase of Mts. trichosporium   *Mts. trichosporium* is the only Type II methanotroph in which methane monooxygenase has been studied and at the time of this writing it is the only organism from which the enzyme has been highly purified. Initial studies by Tonge, Higgins, and their colleagues (115, 116) showed the presence of a NADH-dependent methane monooxygenase in cell-free extracts of this bacterium. However, the activity appeared to be unpredictable from extract to extract, reminiscent of the situation in *M. capsulatus* (Texas). In addition, activity was apparently lost quite rapidly on storage at 0–4 °C. On finding that high phosphate concentrations would inhibit the further oxidation of methanol, the enzyme activity could be measured by methanol formation as well as or instead of oxygen or NADH consumption. It was then found that preparations that did not show stimulation of endogenous NADH oxidase activity in the presence of methane (i.e., they were apparently inactive) often showed enzyme activity when methanol formation was assayed. In addition, the specific activities of preparations showing methane-stimulated NADH-dependent oxygen consumption were approximately an order of magnitude higher on the basis of methanol formation as compared with oxygen or methane consumption. These apparent anomalies were explained in terms of methane addition causing a redirection of endogenous electron flow from NADH to oxygen into the monooxygenase function. In other words, it was erroneous to regard the endogenous NADH oxidase activity as the no-substrate control. This was an important finding because it was subsequently found that stored preparations were active on the basis of methanol formation; indeed, extracts stored for 14 days at 4 °C did not show appreciable loss of activity. Specific methane monooxygenase activities in this bacterium are approximately an order of magnitude greater than figures published for other preparations.

The methane-oxidizing activity of crude cell-free extracts of this methanotroph is not obligatorily NADH-dependent. Ascorbate and, in the presence of low concentrations of phosphate, methanol were also effective electron donors. This suggests that electrons derived from the further oxidation of methanol can be recycled into the monooxygenase reaction (Figure 8). Also, Amytal, a mid-chain electron transport inhibitor, completely blocked NADH-dependent methane oxidation without having any effect on ascorbate-dependent activity, suggesting that the enzyme cannot link directly to NADH but rather uses electrons derived from NADH via an electron transport chain (115, 116).

Fractionation of cell-free extracts revealed that the monooxygenase involves at least two components, one particle-bound and the other mainly soluble or at least solubilized on disruption. It soon became clear that the latter is an unusual carbon monoxide–binding cytochrome *c* with a redox potential of +310 mV, remarkably high for a *c*-type cytochrome. Cytochromes of this type have been found in several methylotrophs and their role in electron transport and especially in methanol and formaldehyde oxidation is discussed later in this section. Methane monooxygenase activity that had been lost completely from thoroughly washed particle preparations could be restored fully by addition of purified samples of this cytochrome (115, 116). The membrane-bound portion of the monooxygenase has been solubilized using a variety of procedures, phospholipase-D treatment being the preferred one because it is rather selective. Ultrafiltration of this solubilized material revealed that two protein components in addition to the

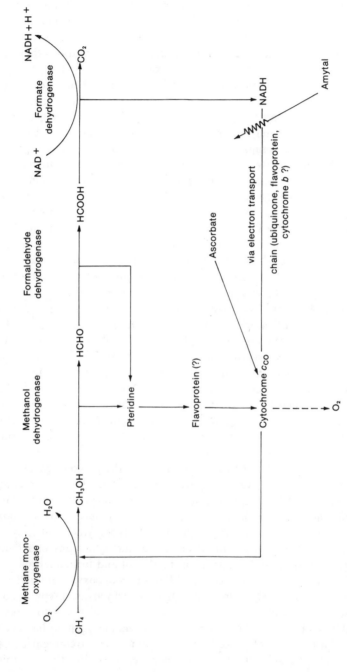

Figure 8. Probable mechanism for supply of reducing power to the methane monooxygenase of *Methylosinus trichosporium*.

cytochrome are required for activity. All three components have now been obtained in pure form (117). Some of the properties of the enzyme system are summarized in Table 4. Although cytochrome c is found mainly in the soluble fraction of disrupted organisms, some remains in the membrane fractions unless they are thoroughly washed. It seems likely that the cytochrome is associated with the membranes in vivo.

The artificial electron donor ascorbate is the only reducing agent that functions with the pure enzyme system. However, it can be replaced by methanol if a partially-purified preparation of methanol dehydrogenase from *Mts. trichosporium* is also included in the assay system. This substantiates the findings with crude cell-free systems that electrons derived from the further oxidation of methanol can be used to drive the monooxygenase reaction (Figure 8).

The pure methane monooxygenase has a high specific activity and is fairly stable on storage at 0-4 °C, but is inactivated by freezing. In the absence of methane, but in the presence of ascorbate, the cytochrome component shows high ascorbate oxidase activity and it is thought that even in the presence of methane some of this oxidase activity persists. This explains the stoichiometry data, which is imperfect for a monooxygenase; more oxygen is consumed than would be expected for such an enzyme system (Table 4). The system is absolutely dependent on cytochrome c_{co} from *Mts. trichosporium*. Although similar cytochromes are present in other methylotrophs (135, 136), those from facultative methylotrophs probably dif-

Table 4. Properties of the methane monooxygenase enzyme system purified from *Mts. trichosporium* (117)

	Cytochrome c_{co}	Protein 1	Protein 2
Molecular weight	13,000	47,000	9,400
Metal ion content	1 atom iron/molecule. Variable amount of copper (0.3-0.8 atoms/molecule)	1 atom copper/molecule	none

Molar ratio of components for optimum activity: 1:1:1
Specific activity: 6 μmol/min/mg of protein
pH optimum: 6.9-7.0
Stoichiometry: 5:4:7 (methane utilization: methanol formation: oxygen consumption)
K_m for methane: 66 μM
Substrate specificity: active with CH$_4$, CO, ethane, *n*-propane, *n*-butane

Electron donor specificity: ascorbate only for pure system
Effect of inhibitors: sensitive to inhibition by CN$^-$, chelating agents (especially effective copper chelators) SKF 525A, Lilly compounds 18947 and 53325, 2-mercaptoethanol, and dithiothreitol.

fer somewhat from the *Mts. trichosporium* cytochrome. This component of the oxygenase system cannot be replaced by a closely similar cytochrome purified from the facultative methylotroph *Pseudomonas extorquens* or by horse-heart cytochrome *c*. Because the monooxygenase is a copper-containing enzyme, inhibition by chelating agents is not surprising, while inhibition by 2-mercaptoethanol and dithiothreitol is rather unusual but could also be due to interaction with copper. Details of the enzymic mechanism are awaited. Perhaps the closest analogy can be drawn with the cytochrome P_{450} *n*-alkane monooxygenase isolated from a corynebacterium (137). However, the methane monooxygenase differs in not being directly NAD(P)H-linked. There is evidence that cytochrome c_{co} is responsible for methane-binding while protein 1 may bind oxygen. In this way methane would inhibit the oxidase activity of the cytochrome (117, 138). The role of protein 2 is not known.

Comparison of Methane Monooxygenases from Different Sources The most detailed information is available for the purified enzyme system from *Mts. trichosporium*, but there is insufficient knowledge of the other systems to allow detailed comparison. One obvious question is whether or not the *Mts. trichosporium* system is unique in not being directly NAD(P)H-linked. In this case it is likely that the cytochrome *c* component is reduced by electrons derived from the oxidation of methanol and formaldehyde. There is good evidence (discussed later in this section) that cytochrome *c* is the terminal oxidase for in vitro methanol oxidation and is most likely to be the in vivo indirect electron acceptor for methanol and formaldehyde dehydrogenase. However, NAD(P)H clearly must be capable of supplying reducing power derived from endogenous substrates to the monooxygenase in order to initiate growth on methane (116) (Figure 8).

Without this mechanism for recycling electrons from methanol and formaldehyde oxidation, there is some difficulty in accounting for an adequate supply of NAD(P)H to drive the monooxygenase reaction because only one NADH molecule is generated during the oxidation of methane to carbon dioxide via formate (the formate dehydrogenase reaction) by the normally accepted route. Because carbon is incorporated at the level of formaldehyde, substantially less than one NADH can be generated per methane molecule oxidized. It would seem, therefore, that an obligatorily NAD(P)H-linked monooxygenase would necessitate reversed electron transport, as pointed out by Van Dijken and Harder (139). Such a process might be energetically expensive and probably inconsistent with similar molar growth yields on methane and methanol. However, in this context the recent proposal that a cyclic mechanism may operate for formaldehyde oxidation in Type I methanotrophs is very interesting (53) (Figure 4) especially because it is now known to be the major mechanism for formaldehyde oxidation in a facultative methylotroph (70). This mechanism yields 2 molecules of NADPH per molecule of formaldehyde oxidized. If this scheme were to operate in Type I methanotrophs there would be sufficient NADPH formed to drive the monooxygenase without resorting to reversed electron transport.

Returning to the question of whether or not this newly discovered electron recycling process for the *Mts. trichosporium* monooxygenase occurs in other methanotrophs, to date there has not been any direct demonstration that cell-free methane monooxygenase either from *M. capsulatus* or *P. methanica* can use any alternative electron donors. Ethanol, methanol, and sodium ascorbate were tested without success with the *M. capsulatus* (Bath) system (112). It would therefore be interesting to determine the relative molar growth yields for methane and methanol for this particular strain. Similarly there is no evidence for an electron recycling system in the particulate preparations from *M. capsulatus* (Texas) (110). The relationship between these two *Methylococcus* systems is not clear but, in addition to one being soluble and the other particulate, there are marked differences in their sensitivities to inhibitors (110, 112). It is perhaps surprising to find these apparently different systems in such closely related bacteria. From the current state of knowledge, both systems are probably also different from that in *Methylosinus*. For example, the methane-independent NADH oxidase activity of *M. capsulatus* (Texas) is not sensitive to several common electron transport inhibitors, while the methane-dependent activity is quite sensitive. In *Mts. trichosporium*, both activities show similar sensitivities to inhibitors (see below), with the exception that the oxygenase is more sensitive to cyanide.

Interestingly, indirect evidence suggests that there are closer similarities between the *P. methanica* enzyme and that in *Mts. trichosporium*, which is all the more surprising because these bacteria are representative of the two main groups of methanotrophs. There is evidence from whole organism studies that the *P. methanica* system may not be obligatorily NAD(P)H-linked because ethanol stimulates carbon monoxide oxidation and the alcohol is oxidized to acetate by $NAD(P)^+$-independent enzymes (28, 29). This led Ferenci, Strøm, and Quayle (29) to suggest the two possible mechanisms for electron supply to the carbon monoxide (methane) monooxygenase shown in Figure 9. Scheme (a) involves reversed electron transport and scheme (b) is formally analogous to the one proposed independently for the *Mts. trichosporium* system (Figure 8). However, ethanol would not serve as the reductant in cell-free systems, perhaps due to the partial disruption of the electron transport system. It has been pointed out that the alternative reversed electron transport scheme (Figure 9a) may not necessarily require ATP; it depends on the half reduction potential of the X/XH_2 couple (29).

The sensitivity to inhibitors of the *P. methanica* system (29, 31) was almost identical to that of the pure *Mts. trichosporium* enzyme (117). Various chelating agents, dithiothreitol, reduced glutathione, and cyanide were potent inhibitors and the oxygenase was more sensitive to cyanide than the NADH oxidase activity. There is a major difference in the inhibitor pro-

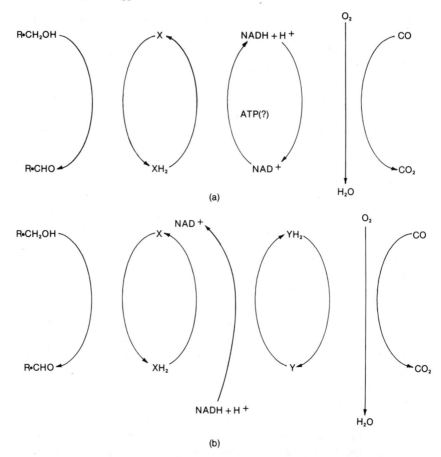

Figure 9. Two possible mechanisms for channeling of reductant from alcohol oxidation to *Pseudomonas methanica* carbon monoxide (methane) monooxygenase. X represents an unknown physiological electron acceptor for the reactions catalyzed by the primary alcohol dehydrogenase and Y represents an electron carrier (or sequence of carriers) between NADH and the monooxygenase. [From Ferenci, Strøm, and Quayle (29) by courtesy of Journal of General Microbiology.]

file of the *M. capsulatus* (Bath) soluble enzyme system and those of the other monooxygenase systems. While methane oxidation by washed suspensions of *Mts. trichosporium* (140), pure *Mts. trichosporium* methane monooxygenase (117), and active particulate preparations from *P. methanica* (31) and *M. capsulatus* (Texas) (110) were all potently inhibited by chelating agents, the *M. capsulatus* (Bath) system remained unaffected (112).

The Possible Role of Free Radicals in Methane Oxidation   In spite of the accumulation of a considerable body of evidence in favor of monooxygenases being responsible for methane oxidation in several methanotrophs, it has been argued that there are some findings that are not entirely consistent with the operation of such enzyme systems. Hutchinson, Whittenbury, and Dalton (141) have proposed that, at least in *M. capsulatus,* a mechanism involving free radicals rather than a monooxygenase system may be involved (Figure 10). The process would be initiated by removal of a hydrogen atom from methane by an iron (III) hydroperoxide. This would generate a methyl radical that could be stabilized by coordination to an iron atom. The iron-methyl could then react with molecular oxygen to form a methylperoxy iron derivative (Figure 11). Methanol, formaldehyde, and formate could then be formed subsequently by the processes shown in Figures 10 and 11.

There is at present far less experimental evidence for a free radical mechanism of this particular type than for a monooxygenase mechanism. However, it might be easier to rationalize the extraordinarily broad substrate specificity of the soluble *M. capsulatus* (Bath) system if free radicals were involved. While this appears to be a monooxygenase, it seems reasonable to suppose that free radicals may be involved in the reaction mechanism and that free radical mechanisms and monooxygenases are not necessarily mutually exclusive.

Methanol Dehydrogenase   Bacterial methanol dehydrogenase was first studied in detail in a facultative methylotroph (*Pseudomonas* M27) by Anthony and Zatman (142-145). Methanol oxidation in vitro can be coupled via phenazine methosulfate to oxygen or to the artificial electron acceptors 2,6-dichlorophenol indophenol or cytochrome *c*. Ammonia or methylamine are required as activators and the enzyme has a wide substrate specificity, oxidizing many primary alcohols, the rate decreasing with increasing chain length. The enzyme does not contain metal ions, has a high isoelectric point, a molecular weight of 120,000–146,000 is made up of two subunits with molecular weights of about 60,000, and possibly contains a pteridine prosthetic group. Details of the enzymic mechanism are not clear, but two possibilities were suggested by Anthony and Zatman (146). A pteridine might act directly as an electron acceptor or, alternatively, methanol might condense with it before oxidation of the bound C_1 unit. It has been pointed out that the standard electrode potentials of pteridines are rather low to permit the former mechanism and that the substrate specificity is consistent with a mechanism involving a $N^{5,10}$-methylene pteridine derivative. This could then either be oxidized to formate or, in a serine pathway organism, condense with glycine to form serine (41). There is some evidence that the in vivo electron acceptor for the dehydrogenase is a flavoprotein (144, 147). The enzyme seems to be widespread among methylotrophs and its importance is demonstrated by mutants lacking it that are unable to grow on methanol (85, 147, 148).

A closely similar enzyme has been extensively purified from *M. capsulatus* (Texas) (101, 120–121). It catalyzes the oxidation of methanol and formaldehyde to formate and is made up of two subunits, each of molecular weight 62,000, into which it dissociates at acid pH. In this bacterium no other mechanism has been demonstrated for oxidation of formaldehyde to

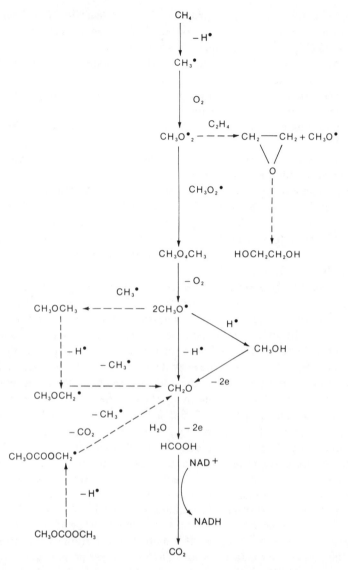

Figure 10. Pathways that might be involved in a free radical oxidation of methane: _____, main pathways; ———, ancillary pathways [From Hutchinson, Whittenbury, and Dalton (141) by courtesy of Journal of Theoretical Biology.]

formate, so presumably in this case the enzyme fulfills a dual role for oxidation of both methanol and formaldehyde. The enzymes from both *M. capsulatus* and *P.* M27 have closely similar amino acid compositions but differ somewhat in charged amino acids and therefore have different electrophoretic mobilities. Wadzinski and Ribbons (122) have also purified the *M. capsulatus* enzyme, and have shown that it is probably membrane

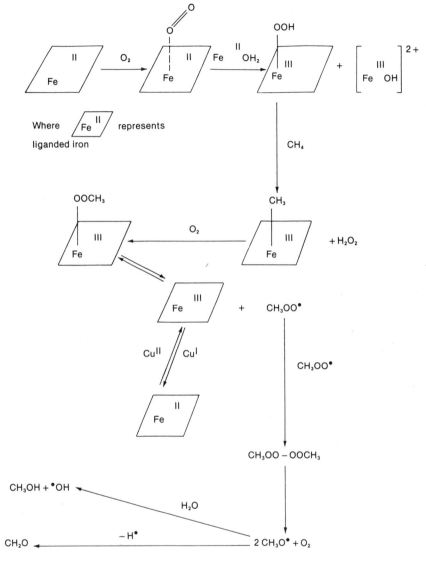

Figure 11. Possible role of metal ions in stabilization by coordination of free radicals that could be involved in methane oxidation. [From Hutchinson, Whittenbury, and Dalton (141) by courtesy of Journal of Theoretical Biology.]

bound in vivo and that the major site for methanol and formaldehyde oxidation lies on the intracytoplasmic membranes, where it is coupled to oxygen reduction. Under the breakage conditions used by these authors only about 40% of the total methanol dehydrogenase activity was found in the soluble fraction of cell-free extracts. The membrane-bound enzyme, after subsequent solubilization by phospholipase or detergent treatment, proved to be indistinguishable from the soluble form and there was concomitant loss of methanol oxidase activity from the membranes on solubilizing the dehydrogenase. Methanol oxidase activity was not solubilized, presumably because either the appropriate electron transport components were not solubilized or the electron transport system had been disorganized.

Patel and Felix (119) have recently examined several species of both Type I and Type II methanotrophs for distribution of methanol dehydrogenase and methanol oxidase activities, using essentially the same disruption and fractionation procedures as Wadzinski and Ribbons (122), namely, treatment in a French pressure cell followed by differential centrifugation. While Type I organisms showed a distribution of methanol dehydrogenase between soluble and particulate fractions and methanol oxidase activity resided exclusively in particulate fractions, Type II species showed only soluble methanol dehydrogenase and no oxidase activity. However, Higgins, Tonge, and co-workers (115, 117) found methanol oxidase activity distributed between particulate and supernatant fractions obtained from sonicated *Mts. trichosporium*. The reason for this discrepancy probably lies in the different breakage techniques used, but it is likely that the methanol dehydrogenase of Type II methanotrophs is also membrane-bound in vivo.

The methanol dehydrogenase of *Mts. sporium* has been crystallized and is closely similar to those studied in other methylotrophs (119). Again formate is the reaction product from methanol. There are some minor differences in properties as compared with the enzymes from other species. For example, the *Mts. sporium* enzyme oxidizes various primary alcohols at similar rates while in other cases the rate decreases as the alkyl chain length of the alcohol increases (120, 145, 149). In addition, the *Mts. sporium* enzyme gives only weak precipitin bands with enzymes from Type I organisms but much stronger bands with enzymes from Type II organisms, showing that the Type I organisms lack some antigenic determinants present in *Mts. sporium* and other Type II methanotrophs (119). Immunochemically, the methanol dehydrogenases from Type II methanotrophs are more closely related to those from facultative methanol utilizers than to those from Type I methanotrophs (120).

Progress in our understanding of the mechanism of action of methanol dehydrogenase depends in part on the elucidation of the exact structure of its prosthetic group. It seems clear that it is not a folic acid derivative, while various pteridines that may be derived from its prosthetic group have been isolated from methylotrophs (150, 151); it has been considered to most likely be a pteridine (146), probably a 2,4-dihydroxypteridine (152).

Formaldehyde Dehydrogenase   Methanol dehydrogenase also oxidizes formaldehyde to formate and, as discussed above, there is no evidence in *M. capsulatus* that any other enzyme catalyzes formaldehyde oxidation.

However, the possibility of a cyclic scheme for formaldehyde oxidation in a Type I methanotroph as described in *Pseudomonas* C (70) should be borne in mind. Two other types of formaldehyde dehydrogenase activity have been detected in methanotrophs; one found in a strain of *P. methanica* by Harrington and Kallio (123) catalyzes the reaction:

$$HCHO + NAD^+ + H_2O \xrightarrow[\text{glutathione}]{\text{reduced}} HCOOH + NADH + H^+$$

The second enzyme was found in another strain of *P. methanica* by Johnson and Quayle (118) and oxidizes formaldehyde in the presence of the artificial electron acceptor dichlorophenol indophenol:

$$HCHO + DCPIP + H_2O \rightarrow HCOOH + DCPIPH_2$$

It seems in the case of *P. methanica,* therefore, that there may be three enzymes capable of oxidizing formaldehyde but their relative importance in vivo remains to be elucidated.

Formate Dehydrogenase   While NAD-linked formate dehydrogenase activity has been detected in a variety of methylotrophs (118, 153) it has only been reported in three methanotrophs (101, 111, 118, 215). The partially purified enzyme from *P.* AM1 was specific for formate and was inhibited by cyanide and ferrous and cupric ions.

Electron Transport Systems and Energy Transduction

Ribbons, Harrison, and Wadzinski (41) reported the presence of *a*- and *c*-type cytochromes in *M. capsulatus* but could not detect any *b*-type cytochromes from difference spectra of whole organisms. Carbon monoxide difference spectra of several methanotrophs showed absorption bands at relatively short wavelengths (412 nm), perhaps due to an unusual oxidase or oxygenase. These authors suggested that the obligate dependence of some microorganisms on methane or methanol may be due to their having electron transport and energy transducing systems substantially different from those of common heterotrophs. Since 1970, there have been several more detailed studies of methanotroph electron transport systems. While there is some confusion in the literature, arising mainly from the problems of interpreting whole organism spectral data, it is clear that there is a remarkable similarity in the cytochrome complements of different methanotrophs and also between those of methanotrophs and methylotrophically grown methylotrophs.

Davey and Mitton (154) examined whole organisms of *Methylomonas albus* and *Mts. trichosporium* and concluded that these bacteria had qualitatively similar cytochrome patterns containing cytochromes *a* and *c*, but there was no clear indication of cytochrome *b* from difference spectra. However, carbon monoxide binding spectra showed, in addition to the cytochrome a_3 band, strong absorbance at 416 nm. This resembled the earlier finding with *M. capsulatus* (41) and the authors concluded that the band might be due to cytochrome *o*. The presence of cytochromes *c* and *a* in *Mts. trichosporium* was confirmed by Weaver and Dugan (75) and by using relatively gentle cell disruption techniques it was shown that the cytochromes are located in the membrane fractions together with the ATPase activity. Monosov and Netrusov (155) obtained evidence for the presence of cytochrome *b* in addition to *a*- and *c*-type cytochromes in both *Mts. trichosporium* and *Methylomonas agile*. These authors also demonstrated the localization of both terminal oxidase and ATPase activities in the intracytoplasmic membranes.

A comparative study of the cytochrome systems of five facultative methylotrophs and methanotrophs, including *Mts. trichosporium* and *P. methanica*, was carried out by Tonge, Higgins, and co-workers (116, 135). After growth on methane or methanol, the cytochrome complements were closely similar in all cases, being somewhat unusual in that the concentrations of particle-bound cytochromes *a, b,* and *c* were low, while in extracts prepared by sonication there was a high concentration of an unusual carbon monoxide–binding cytochrome *c* (cytochrome c_{co}) in soluble fractions. No cytochrome *o* was detected and it is likely that the presumed cytochrome *o* of Davey and Mitton (154) was due to this *c*-type cytochrome, which would also account for the peak at 412 nm in carbon monoxide–binding spectra of methylotrophs observed previously (41). Interestingly, facultative methylotrophs when grown on heterotrophic substrates showed grossly different cytochrome complements from those found after growth on methanol. Growth on heterotrophic substrates resulted in massive increases in the amounts of membrane-bound cytochromes aa_3, *b*, and *c* and the synthesis of some cytochrome a_2, which is not present in methylotrophically grown organisms. This was accompanied by a decrease in the concentration of soluble cytochrome c_{co}. This work indicates a radically different requirement for respiration and energy transduction for growth on methane or methanol than for growth on common heterotrophic substrates.

The cytochrome c_{co} is most likely membrane-bound in vivo and its role as a component of the methane monooxygenase enzyme system of *Mts. trichosporium* was discussed under "Methane Monooxygenase," above. It should be pointed out that soluble CO-binding *c*-type cytochromes are not unique to methylotrophs but are found in a variety of microorganisms (156). In a marine bacterium, *Beneckea natriegens*, there is evidence that the CO-binding properties of this cytochrome reflect an oxidase function (157).

There is little information concerning other electron transport components in methanotrophs, but ubiquinone-10 has been isolated from *Mts. trichosporium* (151) and this bacterium also contains flavoproteins (158). Inhibitor studies with *Mts. trichosporium* and *P. methanica* indicate that there are two terminal oxidases in both bacteria, presumably cytochrome aa_3 and cytochrome c_{co} (115–117, 138, 159). Although there is evidence that cytochrome c_{co} can act as an oxidase in *Mts. trichosporium* and in the facultative methylotroph *P. extorquens* (147), this appears not to be the case in *P.* AM1, in which the cytochrome c_{co} shows some different properties from that examined in other methylotrophs (160).

The exact sites of ATP synthesis in methanotrophs are not known, but measurements of respiration-induced proton extrusion by *Mts. trichosporium* oxidizing methane, methanol, formaldehyde, or formate indicate a P:O ratio of 1 for each substrate (161). While accepting the fact that there are some uncertainties in applying this technique to estimation of P:O ratios in bacteria, a P:O ratio of 1 for formate is unexpected and suggests that two of the three proton-translocating loops normally associated with NADH oxidation may be missing. However, as mentioned previously, the formate dehydrogenase reaction is the major NADH-generating step during the oxidation of methane to carbon dioxide in this serine pathway organism. Because most of this reducing power will be required for carbon dioxide fixation and for biosynthesis, there will be little NADH available for energy transduction and hence little selective advantage in developing more than one coupling site.

The indication of a P:O ratio of 1 for methane is especially interesting because during the oxidation of methane to methanol half the oxygen consumed is incorporated into the product rather than acting as an electron acceptor. This suggests that the true P:O ratio may in this case be 2, implying an extra coupling site for electrons recycled into the oxygenase reaction. However, recent relative molar growth yield data for *M. capsulatus* grown on methane as compared with various pseudomonads grown on methanol indicate only about an 8% variation in molar growth yields; this has been interpreted as indicating that the methane to methanol step is energetically neutral (162). While this may indeed be the case for *M. capsulatus*, it is by no means clear that comparisons from one methylotroph to another are valid in this context. For example, proton extrusion data for the facultative methylotroph *P.* AM1 suggest a maximum P:O ratio of 2 (163, 164); the obligate methanol utilizer *Pseudomonas* EN probably also has a maximum P:O ratio of 2 (38). Interestingly, respiration-induced proton extrusion could not be measured in *M. capsulatus*, which may be due to the sites of ATP synthesis being localized in the stacked internal membranes typical of that organism (162), as opposed to the situation in *Mts. trichosporium*, in which these sites are presumably in the peripheral membrane system of that bacterium.

Schemes for electron transport and energy transduction in *Mts. trichosporium* have been put forward largely on the basis of studies with electron transport inhibitors (115, 116) and the most recent version is shown in Figure 12. In addition to depicting the electron recycling phenomenon discussed above under "Methane Monooxygenase," the nature of electron supply to the oxygenase from NADH is also shown. The scheme allows for the possibility that the bacterium may possess more than one *c*-type

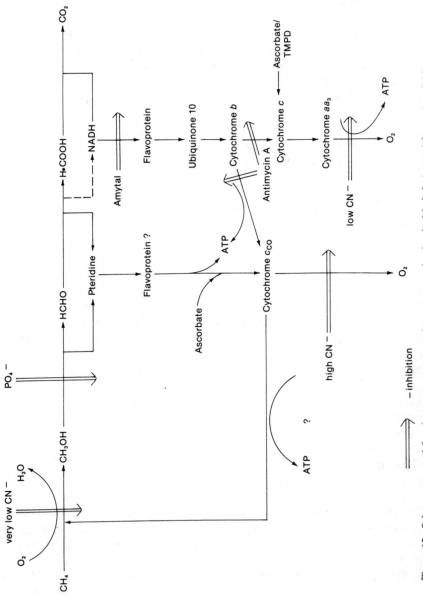

Figure 12. Scheme proposed for electron transport and energy transduction in *Methylosinus trichosporium* (138).

cytochrome or that the CO-binding cytochrome c may exist in more than one form. In facultative methylotrophs so far examined, it is clear from mutant studies that the c-type cytochrome is essential only for growth on alcohols oxidized by methanol dehydrogenase (136, 147, 165). In both facultative methylotrophs and in *Mts. trichosporium* this cytochrome has been shown to be closely linked to the methanol dehydrogenase enzyme (117, 147, 165, 166). Indeed, *Mts. trichosporium* cell-free supernatant preparations in which cytochrome c_{co} is the only cytochrome present show high methanol oxidase activity (116). It seems, therefore, that cytochrome b is not an essential intermediary for reduction of the c-type cytochrome by methanol dehydrogenase but, as mentioned previously, there is evidence for involvement of a flavoprotein.

Metabolic Implications for Molar Growth Yields

In view of the commercial interest in using bacteria that grow on methane or methanol for single-cell protein production, there have been numerous measurements of molar growth yields and considerable theoretical speculation. Some of the difficulties inherent in direct measurement of yields, especially on methane, and the values that have been obtained were discussed under "Methane Monooxygenase." A complementary approach to the problem is to attempt to predict yields from the known biochemistry of these microorganisms and such theoretical analyses have been published, most notably by Van Dijken and Harder (139, 167) and by Drozd and coworkers (38, 162). The problem with the theoretical approach at present lies in our incomplete knowledge of methanotrophic biochemistry. While a detailed analysis is not appropriate here, it is interesting to examine briefly the problems inherent in yield prediction and the biochemical aspects that are most important for these predictions.

The three main factors affecting the yield are the nature of the catabolic reactions involved in methane and methanol oxidation, the P:O ratios, and the ATP requirements for biosynthesis of cell material. While the energy requirements of the biosynthetic pathways are quite well defined, there is less certainty about the energetics of methane monooxygenase (discussed previously), the true P:O ratios, and the yield of NAD(P)H from formaldehyde oxidation in bacteria using the ribulose monophosphate pathway. In spite of these uncertainties, the range of theoretical yields obtained, making reasonable assumptions, generally agrees fairly well with the range obtained from direct yield determinations (38, 139, 167).

The dramatic effect of the nature of methane monooxygenase on yields (discussed above) is emphasized by the calculations of Van Dijken and Harder (139). The maximum theoretical yields on methane for a ribulose monophosphate pathway organism using an energetically neutral reaction for converting methane to methanol as compared with one using a conventional NADH-linked monooxygenase, making otherwise identical assumptions, are 1.46 and 0.91 g of cells/g of methane, respectively. While there is evidence (see "Methane Monooxygenase") for an energetically conservative or even ATP-yielding enzyme in *Mts. trichosporium*, the energetics of the methane to methanol step in *M. capsulatus* and other methanotrophs remain uncertain. Van Dijken and Harder (139) also pointed out that even assuming the maximum possible theoretical value for Y_{ATP} (168), the high yields reported on methane of 1.1 g of cells/g of methane (12) are not possible if a conventional NADH-linked monooxygenase is involved.

In general, the theoretical implications of carbon incorporation pathways for cell yields are more clear. As discussed under "Carbon Metabolism," there are known to be two variants of the serine pathway, the icl-serine pathway and the icl⁻-serine pathway. To date there is no evidence for the existence of the icl-serine pathway in methanotrophs. There are four possible variants of the ribulose monophosphate cycle (Table 5). There is evidence for operation of only two of these in methanotrophs, the FDP- and possibly the KDPG-aldolase variants, with transaldolase-catalyzed rearrangement phase (53), although the SDPase variants seem to operate in some facultative methylotrophs (169). Because the detailed operation of the icl⁻-serine pathway is not known, the exact energy change cannot be predicted, but it is likely to be similar to that for the icl-serine pathway.

From the data in Table 5 it is clear that on the basis of energetics of carbon incorporation alone Type I methanotrophs would be expected to give higher growth yields than Type II organisms. For methanotrophs there is insufficient accurate experimental data to substantiate these conclusions (139). However, recent comparative yield data for a variety of methylotrophs growing on methanol and other 1-carbon substrates (170–172) show that ribulose monophosphate pathway organisms give yields on methanol 17–44% higher than those for serine pathway bacteria (170). This concurs quite well with the predictions of Van Dijken and Harder (139) that yields for serine pathway organisms would be about 20% lower than those for ribulose monophosphate pathway organisms. Anthony has recently discussed in detail growth yield predictions for methylotrophs (242).

The Role of Intracytoplasmic Membranes in Methanotrophs

It is generally accepted that all methane-utilizing bacteria contain extensive intracytoplasmic membrane systems (40, 41, 173–177) and many of the data reviewed above suggest that the enzymes involved in oxidizing methane to formate, the electron transport components, and the energy-transducing systems are located in these membranes, implicating them in the processes of energy generation. This is hardly surprising and localization of such activities in microbial membranes is not unique. For example, cell-free membrane systems showing closely coupled oxidative phosphorylation have been isolated from a facultative methylotroph (166). These preparations were presumably derived from the pericytoplasmic membrane. There is also evidence that in Type I methanotrophs, the hexulose phosphate synthase may be membrane-bound in vivo (40).

Table 5.   Comparison of the bioenergetics of carbon-1 incorporation pathways.[a]

Pathway	Cleavage phase		Rearrangement phase		Reactants	Product	Energy change		
	FDP[b] aldolase	KDPG[c] aldolase	Trans-aldolase	SDPase[d]			ΔNAD(P)H$_2$	ΔFPH$_2$[e]	ΔATP
Ribulose monophos-phate	−	+	−	+	3HCHO	Pyruvate	+1	0	−3
	−	+	+	−	3HCHO	Pyruvate	+1	0	0
	+	−	−	+	3HCHO	Triose-P	0	0	−2
	+	−	+	−	3HCHO	Triose-P	0	0	−1
icl-Serine					2HCHO + 1 CO$_2$	P-glycerate	−2	+1	−3

[a] Adapted from ref. 243.
[b] FDP, fructose 1,6-phosphate.
[c] KDGP, phospho-2-keto-3-deoxygluconate.
[d] SDP, sedoheptulose 1, 7-diphosphate.
[e] FP, flavoprotein.

Intracytoplasmic membranes are clearly not necessary for growth on methanol because most facultative methylotrophs are devoid of them, with the exception of *Hyphomicrobium* (178). Of particular interest is the recent work with *Mtlb. organophilum*, which contains membranes when grown on methane but not when grown on glucose (14) or methanol (179). In addition, this facultative methanotroph increases the amount of membranes present during growth on methane in response to low dissolved oxygen tension (179). Growth on methane is clearly not unique in being associated with intracytoplasmic membranes. Similar membranes are found in photosynthetic bacteria (180), ammonia and nitrite oxidizers (181, 182), blue-green algae (183), and in some higher hydrocarbon utilizers (184, 185). These membranes contain lipids that are unusual in bacteria, such as phosphatidylcholine, squalene, and sterols (179, 186–188). Interestingly, the phospholipids of *Mtlb. organophilum* (179) and *Mts. trichosporium* (188), the only methanotrophs studied in this respect, differ somewhat from each other.

There has been much speculation about the role of these membranes. Clearly it is not necessary to assume that they perform exactly the same type of function in these different groups of microorganisms. However, in view of the startling similarities between methane and ammonia oxidation discussed below it seems likely that similar functions are involved in these cases. Methane is not very soluble in water but it dissolves more readily in lipid. Therefore, one function may be to increase the local substrate concentration for the monooxygenase by a simple solubility effect. The free-energy change for methane oxidation is greater if the gas is present in a hydrophobic environment (189) and the oxygenation reaction may be thermodynamically more favorable if the oxygen involved is also present in the lipid phase. In addition, it is known that the kinetic parameters of enzymes can be changed quite dramatically as a result of being incorporated into membrane structures (190, 191); in this respect the finding of lower K_m values for methane in an intact methanotroph (192) than for solubilized pure methane monooxygenase (117) may be significant. On the other hand, the need to synthesize intracytoplasmic membranes may reflect a problem of physical accomodation of all the protein components involved in methane oxidation in the limited area of the pericytoplasmic membrane.

In conclusion, while it is clear that the intracytoplasmic membranes are involved in oxidation processes, energy transduction, and perhaps also some biosynthetic processes, the reason for their presence in addition to the periplasmic membrane remains uncertain. Indeed, very recent work suggests that aerobic methane oxidation may not be obligatorily associated with intracytoplasmic membranes after all (193). *Mts. trichosporium* only contains these membranes when growing under conditions of oxygen limitation (Figure 2b is an electron micrograph of *Mts. trichosporium* grown under these conditions). Typically the membranes are somewhat closer to the pericytoplasmic membrane (40, 41) than shown in that particular photograph, but clearly there is some variability. Cultures growing rapidly under conditions of efficient gas transfer do not contain membranes. This is reminiscent of the inverse relationship between membrane and associated chlorophyll content of photoautotrophs and light intensity (194, 195). It would be of interest to determine whether or not the methane monooxygenase activity of methanotrophs varies with growth conditions.

Anaerobic Methane Oxidation

The recent isolation of bacteria from lake sediment surfaces that couple the anaerobic oxidation of methane to carbon dioxide with reduction of sulfate to hydrogen sulfide was mentioned earlier (18). These bacteria do not incorporate label from $^{14}CH_4$ into cell material but they incorporate carbon from acetate, lactate, or yeast extract; acetate is not oxidized to carbon dioxide. While nothing is known about the biochemistry of this process of methane oxidation, the free-energy change for the overall reaction has been calculated as -4.5 kcal/mol (196). Assuming that these anaerobes have a similar biochemical efficiency to aerobic methane utilizers, cell yields of about 20 mg of organisms/g of methane oxidized are possible.

NITROGEN METABOLISM

All methanotrophs utilize ammonia as nitrogen source, most will use nitrate and nitrite, and some use urea, casamino acids, or yeast extract (12). Most of them therefore have the capacity to reduce nitrate and nitrite to ammonia. *M. capsulatus* incorporates ammonia via the glutamine synthase and L-glutamine(amide)-2-oxoglutarate aminotransferase (GOGAT) system. The organism is constitutive for these two enzymes, but does not produce glutamate dehydrogenase even in the presence of high ammonia concentrations (197).

Of particular interest are the numerous reports that methanotrophs cooxidize ammonia to nitrite and/or nitrate (12, 30, 33, 198). Ferenci, Strøm, and Quayle (29) showed that ammonia is a competitive inhibitor of both methane and carbon monoxide oxidation by *P. methanica,* suggesting that ammonia is probably oxidized by the methane monooxygenase system. These authors pointed out that ammonia oxidation by *Nitrosomonas europaea* is strikingly similar to methane oxidation by methanotrophs, involving the route:

$$NH_3 \rightarrow NH_2OH \rightarrow (NOH) \rightarrow NO_2^-$$

Although cell-free systems that oxidize ammonia have been isolated from *Nitrosomonas* and it has been shown that a monooxygenase is most probably responsible for this activity (199–203), the enzyme has not yet been purified, so that a detailed comparison with methane monooxygenase is not yet possible. It is clear, however, that ammonia monooxygenase is closely similar to the methane system, especially the one found in *Mts. trichosporium*. Ammonia oxidation shows a similar pattern of sensitivities to inhibitors (203, 204) and is stimulated by hydroxylamine (199, 205). Although the evidence is indirect, it is possible that an electron recycling system similar to the one found in *Mts. trichosporium* may be involved in ammonia oxidation and it has been suggested that the further oxidation of nitroxyl may supply the reducing power for ammonia monooxygenase (205). Hydroxylamine oxidation is linked to cytochrome *c* (206–209), which exists in multiple forms in *Nitrosomonas* (210, 211), one form having a high redox potential and reacting with carbon monoxide (212). In addition, proton extrusion studies with *N. europaea* suggest a P:O ratio of 1 for both ammonia and hydroxylamine oxidation (213).

Recently O'Neill and Wilkinson (214) have studied ammonia oxidation by methanotrophs in some detail. They suggest, on the basis of pH effects, that NH_3 rather than NH_4^+ interacts with the enzyme. This is also thought to be the case in *Nitrosomonas* (202). In addition, there is kinetic evidence that ammonia and methane bind to the same site in methane monooxygenase (214). It is noteworthy that hydroxylamine is oxidized more rapidly by cell-free extracts of *Mts. trichosporium* than the analog, methanol (215). The relationship between ammonia and methane oxidizers is discussed further under "Evolutionary Considerations."

In addition to their capacity to oxidize ammonia to nitrate and reduce nitrate to ammonia, many methanotrophs are capable of fixing atmospheric nitrogen (7, 12, 35–37). The first demonstration that they can fix ^{15}N was described by Coty (37) and de Bont and Mulder (24) later confirmed this finding with a pure culture of a methanotroph. Because ^{15}N work is expensive and relatively tedious, the acetylene reduction test is commonly used to measure the nitrogen-fixing ability of cultures (216–219). However, attempts to detect nitrogenase activity in methane utilizers by measuring ethylene formation from acetylene were generally unsuccessful in cultures growing in the absence of a nitrogen source other than dinitrogen and therefore presumably fixing nitrogen. Nevertheless, on prolonged incubation a strain of *Mts. trichosporium* did catalyze the reaction (12) and in the absence of methane and at low partial pressures of oxygen samples of *M. capsulatus* (Bath) from a culture growing on methane and dinitrogen did reduce acetylene in the presence of a suitable electron donor (220).

A possible explanation for the difficulties involved in using the acetylene reduction assay with methanotrophs was put forward by de Bont and Mulder (24), who showed that ethylene, the reduction product from acetylene, was cooxidized in the presence of methane, and these authors suggested that this may be the reason that the product cannot be measured. There was, however, no difficulty in measuring acetylene reduction by organisms growing on methanol. Dalton and Whittenbury (220) have pointed out that this is not entirely consistent with the cooxidation hypothesis because methanol-grown organisms contain methane monooxygenase. However, de Bont and

Mulder (221, 222) and Colby, Dalton, and Whittenbury (220, 223) showed independently that acetylene is a potent inhibitor of methane monooxygenase and its presence may therefore preclude methane oxidation from supplying the reducing power necessary for nitrogenase activity. These researchers now all believe this inhibitory effect to be the cause of failure of the acetylene reduction test (220, 221, 223).

It is now clear that nitrogenase activity can be measured reliably in methanotrophs using the acetylene reduction assay provided the oxygen tension is low and a source of reducing power for the enzyme, such as methanol, ethanol, hydrogen, formaldehyde, or formate, but not methane, is present (220, 223, 224). It is also possible to measure activity in the presence of methane using the reduction of nitrous oxide to dinitrogen as the assay method (220, 223). The nitrogenase system of methanotrophs shares with those of other microorganisms the property of oxygen sensitivity. It has been suggested that the methane monooxygenase, by consuming excess oxygen, serves as a protective mechanism for nitrogenase (220).

Preliminary studies of the activities of glutamate dehydrogenase, glutamine synthase, and nitrogenase under different conditions (220) suggest that the nitrogenase system in *M. capsulatus* is controlled in a way similar to those in other nitrogen-fixing bacteria, being regulated by glutamine synthase (225, 226).

GENETICS OF METHANOTROPHS

There is little information concerning the genetics of methanotrophs and most genetic studies with the obligate bacteria have involved *M. capsulatus*, although other species show similar properties. Spontaneous mutants of this bacterium have been detected at frequencies of 1 in 10^7 to 1 in 4×10^8; these are resistant to antibiotics, amino acid analogs, or some other compounds (227). Attempts to induce mutation to auxotrophy using a variety of mutagens have been rather unsuccessful, with only one mutant (requiring *p*-aminobenzoic acid) being isolated from 35,000 colonies (227, 228). Some unstable auxotrophs were obtained but these rapidly reverted to the wild-type phenotype. Other methanotrophs also seem to be stable to mutagens. Because DNA repair mechanisms are necessary for production of stable mutations (229), these mechanisms have been studied in *M. capsulatus*. Filament formation and photoreactivation are absent while excision repair and recombination repair are present. The recombination repair system may be inefficient and of an accurate type that would not favor formation of stable mutants (230).

Transformation clearly occurs in *M. capsulatus* because the *p*-aminobenzoic acid-requiring auxotroph can be transformed at a frequency of 0.17% using DNA extracted from wild-type organisms (231). This transformation frequency is similar to that found in other bacteria (232–234).

Obligate methanotrophs that have been examined do not appear to contain plasmids, while the facultative methylotroph *P.* AMl contains three plasmids (235) and the facultative methanotroph *Mtlb. organophilum* also contains plasmid DNA (236). This organism behaves quite differently from the obligate species when exposed either to chemical mutagens or ultraviolet radiation. In this case, these agents do cause mutation although at a rather low frequency, and about fifty mutants have been obtained; some of these are drug-resistant, some are auxotrophic, and others have lost the ability to grow on methane or methanol (237). A glutamate auxotroph of this organism lacking isocitrate dehydrogenase was transformed to phototrophy using wild-type DNA with a frequency of 0.5% (237). The relative ease with which mutants of this bacterium can be obtained should facilitate studies of its biochemistry and of the interesting problems of metabolic regulation that it presents.

EVOLUTION OF METHANOTROPHS

The inevitable lack of fossil records makes bacterial paleontology inherently highly speculative. This final section is therefore brief. The increasingly apparent close similarities between obligate methanotrophs (especially Type I organisms) and the ammonia-oxidizing chemolithotrophs already mentioned tempts speculation, and indeed Whittenbury and Kelly in a recent review go so far as to propose a new definition of the term autotrophy that would include both Type I and Type II methanotrophs (238). The similarities between Type I methane utilizers and ammonia oxidizers include complex intracytoplasmic membranes, possession of an incomplete tricarboxylic acid cycle, assimilation of carbon-1 compounds by closely similar pathways initiated by a $C_1 + C_5$ condensation, and ability to oxidize ammonia. It is both tempting and satisfying to regard this as evidence for common ancestry, but of course it could be a case of convergent evolution. Apart from the carbon incorporation pathways these similarities also exist between ammonia oxidizers and Type II methanotrophs. Yet the serine pathway is heterotrophic in nature and unrelated to carbon incorporation pathways in classical autotrophs. It may be that Type I methanotrophs and ammonia oxidizers share a common ancestor while Type II methanotrophs evolved separately, but nevertheless convergently, from heterotrophs. However, this theory may be disturbed by determinations of G + C base ratios. While *M. capsulatus* has the same G + C ratio as Type II methanotrophs, a *Methylomonas* species (another Type I organism) differs quite markedly (1).

It seems likely in view of the reducing conditions and high methane concentration in the primordial atmosphere before the evolution of autotrophs that bacteria capable of utilizing methane would have been represented among early life forms. Indeed, Quayle (239) has suggested that the ribulose monophosphate pathway may well predate the Calvin cycle, which may indeed have evolved from it. Clearly, present-day methanotrophs may bear little resemblance to such anaerobic ancestors and certainly the latter could not have utilized a monooxygenase for methane oxidation. The isolation of pure cultures of anaerobic methane utilizers (18) is obviously of great interest in this regard and detailed biochemical studies with this organism are eagerly awaited. Apparently one strain is capable of incorporating formaldehyde into cell material (238).

A final point concerns the unusual CO-binding soluble cytochrome c found in methanotrophs, methylotrophs, and ammonia oxidizers. It may be that this cytochrome represents a stage in the development of cytochrome aa_3 from primitive cytochrome c. It is therefore of interest to determine the amino acid sequence and structure of this protein in order to relate it to schemes for the evolution of bacterial cytochromes and microbial energy metabolism (240). For a detailed examination of evolutionary aspects of autotrophs and methylotrophs, the recent review by Quayle and Ferenci should be consulted (243).

ACKNOWLEDGMENTS

I would like to thank Drs. H. Dalton, J. W. Drozd, and R. S. Hanson for generously supplying manuscripts in advance of publication.

REFERENCES

1. Whittenbury, R., Colby, J., Dalton, H., and Reed, H. L. (1976). *In* H. G. Schlegel, G. Gottschalk, and N. Pfennig (eds.), Microbial Production and Utilisation of Gases (H_2, CH_4, CO), p. 281. Erich Goltze KG, Göttingen.
2. Ehhalt, D. H. (1976). *In* H. G. Schlegel, G. Gottschalk, and N. Pfennig (edg.), Microbial Production and Utilisation of Gases (H_2, CH_4, CO), p. 13. Erich Goltze KG, Göttingen.
3. Ehhalt, D. H., and Volz, A. (1976). *In* H. G. Schlegel, G. Gottschalk, and N. Pfennig (eds.), Microbial Production and Utilisation of Gases (H_2, CH_4, CO), p. 23. Erich Goltze KG, Göttingen.
4. Kaserer, H. (1906). Zentrabl. Bakteriol. Parasitkde. Abt. II 15:573.
5. Sohngen, N. L. (1906). Zentrabl. Bakteriol. Parasitkde. Abt. II 15:513.
6. Dworkin, M., and Foster, J. W. (1956). J. Bacteriol. 72:646.
7. Davis, J. B., Coty, V. F., and Stanley, J. P. (1964). J. Bacteriol. 88:468.
8. Brown, L. R., Strawinski, R. J., and McCleskey, C. S. (1964). Can. J. Microbiol. 10:791.
9. Stocks, P. K., and McCleskey, C. S. (1964). J. Bacteriol. 88:1071.
10. Foster, J. W., and Davis, R. H. (1966). J. Bacteriol. 91:1924.
11. Whittenbury, R. (1969). Process Biochem. 4:51.
12. Whittenbury, R., Phillips, K. C., and Wilkinson, J. F. (1970). J. Gen. Microbiol. 61:205.
13. Wilkinson, J. F. (1971). *In* D. E. Hughes and A. H. Rose (eds.), Microbes and Biological Productivity, 21st Symposium, Society of General Microbiology, p. 15. Cambridge University Press, Cambridge, England.
14. Patt, T. E., Cole, G. C., Bland, J., and Hanson, R. S. (1974). J. Bacteriol. 120:955.
15. Patt, T. E., Cole, G. C., and Hanson, R. S. (1976). Int. J. System. Bacteriol. 26:226.
16. Enebo, L. (1967). Acta Chem. Scand. 21:625.

17. Zajic, L. E., Volesky, B., and Wellman, A. (1969). Can. J. Microbiol. 15:1231.
18. Panganiban, A. T., Jr., Patt, T. E., Hart, W., and Hanson, R. S. (in preparation)
19. Adamse, A. D., Hoeks, J., de Bont, J. A. M., and van Kessel, J. F. (1972). Arch. Microbiol. 83:32.
20. Zobell, C. E. (1946). Bacteriol. Rev. 10:1.
21. Harwood, J. H., and Pirt, S. J. (1972). J. Appl. Bacteriol. 35:596.
22. Davis, J. B. (1967). Petroleum Microbiology. Elsevier, Amsterdam.
23. Leadbetter, E. R., and Foster, J. W. (1960). Arch. Mikrobiol. 39:92.
24. de Bont, J. A. M., and Mulder, E. G. (1974). J. Gen. Microbiol. 83:113.
25. Hazeu, W. (1975). Antonie van Leeuwenhoek 41:121.
26. Thomson, A. W. (1974). Ph.D. thesis, University of Edinburgh.
27. Hubley, J. H., Mitton, J. R., and Wilkinson, J. F. (1974). Arch. Microbiol. 95:365.
28. Ferenci, T. (1974). FEBS Lett. 41:94.
29. Ferenci, T., Strøm, T., and Quayle, J. R. (1975). J. Gen. Microbiol. 91:79.
30. Hutton, W. E., and Zobell, C. E. (1952). J. Bacteriol. 65:216.
31. Colby, J., Dalton, H., and Whittenbury, R. (1975). Biochem. J. 151:459.
32. Stirling, D. I., and Dalton, H. (1976). Proc. Soc. Gen. Microbiol. 4:31.
33. Drozd, J. W., Bailey, M. L., and Godley, A. (1976). Proc. Soc. Gen. Microbiol. 4:26.
34. Rudd, J. W. M., and Hamilton, R. D. (1975). Arch. Hydrobiol. 75:522.
35. Schollenberger, C. J. (1930). Soil Sci. 29:261.
36. Harper, H. J. (1939). Soil Sci. 48:461.
37. Coty, V. F. (1967). Biotechnol. Bioeng. 9:25.
38. Barnes, L. J., Drozd, J. W., Harrison, D. E. F., and Hamer, G. (1976). In H. G. Schlegel, G. Gottschalk, and N. Pfennig (eds.), Microbial Production and Utilisation of Gases (H₂, CH₄, CO), p. 301. Erich Goltze KG, Göttingen.
39. Quayle, J. R. (1961). Annu. Rev. Microbiol. 15:119.
40. Quayle, J. R. (1972). Adv. Microb. Physiol. 7:119.
41. Ribbons, D. W., Harrison, J. E., and Wadzinski, A. M. (1970). Annu. Rev. Microbiol. 24:135.
42. Anthony, C. (1975). Sci. Prog. (Oxford) 62:167.
43. Wake, L. V., Rickard, P., and Ralph, B. J. (1973). J. Appl. Bacteriol. 36:93.
44. Lawrence, A. J., and Quayle, J. R. (1970). J. Gen. Microbiol. 63:371.
45. Whittenbury, R., Dalton, H., Eccleston, M., and Reed, H. L. (1975). In G. Terui (ed.), Microbial Growth on C₁ Compounds, p. 1. Society of Fermentation Technology, Tokyo, Japan.
46. Hazeu, W., and Steenis, P. J. (1970). Antonie van Leeuwenhoek 36:67.
47. Namsarayev, V. V., and Zavarzin, G. A. (1972). Microbiology 41:999.
48. Malashenko, Yu. R., Romanovskaya, V. A., and Kvasnikov, E. I. (1972). Microbiology 41:871.
49. Morinaga, Y., Yamanaka, S., Otsuka, S., and Hirose, Y. (1976). Agr. Biol. Chem. 40:1539.
50. Malashenko, Yu. R. (1976). In H. G. Schlegel, G. Göttschalk, and N. Pfennig (eds.), Microbial Production and Utilisation of Gases (H₂, CH₄, CO), p. 293. Erich Goltze KG, Gottingen.
51. Davies, J. B., and Yarborough, H. E. (1966). Chem. Geol. 1:137.
52. Bassham, J. A., and Calvin, M. (1957). The Path of Carbon in Photosynthesis. Prentice-Hall, Inc., Englewood Cliffs, New Jersey.
53. Strøm, T., Ferenci, T., and Quayle, J. R. (1974). Biochem. J. 144:465.
54. Leadbetter, E. R., and Foster, J. W. (1958). Arch. Mikrobiol. 30:91.

55. Johnson, P. A., and Quayle, J. R. (1965). Biochem. J. 95:859.
56. Kemp, M. B., and Quayle, J. R. (1967). Biochem. J. 102:94.
57. Lawrence, A. J., Kemp, M. B., and Quayle, J. R. (1970). Biochem. J. 116:631.
58. Kemp, M. B., and Quayle, J. R. (1965). Biochim. Biophys. Acta 107:174.
59. Kemp, M. B., and Quayle, J. R. (1966). Biochem. J. 99:41.
60. Kemp, M. B. (1972). Biochem. J. 127:64P.
61. Kemp, M. B. (1974). Biochem. J. 139:129.
62. Ferenci, T., Strøm, T., and Quayle, J. R. (1974). Biochem. J. 144:477.
63. Dahl, J. S., Mehta, R. J., and Hoare, D. S. (1972). J. Bacteriol. 109:916.
64. Colby, J., and Zatman, L. J. (1972). Biochem. J. 128:1373.
65. Stieglitz, B., and Mateles, R. I. (1973). J. Bacteriol., 114:390.
66. Fujii, T., and Tonomura, K. (1973). Agr. Biol. Chem. 37:447.
67. Diel, F., Held, W., Schlanderer, G., and Dellweg, H. (1974). FEBS Lett. 38:274.
68. Sahm, H., and Wagner, F. (1974). Arch. Microbiol. 97:163.
69. Davey, J. F., Whittenbury, R., and Wilkinson, J. F. (1972). Arch. Microbiol. 87:359.
70. Ben Bassat, A., and Goldberg, I. (1977). Biochim. Biophys. Acta 497:586.
71. Large, P. J., Peel, D., and Quayle, J. R. (1961). Biochem. J. 81:470.
72. Kaneda, T., and Roxburgh, J. M. (1959). Biochim. Biophys. Acta 33:106.
73. Leadbetter, E. R., and Gottlieb, J. A. (1967). Arch. Mikrobiol. 59:211.
74. Huennekens, F. M. (1968). In T. P. Singer (ed.), Biological Oxidations. Interscience, New York.
75. Weaver, T. L. and Dugan, P. R. (1975). J. Bacteriol. 122:433.
76. O'Connor, M. S., and Hanson, R. S. (1975). J. Bacteriol. 124:985.
77. Salem, A. R., Large, P. J., and Quayle, J. R. (1972). Biochem. J. 128:1203.
78. Hersh, L. B., and Bellion, E. (1972). Biochem. Biophys. Res. Commun. 48:712.
79. Salem, A. R., Hacking, A. J., and Quayle, J. R. (1973). Biochem. J. 136:89.
80. Cox, R. B., and Zatman, L. J. (1973). Biochem. Soc. Trans. 1:669.
81. Kohn, L. D., and Jakoby, W. B. (1968). J. Biol. Chem. 243:2465.
82. Kohn, L. D., and Jakoby, W. B. (1968). J. Biol. Chem. 243:2494.
83. Large, P. J., and Quayle, J. R. (1963). Biochem. J. 87:386.
84. Bamforth, C. W., and Quayle, J. R. (1977). J. Gen. Microbiol. 101:259.
85. Heptinstall, J., and Quayle, J. R. (1970). Biochem. J. 117:563.
86. Harder, W., Attwood, M. M., and Quayle, J. R. (1973). J. Gen. Microbiol. 78:155.
87. Dunstan, P. M., Anthony, C., and Drabble, W. T. (1972). Biochem. J. 128:107.
88. Bellion, E., and Hersh, L. B. (1972). Arch. Biochem. Biophys. 153:368.
89. Quayle, J. R. (1975). In G. Terui (ed.), Microbial Growth on C_1 Compounds, p. 59. Society of Fermentation Technology, Tokyo, Japan.
90. Attwood, M. M., and Harder, W. (1977). FEMS Lett. 1:25.
91. Smith, A. J., London, J., and Stanier, R. Y. (1967). J. Bacteriol. 94:972.
92. Patel, R., Hoare, D. S., and Taylor, B. F. (1969). Bacteriol. Proc. 69:128.
93. Davey, J. F., Whittenbury, R., and Wilkinson, J. F. (1972). Arch. Microbiol. 87:359.
94. Eccleston, M., and Kelly, D. P. (1973). J. Gen. Microbiol. 75:211.
95. Wadzinski, A. M., and Ribbons, D. W. (1975). J. Bacteriol. 123:380.
96. Patel, R., Hoare, S. L., Hoare, D. S., and Taylor, B. F. (1975). J. Bacteriol. 123:382.
97. Taylor, I. J., and Anthony, C. (1976). J. Gen. Microbiol. 93:259.

98. Pan, P., and Umbreit, W. W. (1972). J. Bacteriol. 109:149.
99. Carr, N. G. (1973). *In* N. G. Carr and B. A. Whitton (eds.), The Biology of the Blue-Green Algae, p. 39. Blackwell, Oxford.
100. Amemiya, K. (1972). Can. J. Microbiol. 18:1907.
101. Patel, R. N., and Hoare, D. S. (1971). J. Bacteriol. 107:187.
102. Eroshin, V. K., Harwood, J. H., and Pirt, S. J. (1968). J. Appl. Bact. 31:560.
103. Eccleston, M., and Kelly, D. P. (1972). J. Gen. Microbiol. 71:541.
104. Eccleston, M., and Kelly, D. P. (1973). J. Gen. Microbiol. 75:223.
105. Patt, T. E., O'Connor, M., Cole, G. C., Day, R., and Hanson, R. S. (1976). *In* H. G. Schlegel, G. Gottschalk, and N. Pfennig (eds.), Microbial Production and Utilisation of Gases (H_2, CH_4, CO), p. 317. Erich Goltze KG, Göttingen.
106. Wilkinson, J. F. (1975). *In* G. Terui (ed.), Microbial Growth of C_1 Compounds, p. 45. Society of Fermentation Technology, Tokyo, Japan.
107. Ribbons, D. W., and Michalover, J. L. (1970). FEBS Lett. 11:41.
108. Smith, U. S., Ribbons, D. W., and Smith, D. S. (1970). Tissue Cell 2:513.
109. Ribbons, D. W., and Higgins, I. J. (1971). Bacteriol. Proc. 71:107.
110. Ribbons, D. W. (1975). J. Bacteriol. 122:1351.
111. Ribbons, D. W., and Wadzinski, A. M. (1976). *In* H. G. Schlegel, G. Gottschalk, and N. Pfennig (eds.), Microbial Production and Utilisation of Gases (H_2, CH_4, CO), p. 359. Erich Goltze KG, Göttingen.
112. Colby, J., and Dalton, H. (1976). Biochem. J. 157:495.
113. Colby, J., Stirling, D. I., and Dalton, H. (1977). Biochem. J. 165:395.
114. Ferenci, T. (1976). *In* H. G. Schlegel, G. Gottschalk, and N. Pfennig (eds.), Microbial Production and Utilisation of Gases (H_2, CH_4, CO), p. 371. Erich Goltze KG, Göttingen.
115. Tonge, G. M., Harrison, D. E. F., Knowles, C. J., and Higgins, I. J. (1975). FEBS Lett. 58:293.
116. Higgins, I. J., Knowles, C. J., and Tonge, G. M. (1976). *In* H. G. Schlegel, G. Gottschalk, and N. Pfennig (eds.), Microbial Production and Utilisation of Gases (H_2, CH_4, CO), p. 389. Erich Goltze KG, Göttingen.
117. Tonge, G. M., Harrison, D. E. F., and Higgins, I. J. (1977). Biochem. J. 161:333.
118. Johnson, P. A., and Quayle, J. R. (1964). Biochem. J. 93:281.
119. Patel, R. N., and Felix, A. (1976). J. Bacteriol. 128:413.
120. Patel, R. N., Bose, H. R., Mandy, W. J., and Hoare, D. S. (1972). J. Bacteriol. 110:570.
121. Patel, R. N., Mandy, W. J., and Hoare, D. S. (1973). J. Bacteriol. 113:937.
122. Wadzinski, A. M., and Ribbons, D. W. (1975). J. Bacteriol. 122:1364.
123. Harrington, A. A., and Kallio, R. E. (1960). Can. J. Microbiol. 6:1.
124. Higgins, I. J., and Quayle, J. R. (1970). Biochem. J., 118:201.
125. Harwood, J. H. (1970). Ph.D. thesis, University of London.
126. Leadbetter, E. R., and Foster, J. W. (1959). Nature 184:1428.
127. Vary, P. S., and Johnson, M. J. (1967). Appl. Microbiol. 15:1473.
128. Hamer, G., and Norris, J. R. (1971). *In* Proceedings of the Eighth World Petroleum Congress, Vol. VI, p. 133. Applied Science, London.
129. Silverman, M. P., and Oyama, V. I. (1968). Anal. Chem. 40:1833.
130. Harwood, J. H., and Pirt, S. J. (1972). J. Appl. Bact. 35:597.
131. Wilkinson, T. G., Topiwala, H. H., and Hamer, G. (1974). Biotechnol. Bioeng. 16:41.
132. Bewersdorff, M., and Dostálek, M. (1971). Biotechnol. Bioeng. 13:49.
133. Sheehan, B. J., and Johnson, M. J. (1971). Appl. Microbiol. 21:511.
134. Harrison, D. E. F. (1976). Chem. Technol. 6:570.
135. Tonge, G. M., Knowles, C. J., Harrison, D. E. F., and Higgins, I. J. (1974). FEBS Lett. 44:106.
136. Anthony, C. (1975). Biochem. J. 146:289.

137. Cardini, G., and Jurtshuk, P. (1968). J. Biol. Chem. 243:6070.
138. Higgins, I. J., Tonge, G. M., and Hammond, R. C. (1977). Abstracts of the 2nd International Symposium on Microbial Growth on C_1 Compounds, September 12-16th, Pushchino, USSR, p. 65.
139. Van Dikjen, J. P., and Harder, W. (1975). Biotechnol. Bioeng. 17:15.
140. Hubley, J. H., Thomson, A., and Wilkinson, J. F. (1975). Arch. Microbiol. 102:199.
141. Hutchinson, D. W., Whittenbury, R., and Dalton, H. (1976). J. Theor. Biol. 58:325.
142. Anthony, C., and Zatman, L. J. (1964). Biochem. J. 92:609.
143. Anthony, C., and Zatman, L. J. (1964). Biochem. J. 92:614.
144. Anthony, C., and Zatman, L. J. (1965). Biochem. J. 96:808.
145. Anthony, C., and Zatman, L. J. (1967). Biochem. J. 104:953.
146. Anthony, C., and Zatman, L. J. (1967). Biochem. J. 104:960.
147. Higgins, I. J., Taylor, S. C., and Tonge, G. M. (1976). Proc. Soc. Gen. Microbiol. 3:179.
148. Dunstan, P. M., Anthony, C., and Drabble, W. T. (1972). Biochem. J. 128:99.
149. Sperl, G. T., Forrest, H. S., and Gibson, D. T. (1974). J. Bacteriol. 118:541.
150. Urushibara, T., Forrest, H. S., Hoare, D. S., and Patel, R. N. (1972). Experientia 28:392.
151. Tonge, G. M. (1977). Ph.D. thesis, University of Kent.
152. Sperl, G. T., Forrest, H. S., and Gibson, D. T. (1973). Bacteriol. Proc. 73:151.
153. Kaneda, T., and Roxburgh, J. M. (1959). Can. J. Microbiol. 5:187.
154. Davey, J. F., and Mitton, J. R. (1973). FEBS Lett. 37:335.
155. Monosov, E. Z., and Netrusov, A. I. (1975). Microbiology 45:518.
156. Weston, J. A., and Knowles, C. J. (1973). Biochim. Biophys. Acta 305:11.
157. Weston, J. A., and Knowles, C. J. (1974). Biochim. Biophys. Acta 333:228.
158. Hammond, R. C., and Higgins, I. J., unpublished observations.
159. Ferenci, T. (1976). Arch. Microbiol. 108:217.
160. Anthony, C. (1975). Biochem. J. 146:289.
161. Tonge, G. M., Drozd, J. W., and Higgins, I. J. (1977). J. Gen. Microbiol. 99:229.
162. Drozd, J. W., Linton, J. D., Downs, J., Stephenson, R., Bailey, M. L., and Wren, S. J. (1977). Abstracts of the 2nd International Symposium on Microbial Growth on C_1 Compounds, September 12-16th., Pushchino, USSR, p. 91.
163. O'Keefe, D. T., and Anthony, C. (1977). Proc. Soc. Gen. Microbiol. 4:67.
164. Anthony, C., and O'Keefe, D. T. (1977). Proc. Soc. Gen. Microbiol. 4:68.
165. Widdowson, D., and Anthony, C. (1975). Biochem. J. 152:349.
166. Netrusov, A. I., Rodionov, Y. V., and Kondratieva, E. N. (1977). FEBS Lett. 76:56.
167. Harder, W., and Van Dijken, J. P. (1976). In H. G. Schlegel, G. Gottschalk, and N. Pfennig (eds.), Microbial Production and Utilisation of Gases (H_2, CH_4, CO), p. 403. Erich Goltze KG, Göttingen.
168. Bauchop, T., and Elsden, S. R. (1960). J. Gen. Microbiol. 23:457.
169. Colby, J., and Zatman, L. J. (1975). Biochem. J. 148:513.
170. Goldberg, I., Rock, J. S., Ben-Bassat, A., and Mateles, R. I. (1976). Biotechnol. Bioeng. 18:1657.
171. Amano, Y., Sawada, H., Takada, N., and Terui, G. (1975). J. Ferment. Technol. 53:315.
172. MacLennan, D. G., Ousby, J. C., Vassey, R. B., and Cotton, N. T. (1971). J. Gen. Microbiol. 69:395.
173. Proctor, H. M., Norris, J. R., and Ribbons, D. W. (1969). J. Appl. Bact. 32:118.
174. Davies, S. F., and Whittenbury, R. (1970). J. Gen. Microbiol. 61:227.

175. Smith, U., Ribbons, D. W., and Smith, D. S. (1970). Tissue Cell 2:513.
176. Smith, U., and Ribbons, D. W. (1970). Arch. Microbiol. 74:116.
177. de Boer, W. E., and Hazeu, W. (1972). Antonie van Leeuwenhoek 38:33.
178. Conti, S. F., and Hirsch, P. (1965). J. Bacteriol. 89:503.
179. Patt, T. E., and Hanson, R. S., unpublished work.
180. Oelze, J., and Drews, G. (1972). Biochim. Biophys. Acta 265:209.
181. Murray, R. G. E., and Watson, S. W. (1965). J. Bacteriol. 89:1594.
182. Pope, L. M., Hoare, D. S., and Smith, A. J. (1969). J. Bacteriol. 97:936.
183. Gantt, E., and Conti, S. F. (1969). J. Bacteriol. 97:1486.
184. Kennedy, R. S., and Finnerty, W. R. (1975). Arch Microbiol. 102:85.
185. Stirling, L. A., Watkinson, R. J., and Higgins, I. J. (1977). J. Gen. Microbiol. 99:119.
186. Bird, C. W., Lynch, J. M., Pirt, S. J., Reid, W. W., Brooks, C. J. W., and Middleditch, B. S. (1971). Science 230:473.
187. Hagen, P. O., Goldfine, H., and Williams, P. J. C. (1966). Science 151:1543.
188. Weaver, T. L., Patrick, M. A., and Dugan, P. R. (1975). J. Bacteriol. 124:602.
189. Singer, S. J., and Nicolson, G. L. (1972). Science 175:720.
190. Ackrell, B. A. C., Coles, C. J., and Singer, T. P. (1977). FEBS Lett. 75:249.
191. Ackrell, B. A. C., Kearney, E. B., and Singer, T. P. (1977). J. Biol. Chem. 252:1582.
192. Harrison, D. E. F. (1973). J. Appl. Bact. 36:301.
193. Brannan, J., and Higgins, I. J. (1978). Proc. Soc. Gen. Microbiol. 5:69.
194. Holt, S. C., Conti, S. F., and Fuller, R. C. (1966). J. Bacteriol. 91:344.
195. Jones, O. T. G. (1977). In B. A. Haddock and W. A. Hamilton (eds.), Microbial Energetics, 27th Symposium, Society for General Microbiology, p. 151. Cambridge University Press, Cambridge, England.
196. Baas-Becking, L. G. M., Kaplan, I. R., and Moore, D. (1960). J. Geol. 68:233.
197. Drozd, J. W., personal communication.
198. Malashenko, Y. R. (1976). In H. Dellweg (ed.), Abstracts of the 5th International Fermentation Symposium, p. 209. Institute fur Garungsgewerke und Biotechnologie, Berlin.
199. Suzuki, I., and Kwok, S. C. (1969). J. Bacteriol. 99:897.
200. Watson, S. W., Asbell, M. A., and Valois, F. W. (1970). Biochem. Biophys. Res. Commun. 38:1113.
201. Suzuki, I., and Kwok, S. C. (1970). Biochim. Biophys. Acta 222:22.
202. Suzuki, I., Dular, U., and Kwok, S. C. (1974). J. Bacteriol. 120:556.
203. Suzuki, I., Kwok, S. C., and Dular, U. (1976). FEBS Lett. 72:117.
204. Hooper, A. B., and Terry, K. R. (1973). J. Bacteriol. 115:480.
205. Hooper, A. B. (1969). J. Bacteriol. 97:968.
206. Suzuki, I. (1974). Annu. Rev. Microbiol. 28:85.
207. Aleem, M. I. H., and Lees, H. (1963). Can. J. Microbiol. 41:763.
208. Falcone, A. B., Shug, A. L., and Nicholas, D. J. D. (1963). Biochim. Biophys. Acta 77:199.
209. Hooper, A. B., and Nason, A. (1965). J. Biol. Chem. 240:4044.
210. Rees, M. K. (1968). Biochemistry 7:353.
211. Rees, M. K. (1968). Biochemistry 7:366.
212. Tronson, D. A., Ritchie, G. A. F., and Nicholas, D. J. D. (1973). Biochim. Biophys. Acta 310:331.
213. Drozd, J. W. (1976). Arch. Microbiol. 110:257.
214. O'Neill, J. G., and Wilkinson, J. F. (1977). J. Gen. Microbiol. 100:407.
215. Tonge, G. M., and Higgins, I. J., unpublished work.
216. Dilworth, M. J. (1966). Biochim. Biophys. Acta 127:285.
217. Schollhorn, R., and Burris, R. H. (1967). Proc. Natl. Acad. Sci. USA 57:1317.

218. Hardy, R. W. F., Holsten, R. D., Jackson, E. K., and Burns, R. C. (1968). Plant Physiol. 43:1185.
219. Postage, J. R. (1972). *In* J. R. Norris, and D. W. Ribbons (eds.), Methods in Microbiology, Vol. 6B, p. 343. Academic Press, London.
220. Dalton, H., and Whittenbury, R. (1976). *In* H. G. Schlegel, G. Gottschalk, and N. Pfennig (eds.), Microbial Production and Utilisation of Gases (H_2, CH_4, CO), p. 379. Erich Goltze KG, Göttingen.
221. de Bont, J. A. M., and Mulder, E. G. (1976). Appl. Environ. Microbiol. 31:640.
222. de Bont, J. A. M. (1976). Antonie van Leeuwenhoek 42:245.
223. Dalton, H., and Whittenbury, R. (1976). Arch. Microbiol. 109:147.
224. de Bont, J. A. M. (1976). Antonie van Leeuwenhoek 42:255.
225. Tubb, R. S. (1974). Nature 251:481.
226. Streicher, S. L., Shanmugam, K. T., Ausubel, F., Morandi, C., and Goldberg, R. C. (1974). J. Bacteriol. 120:815.
227. Harwood, J. H., Williams, E., and Bainbridge, B. W. (1972). J. Appl. Bact. 35:99.
228. Williams, E. (1973). Ph.D. thesis, University of London.
229. Bridges, B. A. (1969). Rev. Nucl. Sci. 19:139.
230. Williams, E., and Bainbridge, B. W. (1976). *In* K. D. Macdonald (ed.), 2nd International Symposium on the Genetics of Industrial Microorganisms, p. 313. Academic Press, London.
231. Williams, E., and Bainbridge, B. W. (1971). J. Appl. Bact. 34:683.
232. Anagnostopoulos, C., and Spizizen, J. (1961). J. Bacteriol. 81:741.
233. Hotchkiss, R. D. (1954). Proc. Natl. Acad. Sci. USA. 40:49.
234. Spencer, H. T., and Heriott, R. M. (1965). J. Bacteriol. 90:911.
235. Warner, P. J., Higgins, I. J., and Drozd, J. W. (1977). FEMS Microbiol. Lett. 1:339.
236. Hanson, R. S., personal communication.
237. O'Connor, M., Wopat, A., and Hanson, R. S. (1977). J. Gen. Microbiol. 98:265.
238. Whittenbury, R., and Kelly, D. P. (1977). In B. A. Haddock and W. A. Hamilton (eds.), Microbial Energetics, 27th Symposium, Society of General Microbiology, p. 121. Cambridge University Press, Cambridge, England.
239. Quayle, J. R. (1977), personal communication.
240. Dickerson, R. E., Timkovich, R., and Almassy, R. J. (1976). J. Mol. Biol. 100:473.
241. Colby, J., and Dalton, H. (1978). Biochem. J. 171:461.
242. Anthony, C. (1978). J. Gen. Microbiol. 104:91.
243. Quayle, J. R., and Ferenci, T. (1978). Microbiol. Rev. 42:251.

Index

ABS, *see* Alkyl-benzene sulfonates
Acetaldehyde dehydrogenase, CoA-
 dependent, 93, 94
Acetate
 effect on membrane energization, 32
 as energy source, Y_{ATP} value and, 25
 methanogenesis and, 274, 276, 277,
  293
 precursor, of methane, 271
 product, 94, 105–106
  of dark metabolism, 146, 147
  in energy metabolism, 92
 utilization, ATP requirements and,
  4, 5
  in dark metabolism, 153–154
  in energy metabolism, 97
  in green bacteria, 161
  via icl-citrate pathway, 316, 317
  in methanotrophs, 318
  by purple bacteria, 133–134,
   136–137, 143
  in ribulose diphosphate cycle, 126
[1-^{14}C] Acetate, 75
Acetate kinase
 activity, 91, 93, 94
 substrate-level phosphorylation reaction, 88, 91
Acetobacterium, metabolism, 271, 273
Acetobacterium woodii, energy metabolism, 92, 95, 96, 105
Acetoin, product, of dark metabolism,
 146, 147
Acetokinase, 110
Acetyl-CoA, intermediate, in energy
 metabolism, 92, 93, 94, 100, 101
Acetyl-CoA synthetase, in purple bacteria, 133
Acetylene, inhibitor, of methane
 monooxygenase, 345
Acetylene reduction test, 344
Acetyl phosphate, intermediate, in
 energy metabolism, 91, 92, 93
Achromobacter, respiratory chain com-
 position, 53
Achromobacter fischeri, nitrite reduc-
 tase, 257
Acid phosphatase, enzyme evolution
 and, 218

Acinetobacter lwoffi, energy coupling
 sites, 62, 63, 64
Acinetobacter sp., in methane-utilizing
 community, 211, 212
Aconitase, 316
Acrylyl-CoA/propionyl-CoA couple,
 redox potential, 99
Actinomyces israeli, in studies of
 growth yield, 14
Actinomycin sensitivity, 181
Adenine nucleotide translocase, 61, 74
Adenosine diphosphate
 binding site on BF_1, 66
 inhibitor, of nitrogenase, 250
 respiration rate and, 35
 respiratory chain control and, 57
 substrate, of energy metabolism, 86
 see also ADP-
Adenosine triphosphate
 binding site on BF_2, 66
 as energy carrier, 86
 metabolism, 25–26
 methyl reductase and, 285–286, 288
 nitrate reductase activity and, 261
 nitrogenase activity and, 243, 246,
  250
 wastage, 37–39
 see also ATP-; ATPase
Adenosine triphosphate requirement
 for membrane energization, 25–28
 for microbial biomass formation,
  experimental determination,
   13–25
  theoretical calculations, 3–13
 for motility, 28
 for transport processes, 5–7
Adenosine triphosphate synthesis
 in dark metabolism, 146, 151
 during fumarate reduction, 95
 in green bacteria, 157–158
 in methanogens, 291–293
 in methanotrophs, 322–323, 337
 specific rate, 18–19
 substrate-level phosphorylation, reactions for, 89, 91
Adenylate kinase, in dark metabolism,
 151
Adenyl cyclase, 185

355